International Max Planck Re:

at the l

Hamburg Studies on Maritime Affairs
Volume 12

Edited by

Jürgen Basedow
Peter Ehlers
Hartmut Graßl
Hans-Joachim Koch
Rainer Lagoni
Gerhard Lammel
Ulrich Magnus
Peter Mankowski
Marian Paschke
Thomas Pohlmann
Uwe Schneider
Jürgen Sündermann
Rüdiger Wolfrum
Wilfried Zahel

Meltem Deniz Güner-Özbek

The Carriage of Dangerous Goods by Sea

 Springer

Meltem Deniz Güner-Özbek
Istanbul Universitesi
Hukuk Fakultesi
34452 Beyazit Istanbul
Turkey
mdenizguner@hotmail.com

Dissertation zur Erlangung der Doktorwürde
an der Fakultät für Rechtswissenschaft der Universität Hamburg
Vorgelegt von: Meltem Deniz Güner-Özbek
Erstgutachter: Professor Dr. Peter Mankowski
Zweitgutachter: Professor Dr. Ulrich Magnus
Tag der mündlichen Prüfung: 27. Juni 2007

ISBN 978-3-540-75836-5   e-ISBN 978-3-540-75837-2

DOI 10.1007/978-3-540-75837-2

Hamburg Studies on Maritime Affairs ISSN 1614-2462

Library of Congress Control Number: 2007938961

© 2008 Springer-Verlag Berlin Heidelberg

Production: LE-TeX Jelonek, Schmidt & Vöckler GbR, Leipzig
Cover-design: WMX Design GmbH, Heidelberg

Printed on acid-free paper

9 8 7 6 5 4 3 2 1

springer.com

To My Mom, Aysel Güner

Preface

This book represents the Ph.D. study conducted at the Max-Planck Research School for Maritime Affairs at the University of Hamburg between 2004 and 2007. The topic of the study was suggested by my supervisor, Prof. Dr. Peter Mankowski. I want to express my deepest gratitude and thanks to Prof. Dr. Mankowski for his meritorious contribution, invaluable support and encouragement. I would like to extend my appreciation and thanks to second examiner Prof. Dr. Ulrich Magnus for the expeditious submission of the second opinion on my dissertation.

The Max-Planck Research School for Maritime Affairs is distinct in its multidisciplinary post-graduate study on maritime matters. I am deeply honored to be a member of the Max-Planck Research School. I am most grateful to its Directors for admitting me as a scholar, providing me with a generous scholarship and grant for the publication of this book in the Hamburg Studies on Maritime Affairs series. Thanks are also due to the former and current coordinators of the Research School for their guidance and assistance. I owe particular thanks to Dr. Silke Knaut and Ms. Vera Wiedenbeck for their great care. The library of the Max-Planck Institute for Comparative and International Private Law is a treasury of resources and has provided me with a convenient and efficient environment to carry out research for my dissertation. The Hanseatic City of Hamburg with its maritime flair is an excellent place to study maritime matters. I consider myself fortunate to have lived there. This book has been laid out by Ms. Ingeborg Stahl. I am thankful to her.

Further, I feel obliged to Prof. Dr. Tankut Centel (Dean of the Istanbul University Faculty of Law), Prof. Dr. Fehmi Ülgener (Istanbul University Faculty of Law) and Assoc. Prof. Dr. Emine Yazıcıoğlu (Vice Dean of the Istanbul University Faculty of Law) for their constant support and encouragement.

My final words of appreciation are for my family. I am grateful to my parents and my special sister, Özlem Deniz, for giving me the courage to follow my dreams and the strength to make them reality. I wish to express my profound appreciation and indebtedness to my husband, Ural Özbek, whose support, encouragement, understanding, patience, sacrifice and love have contributed immeasurably to my studies at Tulane University in New Orleans, USA, and at the International Max-Planck Research School for Maritime Affairs at the University of Hamburg.

Istanbul, August 2007   Meltem Deniz Güner-Özbek

Summary of Contents

Contents

Abbreviations

ACN	Australia Canada Norway
A.M.C.	American Maritime Cases
Art.	Article
BC Code	Code of Safe Practice for Solid Bulk Cargoes
BGB	Bürgerliches Gesetzbuch
BGH	Bundesgerichtshof
BIMCO	Baltic and International Maritime Council
Cam. L.J.	*Cambridge Law Journal*
CBP	Custom and Border Protection
chap.	chapter
CEFIC	European Chemical Industry Council
CETGD	United Nations Committee of Experts on the Transport of Dangerous Goods
CFR	Code of Federal Regulations
CJS	Corpus Juris Secundum
CLC	International Convention on Civil Liability for Oil Pollution Damage
CMI	Committee Maritime International
COGSA	Carriage of Goods by Sea
Const. L.J.	*Construction Law Journal*
CSI	Container Security Initiative
C-TPAT	Custom Trade Partnership against Terrorism
DGL	Dangerous Goods List
DSC	Dangerous Goods, Solid Cargoes and Containers
ECOSOC	United Nations Economic and Social Council
E.T.L.	*European Transport Law*
ed.	editor
fn.	Footnote
FR	Federal Register
Ga. J. Int'l & Comp. L.	*Georgia Journal of International and Comparative Law*
GGBefG	Gefahrgutbeförderungsgesetz
GGVSee	Verordnung über die Beförderung gefährlicher Güter
GESAMP	Joint Group of Experts on the Scientific Aspects of Marine Environmental
HansGZ	Hansetische Gerichtszeitschrift

HAZMAT	Hazardous Materials
HGB	Handelsgesetzbuch
HMR	Hazardous Materials Regulations
HNS	Hazardous and Noxious Substances
HNS Convention	International Convention on Liability and Compensation for Damage in Connection with the Carriage of Hazardous and Noxious Substances by Sea
Hous. J. Int'l. L	*Houston Journal of International Law*
IAEA	International Atomic Energy
IBC Code	International Code for the Construction and Equipment of Ships Carrying Dangerous Chemicals in Bulk
I.C.C.L.R	*International Company & Commercial Law Review*
I.C.L.Q	*International Comparative Law Quarterly*
IGC Code	International Code for the Construction and Equipment of Ships Carrying Liquefied Gases in Bulk
IJOSL	*International Journal of Shipping Law*
IMO	International Maritime Organization
IMDG Code	International Maritime Dangerous Goods Code
INF Code	International Code for the Safe Carriage of Packaged Nuclear Fuel, Irradiated Nuclear Fuel, Plutonium and High Level Radioactive Wastes on Board Ships
Netherlands YB Int'l L.	Netherlands Yearbook of International Law
IOPC Fund	International Oil Pollution Compensation Fund
Int'l Trade L.J	*International Trade Law Journal*
IntML	*International Maritime Law*
ISM Code	International Safety Management Code
J.B.L.	*Journal of Business Law*
J. Mar. L. & Com.	*Journal of Maritime Law and Commerce*
J. Statist. Sci.	*Journal of Statistical Science*
Law & Pol'y Int't Bus.	*Law and Policy in International Business*
Leg.	Legal
LLMC	Convention on Limitation of Liability for Maritime Claims
LMLN	*Lloyd's Maritime Law Newsletter*
LMCLQ	*Lloyd's Maritime and Commercial Law Quarterly*
L.Q.R.	*Law Quarterly Review*
Mar. Law.	*Maritime Lawyer*
MARPOL	International Convention for the Prevention of Pollution from Ships
Mar. Pol.	*Marine Policy*
McGill L.J.	*McGill Law Journal*

MDR	Monatszeitschrift für Deutsches Recht
MLAANZ	*Maritime Law Association of Australia and New Zealand*
M.R.I.	*Maritime Risk International*
MSA	Merchant Shipping Act
N.O.S.	Not Otherwise Specified
NVOCC	Non Vessel Operation Common Carrier
NYPE	New York Produce Exchange
OLG	Oberlandesgericht
para.	paragraph
P & I	Protection and Indemnity
P & I Int'l.	*P&I International*
RG	Reichsgericht
RGZ	Entscheidungen des Reichsgerichts in Zivilsachen
SDR	Special Drawing Right
SOLAS Convention	Safety of Life at Sea Convention
Stan. L. Rev.	*Stanford Law Review*
S.&T.L.I.	*Shipping and Transport Lawyer International*
Sw. U. L. Rev.	*Southwestern University Law Review*
Syracuse J .Int'l L. & Com.	*Syracuse Journal of International Law and Commerce*
Transp. L.J.	*Transport Law Journal*
TranspR	*Transportrecht*
Tul. L. Rev.	*Tulane Law Review*
Tul. Mar. L.J.	*Tulane Maritime Law Journal*
U.B.C.L.R.	*University of British Columbia Law Review*
ULR	Uniform Law Review
UNCITRAL	United Nations Commission on International Trade
UNCLOS	United Nations Convention on the Law of the Sea
Unif. L. Rev.	*Uniform Law Review*
U. Miami L. Rev.	*University of Miami Law Review*
U.S.F. Mar. L.J.	*University of San Francisco Maritime Law Journal*
Wash. L.J.	*Washburn Law Journal*
Va. J. Int. L.	*Virginia Journal of Law*
VersR	*Versicherungsrecht*
Vol.	volume

Introduction

The history of the carriage of dangerous goods by sea is as old as mankind itself. The dangers inherent in the carriage of such goods have grown with the passage of time and the development of new technologies. In the days of the sailing vessel, the hazards of the sea were so great, due to the smallness of the ship, that the danger to the cargo alone was negligible. In many cases, the operator of the ship was either unaware that he was carrying dangerous goods or was transporting a single commodity and had taken the necessary precautions. The only dangerous goods were rum, brandy and gunpowder. They were dangerous not by virtue of the substance but the combination. Although it might have been slightly uncomfortable to be near an exploding gunpowder cargo, the danger was rather limited.

Today much of this has changed. An ever-increasing number of goods are moved by sea at the present time. For many of these goods, the ship provides the most feasible mode of transport. Moreover, world trade depends to a large extent on the transport of dangerous goods. It is estimated that more than 50% of packed goods and bulk cargoes transported by sea today can be regarded as dangerous, hazardous or harmful to the environment. The cargoes concerned include products which are transported in bulk, such as solid or liquid chemicals, and other materials, gases and products for and from the oil-refinery industry, and wastes. Some of these substances, materials and articles are dangerous or hazardous from a safety point of view and are also harmful to the marine environment; others are only hazardous when carried in bulk and some may be considered as harmful to the marine environment. Between 10%-15% of cargoes transported in packed form, including shipborne barges on barge-carrying ships, freight containers, bulk packaging, portable tanks, tank-containers, road tankers, swap-bodies, vehicles, trailers, intermediate bulk containers (IBCs), unit loads and other cargo transport units, fall under this criterion.

The recent growth registered in the carriage of dangerous goods has been attributed to a number of factors. One factor is that some commodities are no longer available in sufficient quantities and have been replaced by synthetic materials. The production of synthetic materials often requires the use of dangerous substances. This, in turn, leads to an increase in the carriage of dangerous substances both in bulk and in packed form.

A second, yet related factor is the development of new technologies, which has increased the production of dangerous goods which can be carried by sea. The development of different chemicals in increasing amounts has also resulted in the development of new types of packing.

A tendency towards the specialization of ships and their consequent growth in size form a third factor. A few decades ago, the bulk of chemical products were carried in packed form or in general cargo ships. An increasing number of ships are presently dedicated to one type of cargo. There are now in operation chemical product carriers, container ships, vessels carrying tank containers, ro-ro vessels loading tank vehicles, lash ships and bulk carriers.

Dangerous goods as a category makes legal sense only if rules attached to it are different from those governing cargo in general and there are rules based on the concept of "dangerous goods." Worldwide concern with the risks posed by the increased frequency in the carriage of dangerous goods by sea has led to the progressive formulation and adoption of international technical standards to promote maritime safety. As a rule, maritime transport forms part of a chain which may include a number of other modes of transport. Goods may, for instance, be delivered to ports by trains or trucks before being transferred on board ship. At the port of delivery, the goods must be unloaded and delivered to their final destination. This necessitates, at least, that the basic rules applying to packed goods should be identical for all modes of transport.

During recent years there have been comparatively few major accidents at sea involving dangerous goods, with the exception of accidents involving oil tankers and bulk carriers. Nevertheless, dangerous cargoes have been involved in some of the worst disasters in shipping history and relatively small incidents involving dangerous goods occur frequently. The majority of the cases where the misfortune caused by dangerous goods is less momentous in the sense that they are shared only by the parties to the maritime adventure, so they fail to become a public concern. Both situations, nonetheless, highlight the hazards associated with dangerous goods whose shipment is liable to cause loss or damage. Thus, the multiplicity of dangerous goods and the multifarious dangers they present make legal regulation of the carriage of dangerous goods inevitable.

The purpose of this study is to examine the relevant issues regarding the carriage of dangerous goods by sea. The study starts by describing the background and development of dangerous goods regulations. The meaning of dangerous goods is given special attention, as the term in practice is used very widely to cover both dangerous goods within the framework of dangerous goods regulations as well as any goods that may cause damage. The study mainly focuses on the contractual relationship between shipper and carrier and on the most commonly carried dangerous goods. Afterwards, the parties' respective duties with regard to dangerous goods, rights and liabilities are broadly examined. The liability of the shipper is given particular importance, as the basis of his liability has been a controversial subject in common law and is still the subject matter of discussions in the preparation of the CMI/UNCITRAL draft instrument on the carriage of goods by sea. An examination of duties and liabilities are mainly based upon common law, the Hague/Hague-Visby Rules and national enactments thereof, whereby English, American and German Law are compared as far as possible. Moreover, where relevant, the CMI/UNCITRAL draft instrument will be mentioned. After a review of the limitation of liability issue, which was highlighted by recent cases and by discussions in the preparation of the draft instrument on the carriage of goods by sea and the insurance possibilities for charterers and shippers for liability exposures, the Convention on Liability and Compensation for damage in connection with the Carriage of Hazardous and Noxious Substances by Sea, 1996, is briefly examined.

Part 1:   Dangerous goods regulations

A. In general

Dangerous goods as a category make legal sense only if rules are applied which differ in kind from those governing cargo in general and rules do exist which are based on the concept of the term "dangerous".

Since so few dangerous goods were carried by sea until the latter part of the nineteenth century, special regulations had not been considered necessary. The first traceable reference to regulations dealing with dangerous goods appeared in the British Merchant Shipping Act, 1894. Section 301 of the Act entitled "Dangerous Goods and Carriage of Cattle" stipulated that "an emigrant ship could not proceed to sea if she carried an explosive ... or any vitriol, lucifer matches, guano or green hides, or ... any article... which by reason of the nature, quantity or mode of storage... (is) likely to endanger the health or lives of the steerage passengers or the safety of the ship..." Section 446 went on to provide that "every shipper of aquafortis, vitriol, naphtha, benzine, gunpowder, lucifer-matches, nitroglycerin, petroleum, any explosives within the meaning of Explosives Act, 1857 and any other goods of dangerous nature is bound to mark the nature of the goods distinctly on the outside of the package and to give the master or owner of the vessel notice of their nature and of the name and address of the sender or carrier when sending these items to be shipped or taking them on board".

An ever-increasing number of dangerous goods are carried by sea today. The increase in the movement of dangerous goods around the world draws attention to explosions or fire on container ships, spillages, pollution, accidents and potential danger, with the result that public opinion exerts pressure on governments and other bodies to reduce the risk.[1]

---

[1] It is true that there have been comparatively few major accidents at sea involving dangerous goods, with the exception of accidents involving oil tankers and bulk carriers. Nevertheless, dangerous cargoes have been involved in some of the worst disasters in shipping history and small incidents occur frequently. Among the most notable incidents involving dangerous goods are the loss of the *Mont Blanc* in Halifax, Canada, in 1917, causing 3,000 deaths, injuring 9,000 persons and destroying 6,000 homes. The *Mont Blanc* had been overloaded with more than 2,600 tons of explosives at the height of World War I. In 1944, the freighter *Fort Stikine*, carrying 1,400 tons of explosives and a large quantity of cotton exploded in Bombay, killing 1,250 persons and destroying or damaging 15 ships. In 1947, the *Grandcamp*, a freighter loaded with ammonium nitrate, a substance widely used as fertilizer, exploded in the port of Texas City, killing 480 persons, causing USD 700 million damage in today's monetary value

Consequently, dangerous goods are strong evidence of concern at the risks that maritime movement of such cargoes creates. Governments are concerned and so there are rules governing safety carrying penalties for non-compliance. Industry is concerned and so there are special rules in the carriage contracts. Concerns also exist for those suffering from carriage even if entirely unconnected with it. Moreover now there exist security concerns as dangerous goods could be instruments for terrorist attacks. Therefore, the legal framework in which industry must work covers the area of public and private law, of state concern for safety, security and the environment and individual concern for liability.[2]

B. Development of dangerous goods regulations

I. Role of the IMO

Shipping is one of the most international and dangerous industries in the world. It has always been recognized that the best way of improving safety at sea is to formulate international regulations that are followed by all shipping nations. The Convention establishing the International Maritime Organization was adopted on 6 March 1948 by the United Nations Maritime Conference.[3] The Convention entered into force on 17 March 1958. The new Organization was inaugurated on 6 January 1959, when the Assembly held its first session. The purpose of the Organization is defined in Art. 1 of the IMO Convention. This article stipulates that:

> Machinery for cooperation among governments in the field of governmental regulation and practices relating to technical matters of all kinds affecting shipping engaged in international trade; to encourage and facilitate the general adoption of the highest practicable standards in matters concerning maritime safety, efficiency of navigation and the prevention and control of marine pollution from ships; and to deal with administrative and legal matters related to the purpose set out in this Article.

---

and leaving 2,000 persons homeless. In 1989, the chemical tanker *Masquasar*, carrying 257,700 tonnes of chemicals, including 7,000 tonnes of highly toxic acrylonitrile and various amounts of caustic soda, styrene and methanol, exploded, burning 5 days and killing all 23 crew. Between 1998 and 2003 there were nine serious fires onboard of container ships, causing significant danger to ship, crew and cargo and even in one case, the *Hanjin Pennsylvania,* resulting in an insurance claim in excess of USD 100 million.

[2] Jackson, "Dangerous Cargo- A Legal Overview" in *Maritime Movement of Dangerous Cargoes- Public Regulations Private Liability* Papers of One Day Seminar, Southampton University, Faculty of Law (1981), A1.

[3] For detailed information on the role of the IMO, see Henry, *The Carriage of Dangerous Goods by Sea* (1985); "Basic Facts about IMO", <www.imo.org/includes/blastData Only.asp/data_id%3D7983/Basics2000.pdf> (visited 16.10.2006); Johnson, "IMCO: The First Four Years (1959-1962)" [1963] 12 *I.C.L.Q* 31; Simmonds, *The International Maritime Organization,* (1994).

In order to achieve its objectives, IMO has promoted the adoption of about 40 conventions and protocols. In addition to conventions and other treaty instruments, IMO also adopts numerous non-treaty instruments such as codes and recommendations, which are adopted by the Assembly, Maritime Safety Committee and the Marine Environment Protection Committee.

IMO's first task since it came into being in 1959 was to adopt a new version of International Convention for the Safety of Life Sea at Sea (SOLAS).[4] The transport of dangerous goods has been one of its responsibilities since then.

II. Structure of the IMO

The organization of the IMO consists of an Assembly, a Council and four main Committees: the Maritime Safety Committee (MSC), the Marine Environment Protection Committee (MEPC), the Legal Committee and the Technical Co-operation Committee. There is also a Facilitation Committee and a number of sub-committees of the main technical committees.

The highest technical body of the Organization is the MSC. Its functions are to consider any matter within the scope of the Organization concerned with aids to navigation, construction and equipment of vessels, manning from a safety standpoint, rules for the prevention of collisions, handling of dangerous goods, maritime safety procedures and requirements, hydrographic information, log-books and navigational records, marine casualty investigations, salvage and rescue and any other matters directly affecting maritime safety. The Committee is also required to provide machinery for performing any duties assigned to it by the IMO Convention or any duty within its responsibility which may be assigned to it by or under any international instrument and accepted by the Organization. It is also responsible for considering and submitting recommendations and guidelines on safety for possible adoption by the Assembly.

The MEPC is empowered to consider any matter within the scope of the Organization concerned with prevention and control of pollution from ships. In particular, it is concerned with the adoption and amendment of conventions and other regulations and measures to ensure their enforcement.

The MSC and MEPC are assisted in their work by nine sub-committees. Sub-committees deal with: Bulk Liquids and Gases (BLG), Carriage of Dangerous Goods, Solid Cargoes and Containers (DSC), Fire Protection (FP), Radio Communications and Search and Rescue (COMSAR), Safety of Navigation (NAV), Ship Design and Equipment (DE), Stability and Load Lines and Fishing Vessels Safety (SLF), and Flag State Implementation (FSI).

---

[4]   "Focus on IMO, SOLAS: the International Convention for the Safety of Life at Sea, 1974", <www.imo.org/includes/blastDataOnly.asp/data_id%3D7992/SOLAS98final. pdf> (visited 16.10.2006).

The Legal Committee is empowered to deal with any legal matters within the scope of the Organization. It was established in 1967 as a subsidiary body to deal with legal questions which arose in the aftermath of the Torrey Canyon[5] disaster.

The Technical Cooperation Committee is required to consider any matter within the scope of the Organization concerned with the implementation of technical cooperation projects for which the Organization acts as the executing body or cooperating agency and any other matters related to the Organization's activities in the technical cooperation field.

The Facilitation Committee is a subsidiary body of the Council. It deals with the IMO's work in eliminating unnecessary formalities and "red tape" in international shipping.

III. The SOLAS Convention

In the first Convention for the Safety of Life at Sea (SOLAS 1914), "the carriage of goods which by reason of their nature, quantity and mode of stowage" were likely to endanger the lives of the passengers or the safety of the ship was in principle forbidden. However, the decision as to which goods were "dangerous" was left to the Contracting Governments, which were also requested to advice on the precautions which should be taken in the packing and mode of transport, i.e. stowage and segregation of such goods. The latter seems to imply that, if these precautions were followed, the transport of these goods would be permitted and could be regarded as being safe.

Although SOLAS 1914 never entered into force, the principle of relying on national administrations and relevant authorities to decide on the definition and treatment of dangerous goods was established and, unfortunately, resulted in the development of many diversified regulations and practices embedded in national, regional or individual out-of-date port regulations.[6]

The same attitude was maintained in the SOLAS Conference 1929, in Article 24 of which "Dangerous Goods" are mentioned together with "Life-Saving Appliances". It was still forbidden to carry goods which by their nature, quantity and mode of stowage were liable to endanger the lives of the passengers or the safety of the ship. Moreover, it was still left to each administration to determine which goods were to be considered dangerous and to indicate the precautions which had to be taken in their packing and mode of stowage. The 1929 Convention entered into force in 1933. In 1948, a new attempt was made to formulate safety standards

---

[5]   In 1967, the Torrey Canyon struck Pollard's Rock in the Seven Stones reef between the Scilly Isles off Land's End, England. She was the first of the big super-tankers carrying a cargo of 120,000 tons of oil. The oil leaked from the ship and spread along the sea between England and France, killing most of the marine life it touched and blighting the region for many years thereafter. The CLC 1969 was formulated following the Torrey Canyon disaster.

[6]   "Focus on IMO, IMO and Dangerous Goods at Sea", <www.imo.org/includes/blast DataOnly.asp/data_id%3D7999/IMDGdangerousgoodsfocus1997.pdf>, 3, (visited 6.10. 2006).

for the carriage of dangerous goods. Thus up to the time of the 1949 SOLAS Conference it was forbidden to carry dangerous goods in ships unless due precautions were taken.

The lack of enthusiasm in bringing these instruments into force demonstrates what little importance was attached to the question. Indeed, the quantity of dangerous goods carried by sea was relatively small at the time. But by 1948, when the third SOLAS Conference was held in the wake of World War II, the perception of the subject under discussion had visibly changed. Sea traffic had considerably grown and more cargoes were being transported which could be considered dangerous. The War itself had given much impetus to the transport of these substances, and in the post-war years industrial needs and technological innovations contributed to a further increase. This expansion led to rethinking and, as a result, a new Chapter VI was added to the 1948 SOLAS Convention, dealing specifically with the "Carriage of Grain and Dangerous Goods".

However, the Conference recognized that the provisions of the 1948 SOLAS Convention were inadequate.[7] It therefore adopted Recommendation 22 to stress the importance of international uniformity in the safety precautions applied to the transport of dangerous goods by sea and noted that certain countries with an extensive export trade in chemicals had already adopted detailed regulations. In addition to stressing the need for international uniformity in safety precautions, the Conference also established that goods should be considered dangerous on the basis of their properties and characteristics and a labelling system should be developed using distinctive symbols indicating the kind of danger for each class of substances, materials and articles.

Recommendation 22 additionally urged that further study be undertaken with a view to developing uniform international regulations on the subject. However, despite these efforts, there was no common basis for the different modes of transport to work together to develop rules on dangerous goods.[8]

This situation changed in 1956, when the United Nations Committee of Experts on the Transport of Dangerous Goods (CETDG) completed a report which established the minimum requirements applicable for the transport of dangerous goods by all modes. This report, *United Nations Recommendations on the Transport of Dangerous Goods*, offered the general framework within which existing regulations could be adopted and developed, the ultimate aim being world-wide uniformity across all modes of transport.[9] The United Nations Recommendations have been amended and updated by succeeding sessions of the Committee of

---

[7]   *Ibid.*

[8]   In comparison with other modes of transport, the development of regulations for maritime carriage was very late in coming. As regards these other modes, the quantity of dangerous goods moved and their stowage were of very limited importance. Thus, there was no incentive for the various modes of transport to work together, nor was there any evidence of any effort to find basis for their regulation.

[9]   Wardelmann, "Transport by Sea of Dangerous, Hazardous, Harmful and Waste Cargoes" [1991] 26 *E.T.L* 116. See *infra* p. 12 ff.

Experts and published in accordance with subsequent resolutions of the United Nation Economic and Social Council ("ECOSOC").[10]

However, despite the publication of the United Nations Recommendations, as far as maritime transport was concerned, little was done in response to Recommendation 22.[11] In 1960, the conference took place to revise SOLAS 1948. Chapter VII of the revised 1960 SOLAS Convention, which entered into force on 26 May 1965, dealt exclusively with the carriage of dangerous goods. Another conference, held in 1974, further revised the Convention and the 1974 SOLAS version entered into force on 25 May 1980.[12] It has been subsequently modified and amended. Several amendments to SOLAS 1974 concerning the carriage of dangerous goods have also been adopted since then.

The revised Chapter VII now applies to all ships covered by SOLAS and also to cargo ships of less than 500 gross tonnage. Chapter VII of SOLAS 1974 is comprehensive in nature. It deals with dangerous goods in packaged form as well as with bulk cargoes. Part A deals with the carriage of dangerous goods in packaged form. Regulation 2.3 prohibits the carriage of dangerous goods by sea except when they are carried in accordance with the provisions of the SOLAS Convention and Regulation 3 requires the carriage of dangerous goods in packaged form to be in compliance with the relevant provisions of the International Maritime Dangerous Goods Code (IMDG Code). Regulation 2.4 requires each Contracting Government to issue, or cause to be issued, detailed instructions on emergency response and medical first aid relevant to incidents involving dangerous goods in packaged form.[13] Part A-1 is concerned with the carriage of dangerous goods in solid form in bulk. Part B covers construction and equipment of ships carrying liquid chemicals in bulk and requires chemical tankers built after July 1986 to comply with the International Bulk Chemical Code (IBC Code), while Part C covers construction and equipment of ships carrying liquefied gases in bulk and gas carriers constructed after 1 July 1986 to comply with the requirements of the International Gas Carrier Code (IGC Code). Part D includes special requirements for the carriage of packaged irradiated nuclear fuel, plutonium and high-level radioactive waste on board ships and requires ships carrying such products to comply with the International Code for the Safe Carriage of Packaged Irradiated Nuclear Fuel, Plutonium and High-Level Radioactive Waste on Board Ships (INF Code).

---

[10]   The ECOSOC, established by the United Nations Charter, is placed under the authority of the General Assembly. It is responsible for promoting higher standards of living, full employment, conditions of economic and social progress and development as well as solutions to international, social, health and related problems.

[11]   This was largely because the Convention, establishing IMO (then IMCO), adopted in 1948 at the United Nations Maritime Conference, did not enter into force until 1958 and the IMO Assembly did not meet for the first time until the following year.

[12]   156 States are party to the SOLAS 1974.

[13]   Taking into account the guidelines developed by the Organizations: the Emergency Response Procedures for Ships Carrying Dangerous Goods (EMS Guide) (MSC/Cir. 1025) and the Medical First Aid Guide for Use in Accidents Involving Dangerous Goods, (MFAG) (MSC/Cir.857).

IV. The MARPOL Convention

The International Convention for the Prevention of Pollution from Ships (MARPOL Convention) is the main international convention covering the prevention of pollution of the marine environment by ships through operational and accidental causes. It is a combination of two treaties adopted in 1973 and 1978 respectively and updated by amendments throughout the years.[14] The MARPOL Convention covers pollution by oil, chemicals, harmful substances in packaged form, sewage and garbage. The MARPOL Convention includes regulations aimed at preventing and minimizing pollution from ships and includes six technical Annexes.

Annex III to the MARPOL Convention contains general requirements relating to the prevention of pollution by harmful substances carried at sea in packaged form or in freight containers, portable tanks or road and rail tank wagons.[15] The objective behind the regulations contained in Annex III of MARPOL was to identify marine pollutants so that they could be packed and stowed on board ships in such a way as to minimize accidental pollution as well as to aid recovery by using clear marks to distinguish them from other (less harmful) cargoes.[16] The rule on discharging harmful goods was straightforward: "Jettisoning of harmful substances carried in packaged form shall be prohibited, except where necessary for the purpose of securing the safety of the ship or saving life at sea".[17] The Annex requires the issuing of detailed standards on packaging, marking, labelling, documentation, stowage, quantity, limitations, exceptions and notifications, for preventing and minimizing pollution by harmful substances. However, implementation of the Annex was initially hampered by the lack of a clear definition of harmful substances carried in packaged form. This was remedied by amendments to the IMDG Code to include marine pollutants. Annex III of MARPOL was also amended at the same time, to make it clear that "harmful substances" are those substances which are identified as marine pollutants in the IMDG Code.[18]

---

[14] "Focus on IMO, MARPOL 25 Years" <www.imo.org/includes/blastDataOnly.asp/ data_id%3D7993/MARPOL25years1998.pdf> (visited 16.10.2006); "Focus on IMO, MARPOL 73/78", <www.imo.org/includes/blastDataOnly.asp/data_id%3D7575/ MARPOL.1998.pdf> (visited 16.10.2006).

[15] 125 States are party to the Annex III.

[16] "Focus on IMO, MARPOL 25 years" <www.imo.org/includes/blastDataOnly.asp/ data_id%3D7993/MARPOL25years1998.pdf>, 20 (visited 16.10.2006).

[17] MARPOL Annex III.7(1).

[18] In *State of Louisiana, ex. rel. William J. Guste, Jr. v. The M/V Testbank, etc. et. Al.* 564 F.Supp. 729, it was held that non-compliance with the IMDG Code may not have been relied upon where pollution has caused and persons have suffered injury. This was so because the harm suffered must be of the kind of injury that the statute, i.e. the IMDG Code, intended to prevent. The IMDG Code, which originated in the SOLAS Convention, aims at protecting the crew of the vessel at sea and the safety of cargoes; it was not based on environmental concerns. This situation changed as the IMDG Code takes account of pollution aspects.

V. The UN Recommendations on the Transport of Dangerous Goods

1. Development of the UN Recommendations (Model Rules)

After the Second World War the increased pace of industrialization around the world led to a growth in the transport of goods classified as dangerous, including petroleum products, gases, explosives, petrochemicals, acids and radioactive materials. In the early 1950s, the Transport and Communications Commission of the United Nations Economic and Social Council (ECOSOC) noted that at that time the international regulations for the transport of dangerous goods were fragmentary and that the regulations applied to the different means of transport and also that those applied in different parts of the world lacked uniformity.[19] It seemed to the Commission that this was an urgent problem requiring consideration on a world-wide basis and with respect to all forms of transport. In the view of the great importance for the preservation of life and property, the need for the regulation of the transport of dangerous goods by the various means of transport and operations related thereto to be as uniform as possible on a world-wide basis, and the fact that the problem was already under consideration by the Inland Transport Committee of the UN Economic Commission for Europe (UNECE) and that this work would have to be coordinated with any work undertaken by the United Nations on a world-wide basis, the Transport and Communications Commission recommended that the ECOSOC request the Secretary General:

– to examine, in consultation with the competent international and, where appropriate, national bodies, if necessary by convening a meeting, the various aspects of the problem of the transport of dangerous goods, among them classification, packaging and labelling, with a view to determining which of these aspects are appropriate for uniform or approximately uniform regulations with respect to the various means of transport
– to include among the bodies to be consulted the International Civil Aviation Organization, the International Labour Organization, the interim body dealing with the Safety of Life at Sea Convention, the Central Office for International Transport by Rail, the Central Commission for the Navigation of the Rhine, and the Inter-State Commerce Commission of the United States of America.

Following the request by the Transport and Communications Commission, in April 1953 the ECOSOC requested the Secretary-General to appoint a committee to make a study and to present a report to the Transport and Communications Commission. This committee, which subsequently came to be known as the United Nations Committee of Experts on the Transport of Dangerous Goods (CETDG), was charged with the following tasks: (i) recommending and defining

---

[19] Kervella, "IMO Roles on the Transport of Dangerous Goods in Ships and the Work of the International Bodies in the UN system, the Harmonization Issues with Regard to Classification, Criteria, Labelling and Placarding, Data Information, Emergency Response and Training", in *The 11th International Symposium on the Transport of Dangerous Goods by Sea and Inland Waterways* (1992) 74, 75 f.

groupings or classifications of dangerous goods on the basis of the character of risk involved (ii) listing the principal dangerous goods moving in commerce and assigning each to its proper grouping or classification (iii) recommending marks or labels for each grouping or classification which shall identify the risk graphically and without regard to printed text and (iv) recommending the simplest possible requirements for shipping papers covering dangerous goods consignments.

The original committee was composed of "not more than nine qualified experts from countries having a substantial interest in the international transport of dangerous goods.[20] It met for the first time in 1954 and published its first recommendations in October 1956. In April 1957, the ECOSOC approved the first edition of the *UN Recommendations on the Transport of Dangerous Goods*. These recommendations later became known as the "Orange Book" because of the distinctive colour of the book cover.

2. Nature, purpose and significance of the Recommendations

The United Nations Recommendations on the Transport of Dangerous Goods have been developed by the United Nations Economic and Social Council's Committee of Experts on the Transport of Dangerous Goods in the light of technical progress, the advent of new substances and materials, the exigencies of modern transport systems and, above all, the requirement to ensure the safety of people, property and the environment.[21]

The Recommendations provide a basis for the development of harmonized regulations for all modes of transport, in order to facilitate trade and the safe, efficient transport of dangerous goods.[22] They are addressed to governments and international organizations concerned with the regulation of the transport of dangerous goods. However, they do not apply to the transport of dangerous goods in bulk, which in most countries is subject to special regulations.

The Recommendations aim at presenting a basic scheme of provisions that will allow the uniform development of national and international regulations governing the various modes of transport; yet they remain flexible enough to accommodate any special requirements that might have to be met. Since they were published, the Recommendations have gained global acceptance by being adopted as the basis for most international, regional, national and modal transportation regulations. They are the basis for international modal regulations on the transport of dangerous goods prepared by the International Maritime Organization (IMO), the International Civil Aviation Organization (ICAO). The Recommendations are also used as a basis for the development of regional and national regulations such as European Road and Rail Regulations and U.S. Hazardous Materials Regulations.

---

[20] "Radioactive Materials Transport, The International Safety Regime, An Overview of Safety Regulations and the Organizations Responsible for their Development", World Nuclear Transport Institute, Review Serious No: 1 (Revised July 2006), 4.

[21] Recommendations on the Transport of Dangerous Goods, Model Regulations (2005) 1.

[22] "UN Model Regulations the Transport of Dangerous Goods", <http://hazmat.dot.gov/regs/intl/untdg.htm> (visited 16.10.2006).

The Recommendations enhance safety, improve enforcement capability, ease training requirements and enhance global trade and economic development. Safety is enhanced primarily because harmonized requirements simplify the complexity of the regulations, simplify training efforts and decrease the likelihood of non-compliance. The scope of the Recommendations should ensure their value for all who are directly or indirectly concerned with the transport of dangerous goods. They cover all aspects of transportation necessary for providing international uniformity and include a comprehensive criteria-based classification system for substances that pose a significant danger in transportation. Dangers addressed include explosiveness, flammability, toxicity, corrosiveness for human tissue and metal, reactivity, radioactivity, infectious substance hazards and environmental hazards. They prescribe standards for packaging and multimodal tanks used to transport dangerous materials. They also include a system of communicating the dangers of substances in transport through hazard communication requirements, which cover labelling and marking of packages, placarding of tanks, freight containers and vehicles, and documentation and emergency response information that is required to accompany each shipment. With this system of classification, listing, packing, marking, labelling, placarding and documentation in general use, carriers, consignors and inspecting authorities will benefit from simplified transport, handling and control and from a reduction in time-consuming formalities. In general, their task will be facilitated and obstacles to the international transport of such goods reduced accordingly. At the same time, the advantages will become increasingly evident as trade in goods categorized as "dangerous" steadily grows.

3. The re-formatted UN Recommendations

Although the basic provisions for the safe carriage of dangerous goods given in the UN Recommendations provide a basis for the development of harmonized regulations for all modes of transport, not all the modal requirements were aligned in the past.[23] The different structures of the modal regulations have traditionally required consignors of dangerous goods to be familiar with the various sets of applicable provisions. Throughout the 1980s and 1990s, as multimodal transport became more commonplace, much of the revision work on the various sets of modal dangerous goods rules was centred on ironing out the differences between them and bringing the rules into alignment as much as possible. During the early 1990s it was accepted that one final effort was required to achieve the optimum

---

[23]   The regulations laid down in the RID and ADR regimes have their origins in rules developed a century ago for rail movements of dangerous goods in Europe. Similarly, the sea transport rules included in the first edition of the IMDG Code owed much to the pioneering work done by the United Kingdom in the 1920s and 1930s in developing its Report of the Standing Advisory Committee on the Carriage of Dangerous Goods in Ships, commonly known as the "Blue Book".

degree of rule harmonization and that, as part of this exercise, it would be necessary to re-format the UN Recommendations.[24]

Accordingly, the Recommendations recently have been re-formatted in the
form of "Model Rules." Since many national, regional and modal regulations
governing the transport of dangerous goods are now based on the UN Recommendations; some of the regulations were structured differently, requiring consignors
of dangerous goods to be familiar with the unique structure of all applicable regulations. The lack of the structural harmony of regulations can frustrate compliance
and to the extent that it results in non-compliance is detrimental to safety. Furthermore, a Model Regulation can easily be adopted in the national legislations of
countries throughout the world, eliminating the need for countries to reissue the
regulations in the format of their national regulations.[25]

It is expected that governments, intergovernmental organizations and other international organizations, when revising or developing regulations for which they
are responsible, will conform to the principles laid down in these Model Regulations, thus contributing to worldwide harmonization in this field.[26] Furthermore,
the new structure, format and content should be followed to the greatest extent
possible in order to create a more user-friendly approach, to facilitate the work of
enforcement bodies and to reduce the administrative burden. Although only a
recommendation, the Model Regulations have been drafted in the mandatory
sense, i.e. the word "shall" is employed throughout the text rather than "should",
in order to facilitate direct use of the Model Regulations as a basis for national and
international regulations.

4. Status and future development of the Recommendations

The UN Sub-Committee of Experts on the Transport of Dangerous Goods updates
and amends the UN Recommendations every two years. This biennial revision
cycle allows the Experts to keep the Orange Book up to date with the latest developments in dangerous goods transport. The 14th revised edition, published in
2005, is the version which is currently applicable, while the 15th edition will be
agreed in December 2006 and is scheduled for publishing in 2007.

---

[24] "Radioactive Materials Transport, The International Safety Regime, An Overview of
Safety Regulations and the Organizations Responsible for their Development", World
Nuclear Transport Institute, Review Serious No: 1 (Revised July 2006), 12.

[25] In the past, the process of incorporating amendments to the UN Recommendations was
resource intensive. In the case of international organizations, each change to the
Recommendations was re-evaluated before being introduced into the various international regulations. In some cases the amendments had to be re-proposed by governments participating in these meetings. The fact that each of these issues was re-
discussed, re-worded and re-organized by each of the affected regulatory bodies
increased the likelihood of disharmony. The Model Regulations serve to reduce the
necessity for most of these efforts and in turn enhance harmonization.

[26] Recommendations on the Transport of Dangerous Goods, Model Regulations (2005), 1.

5. Content of the Recommendations

The safety regime governing dangerous-goods transport has been established by taking into account the hazards that these goods present during transport. From the outset it was recognized that some form of generic grouping by physical and/or chemical properties was needed for purposes of identification, packaging, labelling and documentation.[27] Therefore, the UN Committee of Experts developed a nine-class substance identification and classification system based on hazardous properties.[28]

The UN Recommendations only apply to "packaged" dangerous goods, i.e. ranging from small combination packagings, drums and intermediate bulk containers up to and including portable tanks, bulk containers and railway wagons. The Recommendations do not apply to the transport of dangerous goods in bulk in ships or inland waterway vessels, which is the subject of other specialist sets of international maritime regulations, depending on the cargo being carried.[29]

IV. International Maritime Dangerous Goods (IMDG) Code

1. Development of the IMDG Code

Resolution 56, adopted at the 1960 SOLAS Conference, recommended that governments should adopt a uniform international code for the carriage of dangerous goods by sea which should supplement the SOLAS regulations and cover such matters as packing, container traffic and stowage, with particular reference to the segregation of incompatible substances. Furthermore, it recommended that IMO, in co-operation with the CETDG, should pursue its studies on such an international code, especially in respect of classification, description, labelling, a list of dangerous goods and shipping documents. To fulfill this, in January 1961, IMO's Maritime Safety Committee established a working group on the Carriage of Dangerous Goods (CDG). Governments with considerable experience in the carriage of dangerous goods were invited to nominate experts.[30] The Group agreed to prepare the "unified international maritime code" as indicated by the SOLAS Conference.

---

[27]   "Radioactive Materials Transport the International Safety Regime, An Overview of Safety Regulations and the Organizations Responsible for their Development", World Nuclear Transport Institute, Review Series No:1 (Revised July 2006), 10.

[28]   Class 1: Explosives, Class 2: Gases, Class 3: Flammable Liquids, Class 4: Flammable solids, substances liable to spontaneous combustion and substances which in contact with water emit flammable gases, Class 5: Oxidizing substances and organic peroxides, Class 6: Toxic and infectious substances, Class 7: Radioactive materials; Class 8: Corrosive substances and Class 9: Miscellaneous dangerous substances and articles. See *infra* Part 2.

[29]   "Radioactive Materials Transport, The International Safety Regime, An Overview of Safety Regulations and the Organizations Responsible for their Development", World Nuclear Transport Institute, Review Serious No: 1 (Revised July 2006), 8.

[30]   Henry, *The Carriage of Dangerous Goods by Sea* (1985), 100.

Preliminary drafts for each class were compiled by individual national delegations and then considered by the Group, which took into account the practices and procedures of numerous maritime countries in order to make such a code as widely acceptable as possible. Close co-operation was established with the United Nations Committee of Experts on Transport of Dangerous Goods. The Working Group on the Carriage of Dangerous Goods subsequently became the Sub-Committee on the Carriage of Dangerous Goods and was combined with the Sub-Committee on Containers and Cargoes, to become the Sub-Committee on Dangerous Goods, Solid Cargoes and Containers (DSC).

By 1965 the International Maritime Dangerous Goods Code had been prepared and was adopted by the fourth IMO Assembly in the same year. Since its introduction in 1965, the IMDG Code has undergone many changes to keep pace with the ever-changing needs of industry.[31] In 1985, the IMO decided to extend the IMDG Code to marine pollutants. The reason was to assist the implementation of Annex III of the MARPOL 73/78 through the IMDG Code. Moreover, the IMDG Code was re-formatted in 2001 to align with the UN Model Regulations on the Transport of Dangerous Goods as well as the other modal regulations governing the movement of dangerous goods by air, road, rail and inland waterway, which have also recently been harmonized with the UN requirements. Alignment of the various modal rules is seen as an important step in the drive to simplify the carriage of dangerous goods and streamline international multimodal transport operations.

2. Statutes of the IMDG Code

a) Sources of international law

The international community lacks a central law-making authority; thus the creation of new law must be through consensual process. Historically, there are two main sources of international law: customary law and treaties. Customary law evolves over time, becoming universally accepted through continuous practice, whereas treaties take the form of documents signed by governments that agree to be bound by their contents. Treaties and custom constitute "hard law", law that nation states are obliged to follow under pain of sanction by the international legal system and community. Another category of law, in contrast, is termed "soft law".

Soft law is a rapidly developing, though controversial, source of international law.[32] In practice, soft law refers to a great deal of instruments: declarations of principles, codes of practice, recommendations, guidelines, standards, charters,

---

[31] Amendments to the IMDG Code originate from two sources; proposals submitted directly to the IMO by Member States and amendments required to take account of changes to the United Nations Recommendations on the Transport of Dangerous Goods.

[32] Strictly speaking, it is not law at all. Shaw, *International Law* (1997), 92; Gruchulla-Wesierski, "A framework for Understanding 'Soft Law'" [1985] 30 *McGill L.J* 37, 44; Chinkin, "The Challenge of Soft Law: Development and Change in International Law" [1989] 8 *I.C.L.Q* 850.

resolutions etc. Although all these kinds of documents lack legal status, i.e. are not binding, there is a strong expectation that their provisions will be respected and followed by the international community.[33]

b) The formerly recommendatory IMDG Code

The IMO addressed the lack of international uniformity in the field of dangerous goods by producing the IMDG Code. The Code attempted to pull together the many varying customary rules and procedures related to the carriage of dangerous goods at sea. Other transport modes were also consulted. For the first time a uniform system of labelling and presenting other information on the nature of dangerous goods was developed. A comprehensive list of dangerous, hazardous and noxious substances was also created. Underwriters quickly noted the benefits of the Code and shippers stipulated carriage according to the Code.

However, from the legal point of view the IMDG Code was of only limited benefit.[34] IMO Member States are entitled to participate in the adoption of resolutions by the IMO Assembly, the MSC and the MEPC, which recommend the implementation of technical rules and standards not included in IMO treaties. Parties to the UNCLOS are expected to conform to these rules and standards, obviously bearing in mind the need to adapt them to the particular circumstances of each case.[35] Accordingly, the IMDG Code was basically a recommendation to governments for adoption or use as the basis for national regulations in pursuance of their obligations under SOLAS 1974 and MARPOL 73/78.[36] If a state had decided to adopt the Code into its own regulatory system, it could have done so, but that mechanism would then have been applicable only to that state's vessel's and/or ports.

c) The IMDG Code as a part of national legislations

Technical codes and guidelines included in the resolutions are frequently made mandatory by incorporation into national legislation. Many states had incorporated the IMDG Code and given it binding effect.[37] Why have so many states introduced

---

[33] A non-legal commitment is often much easier for a state to accept than a legal one. In all probability, here lies the reason why states do not reject resolutions the terms of which they would by no means accept if they were in a treaty. This presents both an opportunity and a danger. As resolutions also give rise to expectations, they trigger a certain pressure for compliance that is often, as has been shown, effective in the long run. They influence practice, and practice influences law. Bothe, "Legal and Non-Legal Norms: a Meaningful Distinction in International Relations" [1980] (11) *Netherlands Y.B. Int'l L.* 65.

[34] Gold, "Legal Aspects of the Transportation of Dangerous Goods at Sea" [1986] (10) *Mar. Pol.* 185, 186.

[35] "Implications of the United Nations Convention on the Law of the Sea for the International Maritime Organization" (IMO) LEG/MISC/4. 5.

[36] The IMDG Code was adopted as a substantive resolution of the IMO, that is, in the terms of Article 15(j) of the IMO Convention, as a recommendation to members. Henry, *The Carriage of Dangerous Goods by Sea* (1985), 108.

[37] "The Safe Transport of Dangerous, Hazardous and Harmful Cargoes by Sea" [1990] 25 *E.T.L* 747, 784.

the Code into domestic law without being under any legal obligation to do so? Part of the answer lies in the practical value of the IMDG Code. This uncontested value relegated to a secondary position the fact that the Code as such imposed no legal obligations.[38]

The basic reason for adopting the Code as a recommendation was that it was not intended to remain static, as new substances were increasingly being carried by sea. Naturally, a system had to be devised by which changes to the Code could be made as quickly as possible to issue regulations on these new substances or on new methods of packaging.[39] As the Code had overriding practical value, states have adopted its provisions in a form most convenient to them and have avoided constitutional difficulties which usually present themselves when the instrument to be adopted as a treaty.

d) The mandatory IMDG Code

In several cases, codes and guidelines initially contained in non-mandatory IMO resolutions are incorporated at a later stage into the IMO treaties. The IMDG Code is an example of this process.[40] It is no longer recommendatory. The majority of the Code is mandatory. The MSC at its seventy-fifth session agreed to make the IMDG Code mandatory as of 2004 due to the need to provide a mandatory application of the agreed international standards in order to facilitate the multimodal transport of dangerous goods and incorporated the IMDG Code into both SOLAS and MARPOL.[41]

The IMDG Code has, among other developments, undergone significant changes, especially in the context of the language used in the Code. The words "shall", "should" and "may", when used in the Code, mean that the relevant provisions are "mandatory", "recommendatory" and "optional" respectively. Although the Code is legally treated as a mandatory instrument, some chapters remain recommendatory.[42] The IMDG Code is of mandatory application for 155 countries.[43]

Technical rules and standards contained in several IMO treaties can be updated through a procedure of tacit acceptance of amendments. This procedure enables

---

[38]   Henry, *The Carriage of Dangerous Goods by Sea* (1985), 109.

[39]   *Ibid.*

[40]   The International Code for the Construction and Equipment of Ships carrying Dangerous Chemicals in Bulk has been incorporated into both SOLAS and MARPOL. Likewise, the International Code for the Construction and Equipment of Ships Carrying Liquefied Gases in Bulk has also been incorporated into SOLAS.

[41]   Resolution MSC. 122 (75).

[42]   Chapter 1.3 (training); Section 2.1.0 of Chapter 2.1 (Class 1 explosives, introductory notes); Section 2.3.3 of Chapter 2.3 (determination of flash point); columns (15) and (17) of the Dangerous Goods List in Chapter 3.2; Chapter 3.5 (transport schedules for Class 7 radioactive material); Section 5.4.5 of Chapter 5.4 (Multimodal Dangerous Goods Form), insofar as the layout of the form is concerned; Chapter 7.3 (special provisions in the event of an incident and fire precautions involving dangerous goods only); and Appendix B.

[43]   <www.unece.org/trans/doc/2006/ac10c4/UN-SCEGHS-11-inf02e.pdf#search=%22 EU%20ImDG%20Code%22> (visited 30.01.2007).

amendments to enter into force on a date selected by the conference or meeting at which they are adopted, unless within a certain period of time after adoption they are explicitly rejected by a specified number of Contracting Parties representing a certain percentage of the gross tonnage of the world's merchant fleet.

3. Application and implementation of the IMDG

The provisions of the IMDG Code are applicable to all ships to which SOLAS 1974 applies and which are carrying dangerous goods as defined in Chap. VII.1 of that Convention.[44] The provisions of Reg. II-2/19 apply to passenger ships and to cargo ships constructed on or after 1 July 2002. For:

- a passenger ship constructed on or after 1 September 1984 but before 1 July 2002
- a cargo ship of 500 gross tons or over constructed on or after 1 September 1984 but before 1 July 2002
- a cargo ship of less than 500 gross tons constructed on or after 1 February 1992 but before 1 July 2002

the requirements of Regulations II-2/54 of SOLAS apply. For cargo ships of less than 500 gross tones constructed on or after 1 September 1984 and before 1 February 1992, it is recommended that contracting governments extend such application to these cargo ships as far as possible. On the other hand, all ships, irrespective of type and size, carrying substances, materials or articles identified in this Code as marine pollutants are subject to the provisions of this Code.

Although primarily designed for mariners, the provisions of the IMDG Code affect a number of industries as well as storage, handling and transport services from manufacturers to consumers.[45] Chemical and packaging manufacturers, packers, shippers, forwarders, carriers and terminal operators are guided by its provisions on classification, terminology, identification, packing and packaging, marking, labelling and placarding, documentation and marine pollution aspects.[46] Feeder services, such as road, harbour and inland water craft are guided by its provisions. Port authorities, as well as terminal and warehousing companies, consult the IMDG Code to segregate and separate dangerous cargoes in loading, discharge and storage areas.

---

[44]   IMDG Code 1.1.1.
[45]   Wardelmann, "Transport by Sea of Dangerous, Hazardous, Harmful and Waste Cargoes" [1991] 26 *E.T.L* 116, 118.
[46]   *Ibid.* at 120; "Focus on IMO, IMO and Dangerous Goods at Sea", <www.imo.org/ includes/blastDataOnly.asp/data_id%3D7999/IMDGdangerousgoodsfocus1997.pdf>, 5 (visited 16.10.2006).

C. Other instruments

I. The IBC Code

The transportation by sea of liquid chemicals in bulk has developed in line with the increasing number of by-products from petroleum refineries. The carriage of chemicals in bulk is covered by regulations in SOLAS Chapter VII, Carriage of Dangerous Goods, and MARPOL Annex II, Regulation for the Control of Pollution by Noxious Liquid Substances in Bulk.[47] Both Conventions require chemical tankers built after 1 July 1986 to comply with the International Code for the Construction and Equipment of Ships Carrying Dangerous Chemicals in Bulk (IBC Code). Chemical tankers built before 1 July 1986 should comply with the requirements of the Code for the Construction and Equipment of Ships Carrying Dangerous Chemicals in Bulk (BCH Code), the predecessor of the IBC Code.

The purpose of the Code is to provide an international standard for the safe transport by sea in bulk of liquid dangerous chemicals, by prescribing the design and construction standards of ships regardless of tonnage involved in such transport and the equipment they should carry so as to minimize the risks to the ship, its crew and to the environment, having regard to the nature of the products carried. The Code primarily deals with ship design and equipment. In order to ensure the safe transport of the products, the total system must be appraised.

The layout of the IBC Code is in line with the International Code for the Construction and Equipment of Ships Carrying Liquefied Gases in Bulk. Gas carriers may also carry in bulk liquid chemicals covered by the IBC Code. The IBC Code is mandatory both under Annex II of MARPOL 73/78 and Chapter VII, Part B of SOLAS 1974.

II. The IGC Code

Severe collisions or strandings could lead to cargo tank damage and uncontrolled release of the product. Such release could result in evaporation and dispersion of the product and, in some cases, could cause fracture of the ship's hull. The requirements in the code are intended to minimize these risks as far as is practicable, based upon present knowledge and technology.

The Code applies to gas carriers constructed on or after the entry into force of Chap.VII.C of SOLAS 1974 contained in the 1983 amendments to the 1974 SOLAS Convention, i.e. 1 July 1986. Gas carriers constructed before that date should comply with requirements of the Code for the Construction and Equipment of Ships Carrying Liquefied Gases in Bulk (IGC Code) or the Code for Existing Ships Carrying Liquefied Gases in Bulk. The purpose of these codes is to provide

---

[47] This is important, because the Annex itself is concerned only with discharge procedures. The Code was revised to take into account anti-pollution requirements and therefore make the amended Annex more effective in reducing accidental pollution.

an international standard for the safe transport by sea in bulk of liquefied gases and certain other substances. The IGC Code is also mandatory under Annex II of MARPOL and Chap.VII.C of SOLAS 1974.

III. The INF Code

The International Code for the Safe Carriage of Packaged Irradiated Nuclear Fuel, Plutonium and High-Level Radioactive Water on Board of Ships (INF Code) was adopted by Resolution MSC.88 (71) in 1999 and became mandatory on 1 January 2001 by amendments adapted to Chapter VII of SOLAS. Complementing the IMDG Code and IAEA Regulations, the INF Code sets standards above those set by the SOLAS Convention for conventional ships. These enhanced standards apply to the design and operation of ships carrying materials included in the Code. The materials covered by the Code include irradiated nuclear fuel, plutonium and high-level radioactive waste. Specific regulations in the Code cover a number of issues, including damage stability, fire protection, temperature control of cargo spaces, structural considerations, cargo-securing arrangements, electrical supplies, radiological protection equipment and management, training and shipboard emergency plans.

IV. The BC Code

Millions of tonnes of cargoes, such as coal, concentrates, grains, fertilizers, animal foodstuff, minerals and ores, are shipped in bulk by sea every year. While the vast majority of these shipments are made without incident, there have been a number of serious accidents which have resulted not only in the loss of the ship, but also in loss of life.[48]

The problems involved in the transport of solid bulk cargoes were recognized by the 1960 SOLAS Conference, but at that time it was not possible to frame detailed requirements except for the transport of grain cargoes. However, it was agreed that an internationally acceptable Code of Safe Practice for the Shipment of Bulk Cargoes should be drawn up and in 1965 the first Code of Safe Practice for Bulk Cargoes (BC Code) was adopted by the IMO.

The BC Code itself provides guidance to administrations, shipowners, shippers and masters on the standards to be applied in the safe stowage and shipment of solid bulk cargoes excluding grain, which is dealt with under separate rules. The BC Code includes practical guidance on the procedures to be followed and the appropriate precautions to be taken in the loading, trimming, carriage and discharge of bulk cargoes. The primary aim of the BC Code is to promote safe stowage and shipment by:

---

[48]   "Focus on IMO, IMO and the Safety of Bulk Carriers" <www.imo.org/includes/ blastDataOnly.asp/data_id%3D7987/BULK99.FIN.pdf>, 1. (visited 16.10.2006).

(1) highlighting the dangers associated with the shipment of certain types of bulk cargoes;

(2) giving guidance on the procedures to be adopted when the shipment of bulk cargoes is contemplated;

(3) listing typical materials currently shipped in bulk, together with advice on their properties, handling etc.; and

(4) describing test procedures to be employed to determine various characteristics of the materials to be carried.

The list of solid bulk cargoes appearing in the BC Code is by no means exhaustive, and the physical or chemical properties attributed to them are intended only for guidance. Therefore, before loading any bulk cargo it is essential to obtain[49] information on the physical and chemical properties of the cargo. The master has to be provided with loading information sufficiently comprehensive to enable him to arrange the loading of his ship so as not to overstrain the structure and to calculate the stability of the ship for the worst conditions anticipated during the voyage.[50] Other information to assist persons responsible for the loading and unloading of solid bulk cargoes is contained in recommendations published by the Organization.[51] The MSC of the IMO has developed a form for cargo information advice on duties of chief mates and officers of the watch at loading and discharging ports, a cargo operations form and advice on safe practices on board bulk carriers.[52] The BC code also lists certain general precautions.[53]

Currently, the practices contained in the BC Code are intended as recommendations to governments, ship operators and masters, and bring the attention of those concerned to internationally accepted methods of dealing with hazards which may be encountered when carrying a cargo in bulk; the BC Code is expected to become mandatory by 2011.

V. The International Grain Code

Originally grain was transported in sacks, but by the middle of the 20th century the normal procedure was to carry it in bulk. It could be stored, loaded and unloaded easily and the time taken to deliver it from producer to customer was greatly reduced, as were the costs involved. However, grain has one great hazard when carried in bulk. It tends to shift within the cargo space of the ship. Because

---

[49] Normally from the shipper. See *infra* Part 3.

[50] Since valuable information leading to improvements in this Code may come from voyage reports, it is recommended that the master notifies his administration of the behaviour of various types of solid bulk cargoes and, in particular, reports any incidents involving such cargoes.

[51] The Code of Practice for the Safe Loading and Unloading of Bulk Carriers, Resolution A.862(20).

[52] See MSC/Cir.663, 665-667.

[53] Such as the need to protect machinery and the interior of the ship from dust and to ensure that bilges and service lines are in order and not damaged during loading.

of this danger and the great amount of grain transported by sea, special rules governing its carriage in bulk have appeared in various international instruments.

Grain has a tendency to settle during the course of the voyage, as air is forced out when the individual grains sink ("sinkage"). This leads to a gap developing between the top of the cargo and the hatch cover. This in turn enables the cargo to move from side to side as the ship rolls and pitches. This movement causes the ship to list and, although initially the ship's movement will tend to right this, the list can eventually become more severe. In worst cases, the ship can capsize. This problem was well known and when the IMO came into being, one of its first tasks was to consider new measures for improving the safety of bulk carriers. These were incorporated into SOLAS 1960. However, new regulations had some deficiencies as far as safety was concerned, for during a period of four years, six ships loaded under the 1960 rules were lost at sea. The IMO began looking at this problem early in 1963 and asked masters of ships to contribute information to a broad study. Further studies and tests showed that some of the principles on which the 1960 regulations were based were invalid; in particular, it was shown that the 1960 Convention had underestimated the amount of sinkage. As a result, the IMO Assembly in 1969 adopted new grain regulations in Resolution A.184 (VI), which became generally known as the 1969 Equivalent Grain Regulations.

Voyage experience over a three-year period showed that the 1969 Grain Equivalents were not only safer but were also more practical and economical than the 1960 regulations and, being based upon operational experience, they were used as the basis for new international requirements which were subsequently incorporated into the SOLAS Convention.[54]

The International Code for the Safe Carriage of Grain is designed to prevent the particular qualities of grain threatening the stability of ships when carried in bulk. Part A contains special requirements and gives guidance on the stowage of grain and the use of grain fittings. Part B deals with the calculation of heeling moments and general assumptions. Part C of the revised SOLAS Chapter VI, Regulation 9 stipulates the requirements for cargo ships carrying grain.[55]

VI. The radioactive materials regulations

Millions of radioactive materials are sent around the world each year. Each shipment consists of either a single package or a number of packages sent from one location to another at the same time. Some shipments are made for the nuclear fuel

---

[54]   SOLAS Chap. VI.C.

[55]   Regulation 9.1 states that a cargo ship carrying grain shall comply with the requirements of the International Grain Code and shall hold a document of authorization as required by that Code. Furthermore, according to Regulation 9.2, a ship without a document of authorization shall not load grain until the master satisfies the contracting government of the port of loading that the ship complies with the requirements of the Code.

cycle industry, but the vast majority of shipments are for medical,[56] agricultural[57] and industrial[58] use. All transport modes are used. The shipments range from small quantities of radioactive substances for hospitals, which may go through the ordinary postal system, to flasks of highly radioactive spent fuel from nuclear power stations, which travel mainly by rail, ships or road. The transport of radioactive materials has a long history dating back several decades before the advent of nuclear power. Over the years, a comprehensive and stringent regulatory framework has been developed at both international and national levels to ensure safety.

The International Atomic Energy Agency (IAEA), a UN body, was established as an autonomous organization in 1957. The IAEA serves as an intergovernmental forum for scientific and technical co-operation in the peaceful application of nuclear technology, provides international safeguards against its misuse and facilitates the application of safety measures in its use. In 1961, the IAEA published advisory regulations for the safe transport of radioactive materials. The Regulations for the Safe Transport of Radioactive Materials have become recognized throughout the world. The Regulations have been regularly reviewed since their publication to reflect the changes in requirements to the basic safety principles; to protect the general public and the environment; to reflect technological advances; and to reflect lessons learned from regulatory and operational experience. The latest version of the Regulations was published in 1996 and entered into force on 1 January 2002. The IMDG Code includes IAEA standards for Class 7 radioactive materials.

The transport of radioactive materials has an outstanding safety record. Indeed, in terms of its safety record and the strict liability imposed by the international system of regulation under which it operates, the transport of nuclear materials could be regarded as a model for the transport of other classes of dangerous goods.

D. Dangerous goods provisions in the carriage conventions

SOLAS 1974 Convention has both public and private law characteristics and its codes provisions are rather technical. Besides, there are provisions pertaining to dangerous goods which belong exclusively to the realm of the private international law as being primarily commercial.

---

[56] For instance, radioactive chemical tracers provide diagnostic anatomical information and radiotherapy employs radioisotopes in the treatment of illnesses such as cancer, while more powerful gamma sources are used to sterilize equipment and bandages.

[57] Radioisotopes have an important role to play in the growing of crops and the breeding of livestock, as well as in the preservation of food and eradication of disease.

[58] There are a variety of uses, ranging from the use of radioisotopes in steel plants and paper mills to the use of radioactive tracing techniques to check the grounding which fixes offshore oil platforms to the sea bed and the use of radioactive materials in consumer products such as some detectors, luminous watches and emergency exit signs.

I. The Hague/Hague-Visby Rules

Article IV.6 of the Hague/Hague-Visby Rules specifically provides that:

> Goods of an inflammable, explosive or dangerous nature to the shipment whereof the carrier, master or agent of the carrier has not consented with knowledge of their nature and character may at any time before discharge be landed at any place or destroyed or rendered innocuous by the carrier without compensation and the shipper of such goods shall be liable for all damages and expenses directly or indirectly arising out of or resulting from such shipment.
>
> If any such goods shipped with such knowledge and consent shall become a danger to the ship or cargo, they may in like manner be landed at any place, or destroyed or rendered innocuous by the carrier without liability on the part of the carrier, except general average, if any.

However, the provision neither gives a definition of dangerous goods nor refers to any related instrument to help resolve any uncertainty. Therefore, it has been the subject matter of many controversies examined in Part II, III and IV.

II. The Hamburg Rules

Art. 13 of the Hamburg Rules pertains to dangerous goods. The Hamburg Rules make three useful technical changes in respect of dangerous goods and then in different language confirm the meaning of Art. IV.6 of the Hague/Hague-Visby Rules.[59] Art. 13 of Hamburg Rules provides as follows:

1. The shipper must mark or label in a suitable manner dangerous goods as dangerous.
2. Where the shipper hands over dangerous goods to the carrier or an actual carrier, as the case may be, the shipper must inform him of the dangerous character of the goods and, if necessary, of the precautions to be taken. If the shipper fails to do so and such carrier or actual carrier does not otherwise have knowledge of their dangerous character:
   a) The shipper is liable to the carrier and any actual carrier for the loss resulting from the shipment of such goods, and
   b) The goods may at any time be unloaded, destroyed or rendered innocuous, as the circumstances may require, without payment of compensation.
3. The provisions of paragraph 2 of this article may not be invoked by any person if during the carriage he has taken the goods in his charge with knowledge of their dangerous character.
4. If, in cases where the provisions of paragraph 2, subparagraph (b) of this article do not apply or may not be invoked, dangerous goods become an actual danger to life or property, they may be unloaded, destroyed or rendered innocuous, as the circumstances may require, without payment of compensation except where there is an obligation to contribute in general average or where the carrier is liable in accordance with the provisions of article 5.

---

[59] Tetley, "Articles 9 to 13 of the Hamburg Rules", in Mankabady (ed.) *The Hamburg Rules on the Carriage of Goods by Sea* (1978), 197, 202 f.

III. The CMI/ UNCITRAL draft instrument

Over the past few years, Committee Maritime International (CMI) was, at the request of United Nations Commission on International Trade Law (UNCITRAL), engaged in the preparation of a draft of a new convention on the contract of carriage by sea. It is UNCITRAL's intention to create, on the basis of this draft, a successor to the Hague Rules, Hague-Visby Rules and Hamburg Rules.[60] In December 2001, the CMI delivered to UNCITRAL a draft instrument on Transport Law.[61]

In the draft, the distinction between dangerous goods and other goods was removed. Such a distinction was considered as out of date, because the notion "danger" has a more relative nature today.[62] Dangerous cargo that is packed according to all applicable rules and is handled by the carrier in the manner appropriate for that kind of cargo is not inherently more dangerous than ordinary cargo. And ordinary cargo, under certain circumstances, may become a danger to humans or to the environment.

However, ongoing discussions revealed the need for a specific provision in the draft instrument. Accordingly, a provision dealing with dangerous goods was inserted in the draft dealing specifically with dangerous goods.[63]

E. The HNS Convention

The limited regulatory measures which existed were principally aimed at preventing the "danger" from causing loss of vessel, life, goods and personal injury. There was almost no concern about the "danger" affecting third parties. The transportation of dangerous goods at sea considered technical matters best left to those

---

[60] For more information on the draft, see, Debattista, "The CMI/UNCITRAL Cargo Liability Regime: Regulation for the 21st Century?" [2002] *LMCLQ* 304-305; Beare, "Liability Regimes: Where We are, How We got there and Where We are Going" [2002] *LMCLQ* 306-315; Roseg, "The Applicability of Conventions for the Carriage of Goods and for Multimodal Transport" [2002] *LMCLQ* 316-335; Berlingieri, "Basis of Liability and Exclusions of Liability" [2002] *LMCLQ* 336-349; Zunarelli, "The liability of the Shipper" [2002] *LMCLQ* 350-355; Clark, "Transport Documents: Their Transferability as Documents of Title; Electronic Documents" [2002] *LMCLQ* 356-369; Huybrechts, "Limitation of Liability of Actions" [2002] *LMCLQ* 370-381; Asariotis, "Allocation of Liability and Burden of Proof in the Draft Instrument on Transport Law" [2002] *LMCLQ* 382-398; Alcantara, "The New Regime and Multimodal Transport" [2002] *LMCLQ* 399-4004; Herber, "Jurisdiction and Arbitration – Should the New Convention Contain Rules on these Subjects?" [2002] LMCLQ 405-417; van der Ziel, "The UNCITRAL/CMI Draft for a New Convention Relating to the Contract of Carriage by Sea" [2002] *TranspR* 265-277.

[61] Printed in CMI Yearbook 2001 and available at <www.uncitral.org>.

[62] Van der Ziel, "The UNCITRAL/CMI Draft for a New Convention Relating to the Contract of Carriage by Sea" [2002] *TranspR* 265, 272.

[63] See *infra* Part 2.

who shipped, loaded, carried and discharged such goods. For a long time, oil and oil products were also basically regulated within the maritime industry. It was only the advent of maritime accidents which made oil pollution a catalyst of more regulation. The analogy to the carriage of dangerous goods should be apparent. From the legal point of view, the regulatory machinery only becomes necessary when the industrial safeguards do not provide adequate protection for third parties or if damage to the marine environment might result.[64]

In 1977, the Legal Committee of the IMO began work on drafting a uniform liability and compensation scheme in response to the increasing amount of hazardous substances carried by vessels and the resulting concerns of coastal states around the world. In 1984, a diplomatic conference was held to consider and adopt a treaty related to liability and compensation for damage arising out of accidents that occur during the shipment of hazardous cargoes. Among treaties considered at the conference was a proposed HNS Convention. However, agreement could not be reached as the issue proved to be too complex.

Because of the heavy workload of the Legal Committee, it was not until 1996 that the matter could be considered again, but this time the attempt was successful. The International Convention on Liability and Compensation for Damage in Connection with the Carriage of Hazardous and Noxious Substances by Sea (HNS Convention) was adopted on 3 May 1996. The HNS Convention will make it possible for up to 250 million SDR to be paid out to victims of disasters involving HNS, such as chemicals.[65]

F. Dangerous goods regulations for other modes of transport

The purpose of the UN Recommendations on the Transport of Dangerous Goods is to present standards for the transport of all types of dangerous goods and for all modes of transport that allows world-wide uniformity. Accordingly there are dangerous goods regulations for all modes transport being based on the UN Recommendations.

I. IATA Dangerous Goods Regulations

The International Air Transport Association (IATA) recognized in the early 1950s that there was a need to standardize the rules governing the transport of dangerous goods by air. These rules would provide not only for the safe, efficient transportation of these materials but would also help identify and stop undeclared and other potentially hazardous shipments from being carried. Consequently, a team of

---

[64]   Gold, "Legal Aspects of the Transportation of Dangerous Goods at Sea" [1986] (10) *Mar. Pol.* 185, 188.

[65]   See *infra* Part VI on the HNS Convention.

airline and technical experts were given the task of producing the IATA Restricted Article Regulations (RAR) and the first set of regulations governing the international transport of dangerous goods was issued in 1956.

While the RAR was used throughout the industry by all main carriers, they were only applicable to IATA members and their adoption and use were voluntary. Nonetheless, over 80 countries adopted the RAR in their national legislation. However, since they could not be effectively enforced, the regulations could only achieve so much in the relatively small air transport industry of that time. In the early eighties, IATA requested the International Civil Aviation Organization (ICAO) to incorporate the RAR in a set of regulations which would be binding on all states involved in civil aviation and members of the Chicago Convention. As a result, the Air Navigation Commission of ICAO developed in 1981 Annex 18 to the Chicago Convention in response to a need for an internationally agreed set of provisions governing the safe transport of dangerous goods by air. The provisions of this Annex are based on the Recommendations of the United States Committee of Experts on the Transport of Dangerous Goods and the Regulations for the Safe Transport of Radioactive Materials of the IAEA. The provisions of Annex 18 are amplified by the ICAO Technical Instructions for the Safe Transport of Dangerous Goods by Air (ICAO TI's), first published in 1983.

However, IATA recognized that the goal of government legislation was to produce a legal document which could be easily used in court to enforce the rules. This is sometimes at odds with clear and plain instructions on how to comply with the legal requirements. While this is not so important when dealing with taxation, the criminal code and so on, it is a serious concern when considering safety issues such as highway rules and dangerous goods regulations. Consequently, IATA decided to continue publishing its Regulations but with a renewed focus on providing the rules in as clear, precise and unambiguous a fashion as possible.

This is indeed the idea of the current IATA Dangerous Goods Regulations, which replace the RAR and are published annually with the latest rules on the transport of dangerous goods by air from states, operators and the ICAO. It constitutes a manual of industry carrier regulations to be followed by all IATA member airlines. While it has no regulatory basis, it incorporates all ICAO requirements and sets higher criteria for member airlines.

II. European Agreement concerning the International Carriage of Dangerous Goods by Road

The European Agreement concerning the International Carriage of Dangerous Goods by Road (ADR) was concluded in 1957 under the auspices of the UNECE and entered into force in 1968.[66] The ADR requirements have been annexed to EU Directive 94/55/EC. Under this initiative, ADR has applied to the domestic trans-

---

[66] The Agreement was amended by the Protocol amending Article 14(3), concluded at New York on 21 August 1975, which entered into force on 19 April 1985. 42 States are party to the ADR.

port by road of dangerous goods both within and between European Member States since 1997.

The Agreement is short and simple. The key article is the second, which says that apart from some excessively dangerous goods, other dangerous goods may be carried internationally in road vehicles subject to compliance with:

- the conditions laid down in Annex X for the goods in question, in particular as regards their packaging and labelling; and
- the conditions laid down in Annex B, in particular as regards the construction, equipment and operation of the vehicle carrying the goods in question.

Annex A and B have been regularly amended and updated since the entry into force of ADR. The most recent set of new amendments entered into force on 1 January 2005 and consequently a third consolidated restructured version was published as document ECE/TRANS/175. The new structure is consistent with that of the UN Model Regulations, the IMDG Code of the International Maritime Organization, and the Technical Instructions for the Safe Transport of Dangerous Goods by Air of the International Civil Aviation Organization and the Regulations concerning the International Carriage of Dangerous Goods by Rail of the Intergovernmental Organization for Carriage by Rail.

III. Regulations concerning International Carriage of Dangerous Goods by Rail

The international carriage of goods by rail is regulated by the Convention concerning International Carriage by Rail of 9 May 1980 (COTIF).[67] This Convention plays a double role in the field of international rail carriage. Firstly, it creates the Intergovernmental Organization for International Carriage by Rail (OTIF).[68] Secondly, it establishes a uniform system of law applicable to the carriage by rail of passengers, luggage and goods in international traffic between member states.

The Uniform Rules applicable to international carriage by rail are contained in the appendices to COTIF: the Uniform Rules concerning the Contract for International Carriage of Passengers and Luggage by Rail (CIV)[69] and the Uniform Rules concerning the Contract for International Carriage of Goods by Rail (CIM).[70] The "Regulations concerning the International Carriage of Dangerous Goods by Rail (RID)[71] form Annex I of CIM.[72]

---

[67] Convention relative aux Transports Internationaux Ferroviaires.

[68] Organisation Intergouvernementale pour les Transports Internationaux.

[69] Convention Internationale concernant le Transport des Voyageurs et des Bagages par Chemins de Fer.

[70] Convention Internationale concernant le Transport de Marchandises par Chemins de Fer.

[71] Règlement concernant le Transport International Ferroviaire des Marchandises Dangereuses.

[72] Other annexes: Annex II: Regulations concerning the International haulage of Private Owners' Wagons by Rail (RIP), Annex III: Regulations concerning the International

The development of the regulations concerning the carriage of dangerous goods by rail is a main task of OTIF. RID has been revised both technically and legally. The new version of RID is more user-friendly and almost the same as ADR and ADN with the exception of mode-specific parts. The content and structure of the provisions were harmonized, to the greatest extent possible, with the UN Recommendations on the Transport of Dangerous Goods. RID applies to the international transport of dangerous goods by rail between the 42 COTIF signatory states. In addition, the RID requirements have been annexed to EU Directive 96/49/EC, so that they also apply to the rail transport of dangerous goods, including radioactive materials, within and between EU member states.

IV. European Agreement concerning the International Carriage of Dangerous Goods by Inland Waterways

The European Agreement concerning the International Carriage of Dangerous Goods by Inland Waterways (ADN) was adopted on 25 May 2000 on the occasion of the Diplomatic Conference organized jointly by the UNECE and the Central Commission for the Navigation of the Rhine (CCNR).[73] The ADN consists of a main legal text and Regulations annexed thereto.

The annexed Regulations contain provisions concerning dangerous substances and articles, provisions concerning their carriage in package and in bulk on board inland navigation vessels and tank vessels, as well as provisions concerning the construction and operation of such vessels. They also address requirements and procedures for inspections, issue of certificates of approval, recognition of classification societies, monitoring, as well as training and examination of experts.

The ADN Agreement is based on the UN classification system for dangerous goods and has been re-formatted to reflect the layout of the UN Model Regulations, in much the same way as has already been done with the other modal dangerous goods transport regulations.

These regulations, at the time of adoption of the Agreement, are the same as those applicable on the Rhine,[74] and the Agreement is therefore intended to set up the same high level of safety on the entire European inland-waterways network.

V. Convention on Civil Liability for Damage Caused during Carriage of Dangerous Goods by Road, Rail and Inland Navigation Vessels

The Convention on Civil Liability for Damage Caused during Carriage of Dangerous Goods by Road, Rail and Inland Navigation Vessels (CRTD) was prepared by the International Institute for the Unification of Private Law (UNIDROIT) and

---

Carriage of Containers by Rail (RICo), and Annex IV: Regulations concerning the International Carriage of Express Parcels by Rail (RIEx).

[73]   ADN is not in force yet.

[74]   The Regulations of the Carriage of Dangerous Goods in the Rhine (ADNR).

adopted by the Inland Transport Committee of the Economic Commission in 1989. The Convention was opened for signature on 1 February 1990, but has been signed only by Germany and Morocco.[75]

At the request of the Inland Transport Committee, the secretariat sent a questionnaire to all heads of delegations to the Committee with a view to identifying what difficulties would prevent accession to the CRTD. Several delegations replied to the questionnaire.[76]

At its seventy-first session, the Working Party on the Transport of Dangerous Goods, having considered the conclusions of the *ad hoc* group of experts on the basis of the questionnaire on the CRTD Convention, recommended that the Inland Transport Committee call an *ad hoc* meeting of experts on the CRTD with the following mandate:

- to consult experts from all sectors concerned by the CRTD (e.g. legal experts on liability, insurance industry, consignors, carriers) in order to determine how to eliminate the obstacles – such as those related to the limiting of liability and compulsory insurance – to the entry into force of the CRTD;
- to propose, on the basis of these consultations and of proposals by governments, modifications to the existing articles of the CRTD which would provide a better basis for application of the CRTD to the various modes of transport;
- to report to the Inland Transport Committee on progress made and difficulties encountered; and
- to submit to the Inland Transport Committee a reviewed text of the CRTD incorporating the modifications mentioned above for possible adoption as a new convention.

G. Regional and national regulations concerning dangerous goods

I. European Union

Council Directive 67/548/EEC of 27 June 1967 on the approximation harmonization of laws, regulations and administrative provisions relating to the classification, packaging and labelling of dangerous substances is the main source of European Union law concerning chemical safety. The Directive applies to pure chemicals and to mixtures of chemicals which are placed on the market in the European Union. However, the Directive does not apply to the transport of dangerous substances and preparations.

---

[75] For a review of the CRTD, see Evans, "Damage from Goods-From Haesselby to Geneva", in Grönfors (ed.) *Festskrift till Kurt Grönfors* (1991), 125 ff.

[76] Answers to the questionnaire available at <www.unece.org/trans/danger/publi/crtd/crtddoce.html>.

The European Union (EU) maritime safety policy aims at eradicating substandard shipping essentially through a convergent application of internationally agreed rules. Although at Community level a few legislative decisions were taken in the period 1978-1992, the real start of the maritime safety policy occurred only in 1993 through the adoption by the Commission of its first communication dealing with maritime safety: "A Common Policy on Safe Seas".[77] Catalysts for this breakthrough were the accidents with the oil tankers "Aegean Sea"[78] and "Brear"[79] together with the abandoning of the unanimity rule for the maritime decision-making process on 1 January 1993. However, new tragedies that occurred in European waters initiated additional actions focusing on specific shortcomings. After the "Estonia" tragedy,[80] the Community adopted a comprehensive set of rules for the protection of passengers and crew sailing on ferries operating to and from European ports, as well safety standards for passenger ships operating on domestic voyages within the Community. The "Erika"[81] and the "Prestige"[82] accidents obliged the Community to revise its existing rules and to adopt new rules for the prevention of accidents with oil tankers.

The main Directives adopted in this period aim to ensure the implementation of international safety rules by all ships visiting European ports and to ensure that ships flying a flag of an EU member state and their crews comply with the international standards. Special attention has also been given to a better protection of European coasts. Of these hazardous materials (HAZMAT) Directive 2002/59/EC[83] on Establishing Community Vessel Traffic Monitoring and Information System was adopted as part of the Erika II Package and repealed Directive 93/75/EC.[84]

---

[77] In its 1993 Communication COM(93) 66 final, 24.2.1993, the Commission analyses the maritime safety situation in Europe and outlines a framework for a common maritime safety policy based on four pillars: (1) convergent implementation of existing global international rules, (2) uniform enforcement of global international rules by the port states, (3) development of navigational aids and traffic surveillance in infrastructures and (4) reinforcement of the EU's role as a driving force for global international rule-making.

[78] The vessel broke up, exploded and caused more than 70,000 tons of oil to spill in Spain on 3 December 1992.

[79] It went aground off the Shetland Islands on 5 January 1993 and spilled 26 million gallons of oil.

[80] A ro-ro passenger ferry which sank on 28 September 1994.

[81] The 25-year-old single-hull tanker 'Erika' broke in two on 12 December 1999. More than 10,000 tons of heavy fuel oil spilled into the Bay of Biscay and polluted more than 400 km of coastline.

[82] The tanker 'Prestige' broke up on 13 November 2002 and spilled more than 11,000 tons of oil on the Galician coast.

[83] OJ L 208.

[84] The system established by Directive 93/75/EC was not very clear to the rest of the world and was often not fully understood by masters and shipowners, particularly those from outside the European Union. The requirement that the operator of a ship coming from a port loaded outside the Community must give notification of departure was not correctly applied, in many cases notification having been given after the ship's

The purpose of the new Directive is to establish in the Community a vessel traffic monitoring and information system with a view to enhancing the safety and efficiency of maritime traffic, improving the response of authorities to incidents, accidents or potentially dangerous situations at sea, including search and rescue operations, and contributing to a better prevention and detection of pollution by ships.[85] Member states shall monitor and take all necessary and appropriate measures to ensure that the masters, operators or agents of ships, as well as shippers or owners of dangerous or polluting goods carried on board comply with the requirements under the Directive.

Directive 2002/59/EC requires prior notification to be given before entering European ports, compliance with the mandatory reporting system set up by member states and approved by the IMO, as well as the use of the vessel traffic services and ship's routing system approved by the IMO. These measures should ensure that, whether or not they carry dangerous or polluting goods, all ships covered by the Directive which enter European waters or use European ports will be identified, will observe the traffic rules in force and will provide the coastal authorities with information that will be important in the event of an emergency at sea.

Furthermore, the Directive places an obligation on the shipper to notify dangerous or polluting goods prior to shipment.[86] The operator, agent or master of a ship carrying dangerous or polluting goods coming from a port located outside the

---

departure and in some cases having been addressed simply to the port of destination according to the local rules of that port (e.g. 48 or 24 hours in advance). Similarly, it was often the case that notification from a port of departure inside the Community was given late. Furthermore, there were some problems in transmitting information on the cargo, notably because the means of communication were inappropriate. Since the Directive laid down no procedures or standard format for the transmission and exchange of data, this could prevent the proper and efficient working of the whole system. Moreover, there was no clear definition of competent authorities, and as a result responsibilities were not clearly established. The list of competent authorities was not updated regularly and was not easily available. In most cases, information notified under the Directive 93/75/EC was merely stored by the recipients for use in the event of an accident, but otherwise remained unused. Thus the information collected was not put to any good use. For these reasons it was considered that the instruments put in place to cope with accidents or pollution caused by substandard shipping in European waters were still inadequate. As the aims of the Commission were too broad to be covered by a simple amendment to Directive 93/75/EC, a new instrument which incorporated the objectives of Directive 93/75/EC but also covered much broader objectives was needed.

[85] Directive 2002/59/EC Art. 1.

[86] Art. 12.Directive defines dangerous and polluting goods by referring to international regulations. Accordingly "dangerous goods" means goods classified in the IMDG Code, dangerous liquid substances listed in Chapter 17 of the IBC Code, liquefied gases listed in Chapter 19 of the IGC Code, solids referred to in Appendix B of the BC Code. (Art. 3g). "Polluting goods" are oil as defined in Annex I to the MARPOL Convention noxious liquid substances as defined in Annex II to the MARPOL Convention, and harmful substances as defined in Annex III to the MARPOL Convention. Art. 3(h).

Community is also obliged to notify that fact upon departure from the loading port.

Directive 2002/59/EC includes the basic provisions of Directive 93/75/EC but changed them to take into account experience with the Directive and the legal and technological developments at international level. One of the basic difficulties of Directive 93/75/EC was that of managing what may be huge amounts of information on cargoes efficiently and cheaply. To this end, Directive 2002/59/EC is designed to keep information on cargoes as close as possible to the source of that information, so that the information is used only in exceptional circumstances. The Directive also harmonizes methods of transmitting information by requiring ships to notify data electronically, specifically by EDI.[87]

II. Germany

In Germany, dangerous goods legislation is based on international regulations and European legislation. While conventions like SOLAS or EU regulations are directly binding, recommendations or directives of the EU must be implemented in national law to have effect. The principal law is the Gefahrgutbeförderungsgesetz[88] (GGBefG) dated 6 August 1975 amended subsequently. However, the Act does not contain direct rules on the transport of dangerous goods. The Act provides a basis for dangerous goods regulations for all modes of transport.

The transport of dangerous goods by seagoing ships is regulated by the Verordnung über die Beförderung gefährlicher Güter mit Seeschiffen[89] (Gefahrgutverordnung See- GGVSee) The Regulations mainly incorporates the SOLAS Convention and the IMDG Code as well as all other IMO Codes.[90] There are enforcement provisions involving fines in the regulations.[91]

III. United Kingdom

The main enabling statutes are the Merchant Shipping Acts, which lay down the safety regulations for the safe design, construction and operation of the ship, qualifications of the master and the crew and various other matters including the carriage of dangerous goods. The regulations made under these Acts give effect to obligations arising from several international conventions.

---

[87] In the past, operators could use any means to give notification. Cargo information was often transmitted by fax, particularly to smaller ports, which obliged the authorities concerned to store large volumes of paper documents on their premises.

[88] The Carriage of Dangerous Goods Act.

[89] The Regulation on the Carriage of Dangerous Goods by Seagoing Ships.

[90] GGVSee § 2.

[91] GGVSee § 10 and GGBefG § 10.1.1. A minimum penalty €1,000 and a maximum assessment of € 50,000.

The main provisions governing the carriage of dangerous goods in UK-registered ships and other ships loading and unloading in UK ports are contained in the Merchant Shipping (Dangerous Goods and Marine Pollutants) Regulations 1997.[92] These require the provision and maintenance of equipment and other arrangements for ensuring health and safety on board in connection with the stowage, segregation, handling and transport of such goods. Duties are imposed also on the consignor of the goods, the shipowner or employer, master and employee. There are also requirements for the proper classification of the cargoes; where necessary, for their packaging, marking, labelling and documentation to ensure proper stowage and segregation; and for known marine pollutants to be identified.

The regulations do not include detailed prescriptions, but require compliance with specific codes or recommendations, such as the IBC Code or the IMDG Code. UK-registered chemical tankers, gas carriers and other ships carrying noxious liquids in bulk must be surveyed and issued with a certificate of fitness.[93] Legislation to implement the reporting requirements of EC Directive 2002/59/EC on vessels carrying dangerous goods entering and leaving Community ports is also necessary. There are enforcement provisions, involving fines and imprisonment, in the regulations.[94] The "competent authority" who is responsible for the safe transport of dangerous goods by sea is the Maritime and Coastal Guard Agency (MCA).

IV. United States of America

In the United States, the Federal Hazardous Materials Transportation Law 49 U.S.C § 5101 is the basic statute regulating hazardous materials transportation. The purpose of the law is to provide adequate protection against the risks to life and property in transporting hazardous materials in commerce by improving the regulatory and enforcement authority of the Secretary of Transportation. Hazardous Materials Regulations (HMR; 49 CFR Parts 171-180) cover five areas:

- hazardous materials definition/classification (Part 172, Sub-parts A-B, Part 173),
- hazard communication (Part 172, Sub-parts C-G),
- packaging requirements (Parts 173, 178-180),
- operational rules (Parts 171, 173-177) and
- training (Part 172, Sub-part H).

---

[92] 1997 No: 2367.
[93] The "International Pollution Prevention (INLS) Certificate" is valid for five years. During the period of validity, the ship has to undergo annual and intermediate surveys, to ensure that it remains in a satisfactory condition.
[94] Art. 24.

The HMR are issued by the Research and Special Programs Administration (RSPA).[95] The HMR apply to interstate, intrastate and foreign commerce and to transportation in commerce by aircraft, railcars, vessels and by motor vehicles operated by interstate carriers.[96] The HMR apply to persons who:

– offer hazardous materials for interstate, foreign, and interstate transportation in commerce (shippers),
– transport hazardous materials in commerce (common, contract and private carriers),
– offer or transport in commerce by any mode or carrier a hazardous waste, hazardous substance, flammable cryogenic liquid, or marine pollutant or
– manufacture, mark, maintain, recondition, repair or test packaging and their components which are represented as qualified for use for hazardous materials.

In the field of maritime transportation, all regulations applicable to vessel carriers and shippers by water are enforced by the United States Coast Guard (USCG). Under the Comprehensive Environmental Response, Compensation and Liability Act,[97] the USCG has been delegated authority, as the designated on-scene coordinator, to respond to the release of hazardous substances into the environment within the U.S. coastal zone. USCG regulatory functions for hazardous materials include bulk transport by vessel. Under delegation from the Secretary, enforcement authority under the Federal Hazmat Law is shared by the RSPA and four modal administrations: the Federal Highway Administration (FHWA), the USCG, the Federal Railroad Administration (FRA) and the Federal Aviation Administration (FAA). The Federal Hazmat Law provides enforcement sanctions including fines and imprisonment.[98]

The general approach taken in the HMR is to permit compliance with certain provisions of the IMDG Code in lieu of the corresponding domestic requirements, rather than to require compliance with the IMDG Code.[99] Accordingly, goods

---

[95] Some of RSPA's regulatory functions are: issuing rules and regulations governing the safe transportation of hazardous materials; issuing, renewing, modifying and terminating approvals for specific activities; receiving and maintaining important records; and making (or issuing) administrative determinations of whether state or local requirements: (1) are preempted by the Federal Hazmat Law; or (2) may remain in effect, under a waiver preemption; representing the Department of Transport in international organizations and working to ensure the compatibility of domestic regulations with the regulations of international bodies.

[96] For an overview of the federal hazardous materials transportation law see <http://hazmat.dot.gov/regs/overhml.pdf> (visited 17.01.2007).

[97] 42 U.S.C § 960 etc.

[98] A minimum penalty of $250 per violation, a maximum assessment of $25,000 per violation per day. (49 U.S.C. § 5123). Criminal penalties for wilful violations of up to $500,000 and five years' imprisonment. (49 U.S.C.§ 5124).

[99] 49 CFR § 171.12. The USA was very reluctant to discard international regulations since the overwhelming number of shipments of dangerous goods involve domestic transport or transport to neighbouring countries with whose laws its own regulations are mostly consistent. Nonetheless, the USA has recognized the importance of its trade in

which are listed in the IMDG Code but not in the HMR may be transported in the U.S. when described, marked and labelled in accordance with the IMDG Code. However, a good that is designed as hazardous under the HMR but not subject to the IMDG Code is subject to the requirements of the HMR. The carriage of dangerous cargoes in bulk is regulated by 46 CFR Parts 146-154.[100]

---

dangerous goods with other regions of the world and has sought to facilitate such shipments. Had it not done so, inconsistencies between national and international regulations would often give rise to a need to delay cargo at the sea/land interface in order to re-label, re-mark or re-pack dangerous goods, to comply with national regulations.

[100] Except bulk packaging as defined by 49 CFR 171.8

Part 2:   Meaning of dangerous goods

A. In general

The parties to a contract for the carriage of goods by sea, whether this is embodied in a bill of lading or one of the standard charterparty forms, customarily negotiate its terms against a background of both commercial and legal considerations. The parties' respective liability for risks to which the vessel and its cargo might be exposed during the course of the contracted voyage will be of paramount importance. Should they arise, the potential risks might expose the parties, through their insurance, to expensive damage claims. Therefore, the description of the cargo for shipment is one of the crucially important components of the negotiations between the parties to any shipping contract.[1] This requirement becomes especially relevant when it has been estimated that more than 50% of the cargoes transported by sea today may be regarded as dangerous, hazardous, and/or harmful and need to be handled with special care. As a result, a shipper does not have unlimited freedom as to what he may transport by sea. Restrictions on goods which a charterer or shipper may ship are imposed by law, the terms of the contract and statutes. These restrictions are commonly elided into a general proposition that a person sending goods by sea must not ship dangerous goods. Yet it is the case that dangerous goods are often shipped and legitimately so; and the carrier should discharge his normal duties in respect of whatever cargo he has agreed to carry. The real issues, therefore, are to identify what risks are involved in the carriage of the cargo in question, how these risks are allocated between the parties and what the consequences of shipping dangerous goods are.[2]

Although there are no *numerous clausus* of cargoes which may be dangerous, hazardous or harmful, "dangerous good" is a useful notion for practical purposes. Whether the goods indeed cause damage is known with certainty only after the damage has occurred, but the trade needs rules about how to deal with goods whose dangerous characteristics can be anticipated beforehand.[3] Obvious examples are inflammable goods, explosives, petroleum products and chemicals. However, some cargo that appears to be safe can also become dangerous. For example, excess moisture in wool can build up great heat and, although rarely, make it li-

---

[1]   Girvin, "Shipper's Liability for the Carriage of Dangerous Goods by Sea" [1996] *LMCLQ* 487, 488.
[2]   Rose, "Cargo Risks: 'Dangerous' Goods" [1996] 55 *Cam. L.J.* 601.
[3]   Tiberg, "Legal Survey", in Grönfors (ed.), *Damage from Goods* (1978) 9, 11.

able to catch fire. Wheat, rice and other grain cargoes can likewise become dangerous through high moisture content.[4]

It may seem that the common usage of "dangerous" to describe such goods does not mean that the risks presented by them need be extreme or that there is one single category of dangerous goods attracting a uniform applicable set of consequences.[5] However, to allocate liability, determining the meaning of "dangerous", or at least of what is not "dangerous", is of paramount importance, because different standards of liability are applied. This chapter tries to find out what dangerous cargo means.

B. Type of good

I. Charterparties

In many respects, voyage and time charterparties are quite different, and one of the differences between voyage and time charter is that in voyage charters, the cargo is stipulated whereas in time charters it is more common to include details of cargoes which cannot be carried.[6] If the charterparty provides for a specific type of cargo, the obligation of the charterers is to tender for loading goods of such a type and in such a condition as would be considered reasonable in the light of the provisions of the contract and the practice at the loading port.[7] If the charterers have an option as to the cargoes to be shipped, the cargo offered must be a reasonable cargo of the description specified. It must be such a cargo as is reasonable to ask the master to load and carry and in such a state as is reasonably within the contract. But charterers are under no obligation to select a cargo or combination of cargoes suitable for the ship.[8] The state of the particular cargo is of paramount importance as regards the question of whether it is reasonable or not. The question in the light of the terms of the contract is whether owners who have expressly agreed to carry cargo of that description may reasonably be said to have agreed to carry particular cargo tendered to them.[9]

If the cargo is described not specifically but entirely in general terms, the same basic approach still holds good. The charterer must tender a cargo of kind and in a

---

[4]   Bulk Grain Cargoes, "Hot Stuff from the US, but not Enough from Brazil and Argentina" 1995 (139) Gard News 15.

[5]   Rose, "Cargo Risks: 'Dangerous' Goods" [1996] 55 *Cam. L.J.* 601, 602.

[6]   Todd, *Contract for the Carriage of Goods by Sea* (1988), 42.

[7]   Cooke/Young/Taylor, *Voyage Charters* (2001), 146.

[8]   *Stanton v. Richardson* (1872) L. R 7.C.P 421, 430, aff'd. (1874) L.R. 9 C.P. 390 (Ex. Ch.), aff'd (1875).

[9]   In *Atlantic Duchess* [1957] 2 Lloyd's Rep. 55, it was held that the charterers were entitled to ship butanised crude oil under the charterparty providing for carriage of crude oil, since no distinction was drawn in the trade between crude oil and butanized crude oil for the purpose of carriage by sea, the two products being regarded as commercially identical.

condition which is reasonable in all circumstances. The relevant circumstances would include any terms of the charter describing the ship, any characteristics of the ship known to the charterer before the fixture was concluded and the kinds of cargoes customarily shipped at the agreed loading port.[10]

German law also lies down that if the goods are described in the contract not only by type but also by specific description, the shipper must load this specific type of cargo.[11] If the goods are not described specifically in the contract, the carrier is obliged to accept other goods proposed by the shipper for the same destination if loading of other goods neither aggravates his situation nor endangers the ship and the rest of the cargo.[12]

II. Bills of lading

In bills of lading, general expressions such as "merchandise" or "produce" are used to describe intended cargo. According to the Hague/Hague-Visby Rules, "goods include goods, wares, merchandise, and articles of every kind".[13] However, two types of cargo excluded from the application of the rules are live animals and cargo which by contract of carriage is stated as being carried on deck and is so carried.[14] Furthermore, parties are entitled to contract on their own terms in relation to the carriage of "particular" goods, provided that these terms are incorporated into a non-negotiable receipt and no bill of lading is issued.[15]

Some bills of lading forms contain a clause defining the goods.[16] It is, however, not always necessary to have a clause defining the goods covered by the bill, as the face of the bill will contain the detailed description of what has been shipped.[17]

---

[10] If the charter describes the dimensions of the ship's hatches, the capacity of her pumps or gear or other features, the charterers must tender a cargo which is suitable for carriage in a ship so described. In the absence of circumstances which point to a different conclusion, the charterer cannot expect the ship to have any special characteristics or equipment required for a particular type of cargo.

[11] HGB § 562(2).

[12] HGB § 562(1).

[13] Art. I(c).

[14] Art. I(c). In both cases, the parties are free to negotiate their own terms of carriage for such cargoes. The exclusion is justified by the peculiar risk attached to the carriage of both categories of cargo, arising in the first cases from the nature and inherent properties of the animals and in the second from the exposed position in which the cargo is stowed.

[15] Art. VI. Such goods are envisaged as "one-off" cargoes not in the usual course of trade, where either the particular character or condition of the goods or the circumstances in which they are to be carried justifies special contract. Obvious examples are experimental cargoes or contracts for the carriage of nuclear waste.

[16] P&O Nedlloyd Bill Cl.1: "Goods means the whole or any part of the cargo received from the Shipper and includes the packing and any equipment or Container not supplied by or on behalf of the Carrier".

[17] Gaskell/Asariotis/Baatz, *Bills of Lading: Law and Contracts* (2000), 206. "K" Line Bill of Lading Cl.1(e): "Goods mean the cargo described on the face of the this Bill of

C. Restrictions on the goods

I. Lawful merchandise or goods

In the absence of any specific description of the cargo, charterparties or bills of lading generally provide a blanket provision as to which cargoes are permitted. Often it is worded as lawful merchandise or lawful goods.[18] To be "lawful", the cargo must be such that it can be loaded, carried and discharged without breach of the local law in force at the loading and discharging ports as well as the flag state's law.[19]

The German Commercial Code, HGB § 564 (2) likewise provides that cargo to be loaded must not violate any law of the port of loading and unloading.

II. Lawful dangerous goods

The IMDG, IBC, and IGC codes are part of the SOLAS 1974 Convention and mandatory under the Convention.[20] As pointed out in Part 1, these codes were adopted into domestic laws even before the codes were made mandatory. Therefore, goods for the purposes of these codes are considered to be lawful unless transport of a specific good is forbidden.[21] Such goods are *lawful dangerous* goods as long as they are loaded, carried and discharged without breach of local law.

Lading, and, if the cargo is packed into container(s) supplied or furnished by or on behalf of the Merchant, include the container(s) as well." The definitions refer to the packing as well as to the cargo itself. Modern containers can be valuable goods in their own right and are part of the cargo and will be treated as such.

[18]   Gentime Cl.2; NYPE 93 Cl.4; Bermuda Container Line, Cl.3 (a).

[19]   In *Leolga v. Glynn* [1953] 2 Q.B. 374, the charterparty stipulated that "in lawful trades for the conveyance of lawful merchandise". The ship was loaded with ammunition and other explosives and ordered to discharge these at Adabiya, in Egypt, although the charterers knew this was prohibited by the Egyptian authorities. She broke down and her repairs took 26 days, since she was black-listed by Egyptian authorities. It was held that as the goods loaded for Adabiya could not be discharged at the nominated port without breach of local law, they were not lawful merchandise and the charterers were liable to pay damages for the delay caused.

[20]   See *supra* Part 1.

[21]   The IMDG Code Art. 1.1.4.1 stipulates that unless provided otherwise by this Code the following are forbidden from transport: Any substance or article which, as presented for transport, is liable to explode, dangerously react, produce a flame or dangerous evolution of heat or dangerous emission of toxic, corrosive or flammable gases or vapours under normal conditions normally encountered in transport. In Chapter 3.3, special provision lists 900 certain substances which may not be transported.

III. Dangerous good clauses

1. Charterparty forms

As it is in the nature of voyage charter that the cargo for shipment is expressly agreed, these charterparties generally do not contain cargo-restriction clauses. This is contrasted with most time charterparty forms, where the charterer has a greater discretion as to the cargo to be shipped and where it is more usual to find an express dangerous goods clause. Clauses in time charters either prohibit the shipment of dangerous cargo[22] or permit shipment under certain circumstances.[23] Generally, restrictions on dangerous cargoes are described together with the lawful merchandise.

For instance, the 1993 revision of the NYPE form contains a special dangerous cargo clause.[24] The first part permits the carriage of dangerous cargo when in accordance with the requirements of specified national authorities and the second part limits the amount of IMO-classified cargo that may be carried. Clause 4 provides that:

(a) In carrying lawful merchandise excluding any goods of a dangerous goods, injuries, flammable or corrosive nature unless carried in accordance with the requirements or recommendations of the competent authorities of the country of the Vessel's registry and of ports of shipment and discharge and of any intermediate countries or ports through whose waters the Vessel must pass. Without prejudice to the generality of the foregoing, in addition the following are specifically excluded: livestock of any description, arms, ammunition, explosives, nuclear and radioactive materials …

(b) If IMO-classified cargo is agreed to be carried, the amount of such cargo shall be limited to … tons and the Charterers shall provide the Master with any evidence he may reasonably require to show that the cargo is packaged, labelled, loaded and stowed in accordance with IMO regulations, failing which the Master is entitled to refuse such cargo or, if already loaded, to unload it at the Charteres' risk and expense.

---

[22] Baltime 1939 Cl.2 provides "…No live stock nor injuries, inflammable or dangerous goods (such as acids, explosives, calcium carbide, ferro silicon, naphta, motor spirit, tar, or any of their products) shall be shipped."; Shelltime 4, Cl.28 titled "injurious cargoes" provides "no acids, explosives, or cargoes injuries to the vessel shall be shipped and without prejudice to the foregoing any damage to the vessel caused by the shipment of any such cargo, and the time taken to repair such damage, shall be for Charters' account…".

[23] Abdul Hamid, *Loss or Damage from the Shipment of Goods, Rights and Liabilities of the Parties to the Maritime Adventure* (Diss. Southampton 1996), 90.

[24] The NYPE 1946 charterparty contains no provision which forbids the shipment of dangerous cargo and only provides that "… to be employed in carrying lawful merchandise, including petroleum and its products in proper containers, excluding ….". Probably this was a deficiency and rectified.

There are similar but less detailed provisions in some of the more specialized time charterparty forms. But none of the forms purports to be explicit in defining precisely what is meant by "dangerous". This is advantageous for carriers and the converse for shippers, because it clearly brings within the ambit of the shipper's obligations the duty to account for a potentially wider category of cargoes.[25]

2. Bill of lading forms

Bill of lading standard forms commonly contain dangerous goods clauses, either expressly or by means of a paramount clause which purports to incorporate the Hague/Hague-Visby Rules. One of the detailed examples is *Ellerman East Africa/Mauritius Service Bill*, which provides in Cl. 19, entitled "dangerous goods", that:

> No goods which are or may become dangerous, inflammable or damaging (including ra-dio-active materials), or which are or may become liable to damage any property what-soever, shall be rendered to the Carrier for Carriage without his express consent in writ-ing, and without the Container or other covering in which the Goods are to be carried as well as the Goods themselves being distinctly marked on the outside so as to indicate the nature and character of any such Goods and so as to comply with any applicable laws, regulations or requirements. If any such goods are delivered to the Carrier without such written consent and/or marking, or if in the opinion of the Carrier the Goods are or are liable to become of a dangerous, inflammable or damaging nature they may at any time be destroyed, disposed of, abandoned, or rendered harmless without compensation to the Merchant and without prejudice to the Carrier's right to Freight.
>
> 1. The Merchant undertakes that such Goods are packed in a manner adequate to with-stand the risks of Carriage having regard to their nature and in compliance with all laws or regulations which may be applicable during the Carriage. ...
> 2. Whether or not the Merchant was aware of the nature of the Goods, the Merchant shall indemnify the Carrier against all claims, losses, damages or expenses arising in consequence of the Carriage of Such Goods....

Other bill of lading forms contain identical or similar clauses.[26] Those forms which do not have specific clauses concerning the carriage of dangerous goods are frequently drafted in a way which is similar to Art. IV.6 of the Hague/Hague-Visby Rules. As in charterparties, however, none of the forms purports to be explicit in defining precisely what is meant by "dangerous".

---

[25] Girvin, "Shipper's Liability for the Carriage of Dangerous Goods by Sea" [1996] *LMCLQ* 487, 490.

[26] Virtually identical provisions are contained in the ANL Tranztas Bill, cl. 18, P & O Containers Bill cl. 19; See also Mitsui OSK Combined Transport Bill 1992 Cl.22 and 1993 Cl.19 respectively.

D. Relevance of the IMDG Code in carriage contracts

Dangerous goods regulations are, in principal, public law regulations that apply to relationships between the state and to persons who must comply with these regulations. In other words, the focal point of the safety framework is to provide safe ships and safe shipping operations for the benefit of those who may be affected by such operations. Insofar as it operates on the ship, the framework is directed at the owner, but insofar as it imposes civil or criminal liability measured in cash, it may be directed at the owner or any other person connected with the carriage or the goods. However, civil liability in respect of damage or loss as between owner, charterer or shipper will depend on the relevant contract. Then the question will arise of the relevance, if any, to that liability of the obligations laid on the parties in relation to the cargo at issue by legislation focused on "public" responsibilities.[27]

I. Common law

In common law liability may arise from breach of statutory duty. In such a case there will be civil liability if the obligation or prohibition was imposed for the benefit or protection of a particular class of individuals.[28] The breach of statutory duty must be of a kind which the statute intends to prevent.[29] The conventional rule is that unexcused violation of a health or safety regulation is negligent *per se* – that is negligent as a matter of law. In the absence of a legally cognizable excuse that for practical purposes renders a statutory or regulatory standard inapplicable, reasonable prudence requires compliance with these standards. As the classic case on the issue puts it unexcused violation "is negligence" itself. This is the rule that pertains in the ordinary case, constituting the vast majority of civil actions in which a regulatory violation plays a part.[30]

Accordingly a shipper who is in breach of statutory provision in relation to the shipment of dangerous goods may at the same time be civilly liable for breach of this statutory duty. Dangerous goods regulations are specific and detailed so as to raise a prima facie case for civil liability.[31] A civil action against shipper may be

---

[27]   Jackson, "Dangerous Cargo- A Legal Overview" in *Maritime Movement of Dangerous Cargoes- Public Regulations Private Liability*, Papers of One Day Seminar (1981), A6.

[28]   *Lonrho Ltd. v. Shell Petroleum Ltd.* (No:2) [1982] A.C. 173; *Hover & Co. v. Denver & R.G.W.R. Co.* 17 F.(2d) 881; Abdul Hamid, *Loss or Damage from the Shipment of Goods, Rights and Liabilities of the Parties to the Maritime Adventure* (Diss. Southampton 1996), 196.

[29]   *Ibid.*

[30]   Abraham, "The Relation between Civil Liability and Environment Regulation: An Analytical Overview" [2002] 41 *Wash. L.J* 379, 394.

[31]   Abdul Hamid, *Loss or Damage from the Shipment of Goods, Rights and Liabilities of the Parties to the Maritime Adventure* (Diss. Southampton 1996), 198.

brought by the shipowners and crew, such persons being the class of individuals the regulations were designed to protect.[32]

II. German law

In German law, on the other hand, Regulations on the Transport of Dangerous Goods by Sea (GGVSee) incorporate the IMDG Code and other codes – *Schutzgesetz* ("Protective law") – in the sense of § 823(2) of the German Civil Code (BGB).[33] Protective law is a legal norm which guards the certain legally protected rights ("Rechtsgüter") of individuals. Such goods contain an order or forbiddance and according to their personal and objective scope of protection aims at protection of individuals.[34] BGB § 823 [duty to compensate for damage] provides that:

(1) A person who wilfully, or negligently, unlawfully injures the life, body, health, freedom, property or other right of another is bound to compensate him or any damage arising therefrom.
(2) The same obligation is placed upon a person who infringes a statute intended for the protection of others. If, according to the provisions of the statute, an infringement of this is possible even without fault, the duty to make compensation arises only in the event of fault.

The dangerous goods regulations based on GGBefG principally serve to the protection of the public safety and order, particularly of the general public, of important public property, of life and health of human as well as animals and other things from the dangerous goods.[35] All standards of conduct in these regulations, show principally individuals protective character and thus protective law in the sense of BGB § 823.2.[36]

III. Standard of reasonableness

In an action for negligence a shipper of goods which are liable to cause loss or damage must have acted in such a way that is below the standard expected of a reasonable and prudent shipper. If anyone practices a profession or is engaged in a transaction in which he holds himself out as having professional skill, the law expects him to show the amount of competence and prudence associated with that profession or trade. If he is not behaved reasonably, he will be liable for consequences.

Although the IMDG Code and other codes do not directly relate to the distribution of risk, they indirectly affect the relation between carrier and shipper or third parties. These codes are manifestly based on the accumulated experience of those

---

[32] *Ibid.*
[33] Bürgerlichesgesetzbuch.
[34] Bremer, *Die Haftung beim Gefahrguttransport* (1992), 65.
[35] GGBefG §§ 2,3.1.
[36] Bremer, *Die Haftung beim Gefahrguttransport* (1992), 77.

involved in the relevant trades and are kept up-to-date. The purpose of the codes is safety, as is the purpose of the specific rule on the dangerous cargoes in the Hague/Hague-Visby Rules and carriage contracts. They require certain conditions for stowing, temperature control, packing, if necessary, etc. These requirements are minimum standards applied in practice. Accordingly, a shipper and a carrier are imputed with the prevailing standards of the shipping industry. Non-compliance with the dangerous goods standards and regulations raises an inference of negligence.[37]

IV. Incorporation of the IMDG Code into contracts

It is clear that if there is to be a category of dangerous goods that attract special rules, contracts cannot solve the problem of definition by listing. Conventions and statutes may do so and therefore allow contracts to define by reference, provided that the category is restricted to substances requiring special facilities.[38] The IMDG Code, IBC, IGC and BC are based on this principle. Parties to a carriage contract may incorporate convention provisions or codes into the contract, even if the purpose of listing of the substances does not relate directly to the distribution of risk between the parties. By incorporating the regulations into the contracts, the parties undertake to comply with certain requirements as to the shipment and handling of the goods.[39] In such a case, failure to comply with the IMDG Code and other codes will also be a breach of contract. Although it is often argued that the purpose of the IMDG Code is the safety of the ship and crew only, the current edition of the IMDG Code expressly states in its preamble that:

> Transport of dangerous goods by sea is regulated in order to reasonably prevent injury to persons or damage to ships or their cargoes.

---

[37] Tiberg, "Legal Survey", in Grönfors (ed.), *Damage from Goods* (1978), 9, 17.

[38] Jackson, "Dangerous Cargo- A Legal Overview" in *Maritime Movement of Dangerous Cargoes- Public Regulations Private Liability*, Papers of One Day Seminar (1981), A4.

[39] In *Islamic Investment Co.1 S.A . v. Transorient Shipping Ltd. (The "Nour")* [1999] 1 Lloyd's Rep. 1, a cargo of bagged fishmeal became hot and caused damage to cargo and delay in discharge. Cl. 27 of the charterparty provided that "… Fishmeal must be shipped under deck and be loaded/stowed/discharged according to IMO and local regulations". It was argued that transport by sea of dangerous cargo is regulated in order reasonably to prevent injury to persons or damage to the ship and there is no reference to the risks of damage to other goods or to the goods themselves. At the time of the proceedings of the case, the 1992 Edition of the IMDG Code's general introduction stated that transport by sea of dangerous goods is regulated "in order reasonably to prevent injury to persons or damage to the ship." Therefore, it was contended that there was no reference here to the risks of damage to other goods or to the goods themselves. The Court held that the effect of cl. 27 is to make compliance with the regulations a contractual undertaking, not necessarily limited to the reasons why the regulations were adopted by the IMO.

E. Meaning of dangerous goods

I. Terminology: dangerous, hazardous, noxious or harmful

Besides the term "dangerous", "hazardous", "harmful" and "noxious" are used to describe such cargoes.[40] Although harmful and noxious are used in general with regard to environmental effects, the use of dangerous or hazardous and the difference between them, if any, are not clear.[41] In the United States, the equivalent of the IMDG Code is entitled "Hazardous Materials Regulations".[42] In particular, the two terms "hazardous" and "dangerous" have been used in a similar and comparable way in international agreements and national legislations.[43] A clear differentiation is related to the different objectives and definition in the different legislative context.

II. What is meant by "dangerous goods"?

The Shorter Oxford Dictionary defines "dangerous" as "fraught with or causing danger; involving risk; perilous, hazardous; unsafe". Despite its general use and long lists included in the codes, the term "dangerous goods" is imprecise and misleading in practice. The fact that any cargo may cause damage makes the meaning of dangerous complicated. What is meant by dangerous?

An Argentine court once held a cargo of fresh fruit to be dangerous within the meaning of the Hague Rules, a decision which stresses that it is as much the nature of a situation as the nature of a substance which creates "danger".[44] Therefore, it

---

[40] For instance, "Harmful substance means any substance which, if introduced to the sea, is liable to create hazards to human health, to harm living resources and marine life, to damage to amenities or to interfere with other legitimate uses of the sea, and includes any substance subject to control by the present Convention." MARPOL 73/78 Art. 2 (2); "For the purpose of this Annex, "harmful substances are those substances which are identified as marine pollutants in the International Dangerous Goods Code (IMDG Code)." MARPOL 73/78 Annex III.

[41] In Australia, hazardous substances are classified based on health effects, while dangerous goods are classified according to their immediate physical or chemical effects, such as fire, explosion, corrosion and poisoning, affecting property, the environment or people. However, many hazardous substances are also classified as dangerous goods <www.ohsrep.org.au/index.cfm?section=10&Category=44&viewmode=content&content id=171>. (visited 10.11.2006).

[42] See *supra* Part 1.

[43] "Identification of Priority Hazardous Substances", European Commission Working Document (ENV/191000/01 of 16 January 2001), 14; "Tracking Intermodal Shipments of Hazardous Materials Using Intelligent Transportation Systems in a New Security Age", Global Trade, Transportation and Logistics GTTL 502 Term Project, Spring 2002, <http://depts.washington.edu/gttl/StudentPapersAbstracts/2002/GTTL-hazmat%20final.pdf> 5. (visited 31.01.2007).

[44] [1979] *ULR* 202.

may be argued that the "situation" is the focal point and that it is a mistake to start with the nature of a substance.[45] But the label "dangerous" implies perhaps the idea that the substance itself has an inherent dangerous characteristic.[46]

Although the label "dangerous" implies that the substance has to be intrinsically dangerous, the substance of the hazard may simply be the likelihood of economic loss through delay or property loss through damage to the ship or other goods. The idea of "dangerous" may impart the idea of personal injury or, at the most, serious damage to property, but in English law the category has been extended to include "unlawful merchandise", the only defect in which was the lack of a licence to land.[47] However, concern at "dangerous goods" persists, first because of the potential scale of damage which may be caused by them and secondly, precisely because of the "danger' factor – which of itself means that the responsibility on those creating and continuing it must be assessed with that in mind.

It has been said that "less dangerous goods are more dangerous than very dangerous goods", simply because in the case of very dangerous goods, everyone appreciates the danger and special precautions are taken.[48] Under special circumstances, almost any product can be dangerous.

The adoption of a category of cargo which attracts special rules necessarily brings with it a problem of definition. Conventions, statutes and judge-made law all use the phrases. Contractual documents base their clauses on them. As for "dangerous goods", in some cases the generality is qualified by the listing of particular substances, which raises the subordinate question of whether the general is to be governed by the particular.[49]

1. In the contracts

As mentioned above, charterparties and bills of lading contain dangerous cargo clauses, but none of them purports to be explicit in defining precisely what is meant by dangerous. Indeed this is not surprising, because it is clear that if there is a category of dangerous goods which is to attract special rules, contracts cannot solve the problem by definition. Statutes and regulations may do so, provided that the category is restricted to substances requiring the special facilities. Parties to the carriage contract may therefore incorporate regulation provisions into the contract even if the purpose of listing the substance does not relate directly to the distribution of risk between the parties.[50] The interpretation of the term "danger-

---

[45]   Jackson, "Dangerous Cargo- A Legal Overview" in *Maritime Movement of Dangerous Cargoes- Public Regulations Private Liability* (1981), A2.
[46]   *Ibid.*
[47]   *Michell Cotts v. Steel* [1916] 2 K.B. 610.
[48]   Grönfors, "Summarizing a Multi-National Problem", in Grönfors (ed) *Damage from Goods* (1978), 106.
[49]   Jackson, "Dangerous Cargo- A Legal Overview" in *Maritime Movement of Dangerous Cargoes- Public Regulations Private Liability*, Papers of One Day Seminar (1981), A4.
[50]   *Ibid.*

ous" is significant, because both liability in general and any rights given in regard to the goods, such as disposal and indemnity, will depend on this interpretation.

2. In the safety, marine pollution and third-party liability conventions

The concern over the carriage of dangerous substances is shown by the growing number of conventions that deal with safety, carriage and third-party liability. Of these, the SOLAS 1974 Convention does not define dangerous goods. It only states that dangerous goods mean the substances, materials and articles covered by the IMDG Code.[51]

MARPOL 73/78 employs the phrase "harmful" in order to define marine pollutants. Accordingly, harmful substances are any substances which, if introduced into the sea, are liable to create hazards to human health, to harm living resources and marine life, to damage amenities or to interfere with other legitimate uses of the sea, and that includes any substances subject to control by the present Convention.[52] Furthermore, for the purpose of Annex III, entitled "Regulations for the Prevention of Pollution by Harmful Substances Carried by Sea in Packaged Form", harmful substances are those substances which are identified as marine pollutants in the IMDG Code.[53]

Under the HNS Convention, the definition of hazardous and noxious substances (HNS) is largely based on lists of individual substances that have been identified in the codes designed to ensure maritime safety and prevention of pollution.[54]

---

[51]   SOLAS Chap. VII/A.1.2
[52]   MARPOL Art. (2).2.
[53]   MARPOL Annex III Reg. 1(1)(1.1)
[54]   HNS Convention Article 5 provides for
a) any substances, materials and articles carried on board a ship as cargo, referred to in (i) to (vii) below:
   (i) in bulk listed in Appendix I of Annex I to the International Convention for the Prevention of Pollution from Ships, 1973, as modified by Protocol of 1978 relating thereto, as amended;
   (ii) noxious liquid substances carried in bulk referred to in Appendix II of Annex II to the International Convention for the Prevention of Pollution from Ships 1973, as modified by the Protocol 1978 relating thereto, as amended, and those substances and mixtures provisionally categorized as falling in pollution category A,B,C or D in accordance with Regulation 3(4) of the said Annex II
   (iii) dangerous liquid substances carried in bulk listed in Chapter 17 of the International Code for the Construction and Equipment of Ships Carrying Dangerous Chemicals in Bulk, 1983, as amended, and the dangerous products for which the preliminary suitable conditions for the carriage have been prescribed by the Administration and port administrations involved in accordance with Paragraph 1.1.3 of the Code.
   (iv) dangerous, hazardous and harmful substances, materials and articles in packaged form covered by the International Maritime Dangerous Goods Code, as amended
   (v) liquefied gases as listed in Chapter 19 of the International Code for the Construction and Equipment of Ships Carrying Liquefied Gases in Bulk, 1983, as amended, and the products for which preliminary suitable conditions for the carriage have been

HNS are very varied and include both bulk cargoes and packaged goods. Bulk cargoes can be solids or liquids, including oils or liquefied gases. The number of substances included is very large. In practice, however, the numbers of HNS that are shipped in significant quantities are relatively small.[55]

Bulk solids are included if they are covered by Appendix B of the BC Code, i.e. they possess chemical hazards, and if they are also subject to the provision of the IMDG Code when carried in packaged form. This means that many of the major bulk solids are excluded, since they either do not posses chemical hazards[56] or they are classified as materials hazardous only in bulk (MHB).[57] Bulk solids that are covered include some fertilizers, sodium and potassium nitrates, sulphur and some types of fishmeal. Bulk liquids are included if they present safety, pollution or explosion hazards and include organic chemicals,[58] inorganic chemicals,[59] and vegetable and animal oils and fats.[60] Both persistent and non-persistent oils of petroleum origin are also included.[61] Bulk liquids that are not covered include potable alcohol and molasses. All liquefied gases which are transported in bulk are included.[62]

3. In the carriage conventions

The Hague/Hague-Visby Rules do not solve any problem of definition, because they do not provide any definition of "dangerous" but link general rules to the category, stating, in Art. VI.6, that the shipper is liable for all damage and expenses arising out of a shipment of goods "inflammable, explosive, or dangerous" unless the carrier has consented to carry them. Art. 13 of the Hamburg Rules pro-

---

prescribed by the Administration and port administrations involved in accordance with Paragraph 1.1.6. of the Code.

(vi) liquid substances carried in bulk with a flashpoint not exceeding 60° C (measured by a closed-cup test)

(vii) solid bulk materials possessing chemical hazards covered by Appendix B of the Code of Safe Practice for Solid Bulk Cargoes, as amended, to the extent that these substances are also subject to the provisions of the International Maritime Dangerous Goods Code when carried in package form, and

b) residues from the previous carriage in bulk of substances referred to in (a)(i) to (iii) and (v) to (vii).

[55] "IMO Guide for Interested Parties on the Workings of the Hazardous and Noxious Substances Convention 1996, (HNS Convention)" LEG 83/INF/3 Annex, 8. Crude oil is the most carried hazardous cargo which is carried in bulk.

[56] Such as iron ore, grain, bauxite and alumina, phosphate rock, cement and some fertilizers.

[57] Such as coal, reduced iron and woodchip.

[58] Such as methanol, xylenes and styrene.

[59] Such as sulphuric acid, phosphoric acid and caustic soda

[60] Such as palm oil, soybean oil and tallow.

[61] Although the Convention only covers non-pollution damage caused by persistent oil. See *infra* Part 6.

[62] Such as liquefied natural gas (LNG), liquefied petroleum gas (LPG), ammnonia, etyhlene, butadiene, ethane and propylene.

vides only that the shipper must mark or label "dangerous goods" as dangerous and must inform the carrier of their dangerous character. Therefore, definition is crucial both to the application of provisions and consent. Any substance listed in the IMDG Code and other codes as dangerous qualifies, but courts tend to apply rules to dangerous situations as well as to dangerous goods.

4. In the statutes

Apart from statutory applications of conventions, statutory obligations may be imposed on shippers and carriers with regard to dangerous goods. The provisions must be viewed in the context of their purpose. What is dangerous in one context may not be so in another. National statutes implementing international conventions and codes[63] give only a general definition of dangerous goods. The reason is that defining the concept of "dangerous goods" exactly may not be appropriate, as technical development in this field changes rather quickly.[64]

5. In the relevant codes

As we have seen so far, neither the SOLAS Convention nor the Hague/Hague-Visby Rules nor the Hamburg Rules define dangerous goods. They either refer to relevant codes or give only some examples of dangerous cargo, such as "inflammable, explosive and any other dangerous goods". This wording presents the classic issue of whether the general catch-all phrase is ultimately governed by the goods specifically listed. If those can be said to form a group of common definition, it may be argued that the generality might be limited by the common denomination. Is that the case with the IMDG Code listing the goods which are dangerous within the Code?

Even assuming that this was so with the IMDG Code, it does not solve the problem, because the IMDG Code regulates only the shipment of packaged dangerous goods. However, in practice not only packaged but also bulk cargoes cause damage.

a) The IMDG Code

The IMDG Code classifies dangerous goods in different classes, subdivides a number of these classes and defines and describes characteristics and properties of goods which would fall within each class or division. Each class has its definition. These definitions have been devised so as to provide a common pattern which it should prove possible to follow in various national and international regulations. The objective of recommended definitions is to indicate which goods are dangerous and in which class, according to their specific characteristics, they should be

---

[63] See *supra* Part 1.

[64] Segolson, *Damage from Goods in Sea Carriage, the Sender's Liability against the Carrier and the Other Owners of the Cargo on Board* (Diss. Stockholm 2001), 14 fn. 45.

included. The definitions, together with the list of dangerous goods, should provide guidance to those who have to use these regulations. The IMDG Code consists of 9 classes according to type of risk. A number of dangerous substances in the various classes have also been identified as substances harmful to the marine environment in accordance with Annex III of the MARPOL 73/78 Convention.

aa) Dangerous goods specifically listed in the IMDG Code

Dangerous goods are assigned UN numbers and Proper Shipping Names according to their hazard classification and their composition. Dangerous goods that are commonly transported are listed in the Dangerous Goods List in Chap. 3.2 of the IMDG Code. Where an article or substance is specifically listed by name, it shall be identified in transport by the Proper Shipping Name in the Dangerous Goods List. This list also contains relevant information for each entry.[65] Entries in the Dangerous Goods List are of four types: single entries for well-defined substances or articles, generic articles for well-defined groups of substances or articles, specific "Not Otherwise Specified" N.O.S. entries covering a group of substances or articles of a particular chemical or technical nature; and general N.O.S. entries covering a group of substances or articles meeting the criteria of one or more classes.

bb) Dangerous goods not specifically listed in the IMDG Code

Substances or articles which are not specifically listed by name in the Dangerous Goods List shall be classified under a "generic" or "not otherwise specified" (N.O.S.) Proper Shipping Name. The substance or article shall be classified according to the class definitions and test criteria, and the article or substance is classified under the generic or N.O.S. Proper Shipping Name in the Dangerous Goods List which most appropriately describes the article or substance.[66]

cc) Mixtures or solutions

A mixture or solution containing one or more substances identified by name in the IMDG Code or classified under an N.O.S or generic entry and one or more substances not subject to the provisions of the IMDG Code is not subject to the provisions of the IMDG Code if the hazard characteristics of the mixture or solution are such that they do not meet the criteria, including human experience criteria, for any class.[67]

With regard to substances or mixtures with multiple hazards, the table in 2.0.3.6 of the Code shows the class of a substance, mixture or solution having more than one hazard when it is not specifically listed by name in the Code. The

---

[65] Such as hazard class, subsidiary risk(s) (if any), packing group (where assigned), packing and tank transport provisions, EmS, segration and stowage properties and observations, etc.
[66] The IMDG Code 2.0.2.7.
[67] The IMDG Code 2.0.2.9.

priority given in the hazard table indicates which of the hazards shall be regarded as the primary hazard. The Proper Shipping Name of a substance, mixture or solution when classified in accordance with the IMDG Code 2.0.3.1 and 2.0.3.2 shall be the most appropriate N.O.S. entry in the Code for the class shown as the primary hazard.[68]

dd) Classes and divisions of dangerous goods in the IMDG Code

(1) Class 1: Explosives

These are among the most dangerous of all goods carried by sea and the precautions outlined in this class of the Code are particularly stringent. The class is divided into six divisions which present different hazards.[69]

- Division 1.1: substances and articles which have a mass explosion hazard. A mass explosion is one which affects almost the entire load virtually instantaneously.
- Division 1.2: substances and articles which have a projection hazard but not a mass explosion hazard.
- Division 1.3: substances and articles which have a fire hazard and either a minor blast hazard or minor projection hazard or both, but not a mass explosion hazard
- Division 1.4: substances and articles which present no significant hazard. This division comprises substances which present only a small hazard in the event of ignition during transport. The effects are largely confined to the package and no projection of fragments of appreciable size or range is to be expected. An external fire need not cause virtually instantaneous explosion of almost the entire contents of the package.
- Division 1.5: very insensitive substances which have a mass explosion hazard. This division comprises substances which have a mass explosion hazard but are so insensitive that there is very little probability of initiation or of transition from burning to detonation under normal conditions of transport.
- Division 1.6: extremely insensitive articles which do not have a mass explosion hazard.

Class 1 is unique in that the type of packaging and method of packing used frequently has a decisive effect on the hazard and, therefore, on the assignment of an explosive to a particular division and compatibility group. Although the safety of goods in Class 1 can be best assured by stowing them separately, this can rarely be done in practice. To ensure that they are stowed as safely as possible, the goods in the class are arranged in twelve compatibility groups.

---

[68]   The IMDG Code 2.0.3.3.
[69]   The IMDG Code 2.1

(2) Class 2 – Gases

Gases carried on board ships have various chemical properties and come in different states. They may be compressed, liquefied at ambient temperature under high pressure, dissolved under pressure in a solvent, which is then absorbed in a porous material, or liquefied by refrigeration.[70] They may, for example, be non-flammable, non-poisonous, flammable, poisonous and corrosive, support combustion or be a combination of all or some of these. Some gases are lighter than air, while some are heavier.

For the purpose of stowage and segregation, Class 2 is divided into 3 subclasses according to the hazards presented by the gases during transport.[71]

– Class 2.1: flammable gases
– Class 2.2: non-flammable, non-poisonous gases
– Class 2.3: poisonous gases

(3) Class 3 – Flammable liquids

This class deals with liquids which give off flammable (ignitable) vapours at or below 61°C closed cup (c.c.).[72] This is called flash-point.[73] Some flammable liquids are included in other classes[74] because of their other more predominant poisonous or corrosive properties.

Class 3 is divided into three subclasses according to the flashpoints of the liquids: Class 3.1 covers liquids with a low flashpoint (below -18°C c.c. (0°F), such as acetone; Class 3.2. covers liquids with an intermediate flashpoint (-18°C up to but not including 23°C c.c. (73°F), such as benzene; and Class 3.3. covers liquids with a high flashpoint (23°C and above up to 61°C c.c. (141°F), such as certain alcoholic beverages.[75] Generally speaking, water is unsuitable in fighting a fire involving flammable liquids.

(4) Class 4 – Flammable solids or substances

This class is divided into three subclasses which have very different properties. The class includes some commonly known products, many of which seem harmless enough but which can be very dangerous unless properly packed, handled and transported.

---

[70] "Focus on IMO, IMO and Dangerous Goods at Sea" May 1996, <www.imo.org/includes/blastDataOnly.asp/data_id%3D7999/IMDGdangerousgoodsfocus1997.pdf> (visited 30.01.2007), 12.
[71] MDG Code 2.2.
[72] 141°F. IMDG Code 2.3.
[73] The lowest temperature of a liquid at which its vapour forms an ignitable mixture with air. IMDG Code 2.3.3.
[74] Mainly classes 6.1 and 8.
[75] IMDG Code 2.3.2.6; "Focus on IMO, IMO and Dangerous Goods at Sea", May 1996, <www.imo.org/includes/blastDataOnly.asp/data_id%3D7999/IMDGdangerousgoodsfocus1997.pdf> (visited 30.01.2007), 13.

Class 4.1 – Flammable solids

The substances and materials in this class are solids possessing the properties of being easily ignited by external sources, such as sparks and flames, and of being readily combustible or being liable to cause or contribute to fire through friction. This class also covers substances which are self-reactive, i.e. liable to undergo at normal or elevated temperatures a strong exothermic decomposition caused by excessively high transport temperatures or by contamination, and desensitized explosives, which may explode if not diluted sufficiently. Under certain conditions, an explosives subsidiary risk label is required for some substances. This class comprises:

– readily combustible solids and solids which may cause fire through friction;
– self-reactive (solids and liquids) and related substances; and
– desensitized explosives

Some common products covered by this class are celluloid; camphor; dry vegetable fibers such as cotton, flax, jute, hemp, kapok; some metal powders; naphthalene; sisal; hay and straw; matches; rubber scrap and sculpture.

Class 4.2 – Substances liable to spontaneous combustion

Substances in this class are liable to become warm and to ignite spontaneously. Some are more likely to do so when wetted with water or in contact with moist air. Some may also give off toxic gases if involved in a fire. Because of these properties, packing and stowage requirements are important. Although some general information is contained in the introduction to the class, more detailed information is given in the individual schedules.

Common products included in this class are charcoal, celluloid scrap, copra, wet or damp or oily fibers, some nitrocellulose-based plastics, fishmeal and seed cakes.

Class 4.3 – Substances which, in contact with water, emit flammable gases

Since the substances in this class give off gases which are sometimes subject to spontaneous ignition and are also toxic, fire-fighting is a particular problem. The use of water, steam or water-foam extinguishers may make matters worse and even the use of carbon dioxide can do more harm than good in some situations; for small fires, neutralizing powders or sand is recommended.

Common products in this class include calcium carbide, metal powders, ferrosilicon, magnesium and magnesium-based products, potassium and potassium-based products, rubidium and sodium.

(5) Class 5 – Oxidizing substances

This class is divided into two subclasses. Class 5.1 deals with oxidizing substances which, although not necessarily combustible in themselves, may increase the risk and intensity of a fire by giving off oxygen. Class 5.2 covers organic peroxides, most of which are combustible.

Class 5.1 – Oxidizing substances (agents)

The fact that all substances in this class give off oxygen when involved in fire creates obvious fire-fighting difficulties, even though they are not necessarily combustible themselves. Some substances may also be sensitive to impact, friction or a rise in temperature, and some may react vigorously with moisture, so increasing the risk of fire.

Mixtures of these substances with organic and combustible materials are easily ignited and may burn with explosive force. There will also be a violent reaction between most oxidizing substances and strong liquid acids, producing highly toxic gases. One fire-fighting problem is caused by the fact that some substances in this class give off oxygen when involved in a fire, so that the use of steam, carbon dioxide or other inert gas extinguishers may be ineffective.

This class includes ammonium nitrate fertilizers, chlorates, chlorites, and calcium and potassium permanganates.

Class 5.2 – Organic peroxides

In addition to being oxidizing substances (agents), most substances in this class are liable to violent or explosive decomposition. Most will burn rapidly and are sensitive to heat. Some are also sensitive to impact or friction. To reduce this sensitivity to a safe level, they are carried in a solution, such as a paste, wetted with water or with an inert solid. Even so, some may react dangerously with other substances.[76]

Some organic peroxides can be particularly dangerous to the eyes, even after only momentary contact, and immediate rinsing of the eyes lasting at least 10 to 15 minutes is essential, followed by medical attention.

Some substances may begin to decompose when a certain temperature is exceeded, and in some cases this may lead to an explosion. To prevent this, some organic peroxides have to be transported at a controlled temperature.[77] Fire is another problem and may result in explosion. Packages containing organic peroxides should be moved away from the heat of any fire or jettisoned.[78] Organic peroxides are carried by sea "on deck only" and are prohibited for carriage on most passenger ships.

(6) Class 6 – Toxic and infections substances

Class 6 is subdivided into two classes.

Class 6.1 – Toxic substances

Substances in Class 6.1 may cause serious injury or even death if swallowed, inhaled or absorbed by contact through the skin. They are arranged in three pack-

---

[76] Violent decomposition may be caused by traces of impurities such as acids, metallic oxides or amines.

[77] IMDG Code 2.5.3.4 and individual schedules contain information on these aspects.

[78] If this is not possible and even when the fire has been extinguished, packages should be treated with great care, since organic peroxides which have been exposed to high temperatures may start a violent decomposition at any time.

ing groups (I, II and III) in descending order of risk. Fire-fighting measures are basically the same as those given for Class 3, flammable liquids, but because of the high risk of poisoning through fumes, the IMDG Code provides that ships carrying poisonous substances should always carry protective clothing and self-contained breathing apparatus. If leakage or spillage involving toxic substances occurs, decontamination should be carried out by trained staff wearing protective clothing and equipment.

This class covers mainly pesticides and insecticides, but also substances such as chloroform, cyanides, strychnine and tear gas.

Class 6.2 – Infectious substances

These are substances containing viable micro-organisms, including bacterium, virus, rickettsia, parasite, fungus or a recombinant, hybrid or mutant that are known or reasonably believed to cause disease in animals or humans. However, they are not subject to the provisions of this class if the spread of disease to humans or animals exposed to such substances is considered unlikely.

Infectious substances carry a special label. In the case of damage or leakage, public health authorities have to be notified immediately.

(7) Class 7 – Radioactive materials

The provisions of this class are based on the principles of the IAEA's Regulations for the Safe 'Transport of Radioactive Materials. They offer guidance to those involved in the handling and transport of radioactive materials in ports and on ships without necessarily consulting the IAEA safety regulations, although references to the IAEA Regulations have been included in the Class 7 IMDG Code.

Packing, labelling and placarding, stowage, segregation and other requirements vary according to the radioactivity of the material. Radioactive materials are divided into three categories, depending on radiation levels, Category I being the least dangerous.

(8) Class 8 – Corrosives

Substances in this class are solids or liquids; they can damage living tissue and materials, in some cases very severely. Some of them give off irritating, poisonous or harmful vapours and some are flammable or give off flammable gases under certain conditions.

Substances in this class may be corrosive to metals such as aluminium, zinc and tin but not to iron or steel, while others are corrosive to all metals. Some substances even corrode glass. Water can also affect some substances by making them more corrosive, by liberating gases and, in a few cases, by generating heat. Due to these different properties, packing, stowage and segregation are extremely important. The substances are also divided into three packaging groups, packaging Group 1 being the most dangerous. The introduction to Class 8 gives detailed information on the types of packaging to be used.

Most fires involving corrosive substances can be dealt with by any extinguishing agent, including water, although those which are also flammable should be dealt with in the same way as substances in Class 3 of the IMDG Code. Care must also be taken in view of the high risk of poisoning through fumes. This class includes battery acid, formic acid, caustic soda and sulphuric acid.

(9) Class 9-Miscellaneous

This class includes substances, materials and articles which, for various reasons, do not come within any of the other classes. As their properties and characteristics are so varied, the individual schedules usually include detailed information on stowage and segregation, packing and further observation.

(10) Marine pollutants

Marine pollutants are substances which, because of their potential to accumulate in seafood or because of their high toxicity to aquatic life, are subject to the provisions of Annex III of MARPOL 73/78. Marine pollutants are not a separate class, although they are regulated under a special title. Many of the substances in Class 1 to 9 are, in fact, considered to be marine pollutants.

b) Dangerous goods in other codes

The IBC, IGC, and BC Codes do not define dangerous goods either. Instead, they list certain substances according to their hazards.

aa) The IBC Code

Products covered by the IBC Code mainly have a significant fire hazard exceeding that of petroleum products. Additionally, those products may have a health hazard, water-pollution hazard, air-pollution hazard, reactivity hazard and marine pollution hazard.[79] Chapter 17 of the Code lists around 500 products.

bb) The IGC Code

The products that the IGC Code comprises may have one or more hazard properties which include flammability, toxicity, corrosivity and reactivity. A further possible hazard may arise due to the products being transported under cryogenic or pressure conditions. The cargoes that fall under the Code are listed in Chapter 19.[80] Some of the cargoes are also covered by the IBC Code and this mentioned in the list.

cc) The BC Code

One of the main dangers with regard to solid bulk cargoes is their effect on the stability of the ship. In the worst cases, they may cause the ship to capsize. The

---

[79] IBC Code Art. 1.2.
[80] Such as ammonia, butane, chlorine, ethane, methane (LNG), propane, sulphur dioxide. The list comprises about 30 products.

BC code deals with three basic types of cargo: materials which may liquefy, materials which possess chemical hazards and materials which do not fall under either of these categories but which may nevertheless pose other dangers.

When loaded, all bulk cargoes tend to form a cone. The angle formed between the slope of the cone and the bottom of the hold, which varies according to the cargo, is known as the angle of repose. Cargoes with a low angle of repose are much more prone to shift during the voyage and special precautions have to be taken to ensure that the cargo movement does not affect the ship's stability. Some cargoes affect stability by liquefying. Moreover, some other cargoes are liable to oxidation, which may result in the reduction of the oxygen supply, the emission of toxic fumes and warming up. Other may emit toxic fumes without oxidation or when wet.

A list of cargoes which may liquefy is contained in Appendix A to the Code. The stowage factor is generally low and it is emphasized that the list of materials is not exhaustive. Appendix A includes concentrates derived from copper, iron, lead, manganese, nickel and zinc ores, various pyrites, fine-particulate coal, coal slurry and various other substances.

Appendix B gives an extensive list of materials which constitute chemical hazards, ranging from aluminium dross to zinc ashes. Some of the classified materials listed also appear in the IMDG Code when carried in packaged form, but others become hazardous only when they are carried in bulk. Because they might reduce the oxygen content of cargo space or are prone to warming up. Examples are woodchips, coal and direct reduced iron (DRI). The various types listed include flammable solids; flammable solids or substances liable to spontaneous combustion; flammable solids or substances which, in contact with water, emit flammable gases; oxidizing substances; poisonous substances; radioactive substances; or corrosives.[81]

Bulk cargoes which are neither liable to liquefy nor constitute chemical hazards are not normally seen as special hazards and are covered in Appendix C. The list ranges from alumina to zircon sand, as well as many of the more commonly carried bulk cargoes such as clay, cement, iron ore, pig iron, sand and sugar. The list includes the angle of repose of each material, its approximate stowage factor and the properties and special requirements connected with each one.

6. Meaning of dangerous in rules of law

From the point of view of the carrier, dangerous goods are goods which pose unforeseeable hazards to the ship and to the other cargo. As will be discussed later, knowledge of the nature of the goods and proper handling methods are indispensable in order to prevent danger. According to trade and industry, dangerous goods are that about which something must be known so as to prevent danger.[82] That means that trade and industry is primarily concerned with rules whose dangerous characteristics can be anticipated beforehand.

---

[81] Such materials should be carefully segregated from other dangerous goods carried in packaged or unitized from. The Code describes how this should be done.

[82] Tiberg, "Legal Survey", in Grönfors (ed.), *Damage from Goods* (1978), 9, 11.

Conventions, statutes or regulations do not give a definition of dangerous goods, but instead list dangerous goods. The cargoes listed in the IMDG Code are deemed dangerous; however, it can be deduced from case law that whether the goods in the particular case are dangerous is subject to individual evaluation of the event leading to the damage. The courts look at many different factors and it is not enough to establish that the goods in question are normally regarded as dangerous according to law, regulations or any other relevant source. A particular cargo may be "dangerous" despite the fact that cargoes of its type are not usually so regarded.[83] Conversely, a cargo listed in dangerous cargo regulations may be said to be not dangerous under the special set of circumstances. Thus numerous cargoes cannot be classified in advance simply by reference to their type as "safe" or "dangerous". Bulk cargoes affecting the stability of the vessel have often been the subject-matter of dangerous cargo cases. Although the expression "dangerous goods" is a convenient description of the category of goods to which the obligation to give notice and a different liability regime apply, the term "dangerous" has been a puzzle in case law.

a) English law

The review of English Common Law below will show that damage by cargo is a field which is particularly vulnerable to reasoning by false analogy.[84] For instance, a dangerous cargo of type X is being carried on board a vessel and becomes involved in a situation in which damage is caused to the goods or properties of a person other than the owner of cargo X. A court analyses the legal situation and concludes that the owner of the cargo, or of the carrying vessel, is or is not liable to the injured party. Subsequently, if cargo Y becomes involved in a similar situation with the result that similar damage ensues, it is tempting to treat this as another example of the carriage of "dangerous" cargo. This process of reasoning, which can be seen in some of the reported cases, leads to unsound results.[85] It ignores the fact that certain types of cargo which may become involved in an incident causing damage to other persons or property may not be dangerous at all.

English courts generally focus on the *situation* in which the damage occurred and they seem to be more concerned with dangerous *situation* rather than dangerous *nature*.[86] Therefore, the word dangerous extends beyond matters likely to

---

[83] Wilford/Coghlin/Kimball, *Time Charters* (2003), 179.
[84] Mustill, "Carriers' Liabilities and Insurance" in Grönfors (ed.) *Damage from* Good (1978), 69, 75.
[85] *Ibid.*
[86] *Ibid.* at 76 f. Furthermore, the distinction between the cargo and the situation also emphasizes the different responsibilities of cargo owner, shippers, consignee, carrier and operator of facilities used in carriage of goods. Responsibility for a substance, the danger of which lies in its make-up, is linked to its manufacturer and user, whereas responsibility for a substance, the danger of which lies in its escape, must be at least linked to the carrier or the operator of facilities, for it is the carriage which lies at the root of the danger. And this is more so when a substance becomes "hazardous" only if certain condition exist.

cause physical loss of or harm to the ship,[87] crew,[88] other cargo,[89] or cleaning expenses and delay,[90] and covers all features of the goods which might lead to detention of the ship.[91]

aa) Physically dangerous

It has been said that the word dangerous in common law obligation is not a term of art but is to be given its ordinary meaning.[92] The Oxford Dictionary defines a "danger" as a thing that causes or is likely to cause harm. Reference to the thing itself and not to the circumstances surrounding it is said to imply that the substance has to be intrinsically dangerous.[93] However, there are cargoes which are not intrinsically dangerous yet are capable of causing loss or damage. Several categories of cargo have been suggested which may cause danger:[94]

1. cargo which is known to be always dangerous, however carried (nitroglycerine)
2. cargo which is known to be capable of safe carriage if exceptional facilities, skill and care are employed (liquefied natural gases)
3. cargo which is not dangerous in itself but which can become dangerous if brought into proximity to other cargoes
4. cargo which is not dangerous in itself but which can damage other goods of a particular type
5. cargo whose characteristics are different than those indicated by its appearance or description
6. cargo which is safe in normal circumstances but unsafe in abnormal circumstances
7. cargo which, although harmless if packed in a proper and recognized manner, becomes potentially deleterious if not so packed.

---

[87] *Mediterranean Freight Services Ltd. v. BP Oil International Ltd. (The "Fiona")* [1994] 2 Lloyd's Rep. 506 fuel oil caused an explosion; *Northern Shipping Co. v. Deutsche Seereederei (The "Kapitan Sakharov")* [2000] 2 Lloyd's Rep. 255, explosion and fire by undeclared dangerous cargo and cargo of isopentane; *General Feeds Inc. v. Burnham Shipping Corp. (The "Amphion")* [1991] 2 Lloyd's Rep. 101, cargo of fish-meal, listed in the IMDG Code, caused a fire.

[88] *Bamfield v. Goole and Sheffield Transport* Company [1910] 2 K.B. 94, ferro silicon caused the death of a seaman.

[89] *Bras v. Maitland* (1856) 26 L.J.Q.B 49; 6 E. & B. 470 corrosive effect of lime caused damage to other cargo.

[90] *Deutsche Ost-Afrika v. Legent* [1998] 2 Lloyd's Rep. 71 IMDG Code Class 1.1 explosive cargo caused off-hire, damages and expenses for port of refuge; *Losinjska Plovidba v. Transco Overseas Ltd. (The Orjula)* [1995] 2 Lloyd's Rep. 395, drums of hydrochloric acid leaked.

[91] Treitel/Reynolds, *Carver on Bills of Lading* (2005), 512.

[92] *Sig. Bergesen D.Y. Co. v. Mobil Shipping (The "Berge Sund")* [1992] 1 Lloyd's Rep. 460, 466.

[93] Jackson, "Dangerous Cargo- A Legal Overview" in *Maritime Movement of Dangerous Cargoes- Public Regulations Private Liability* Papers of One Day Seminar (1981), A2.

[94] Mustill, "Carriers' Liabilities and Insurance" in Grönfors (ed.) *Damage from Goods* (1978), 69, 75 f.

If there is one common strain among all the categories, it is the fact that most of these are capable of causing physical damage to the ship or other cargo or personal injury.[95] However, intrinsic danger is not a common trait among them.

bb) Intrinsic danger not necessary

Certain types of cargoes are capable of causing damage only in certain factual contexts.[96] There are cargoes which are not intrinsically dangerous, but may be subject to special rules. These goods are potentially dangerous and in the course of transit they actually become dangerous.[97] For instance, the danger may have resulted from improper stowage. As a result, the question arises of why the potentiality has been translated into an actuality.[98] Cotton is a rather good example. When it is wet, it ignites easily. Cotton is in Class.4.1 of the IMDG Code.

cc) Relevance of the carrier's knowledge to the meaning of dangerous goods

It has been said that it is often impossible to say in theory whether goods are dangerous or not, since the question often depends on the knowledge of the carrier as to the characteristics of the goods.[99] That is to say that the element of the carrier's knowledge is relevant to the question of whether the goods are in fact "dangerous".

Safe carriage of many types of cargo depends on the knowledge on the part of the carrier in relation to the character of the goods and the necessary precautions to be taken. Absence of knowledge, therefore, may transform the cargo from one which is innocuous but potentially dangerous into one which is actually dangerous. Such an approach appreciates the multi- dimensional nature of the problem but covers all type of goods, whatever hazards they may create, and treats the category of "dangerous cargo" as at most non-existent and at least immaterial.[100] The knowledge of the carrier is of fundamental importance. However, it is relevant to the question of the determination of liability for loss or damage arising from the cargo rather than to the determination of the type of the cargo, e.g.

---

[95] "The normal meaning of the word 'dangerous' in relation to goods does seem to ... imply that the goods are such as to be liable to cause physical damage to some other object other than themselves". *Effort Shipping Co. Ltd. Linden Management SA (The Giannis NK* [1994] 2 Lloyd's Rep. 171, 180. Although the case concerned the Hague-Visby Rules, it is submitted that the observation is equally applicable to the common law obligation. Cooke/Young/Taylor, *Voyage Charters* (2001), 150.

[96] Mustill, "Carriers' Liabilities and Insurance" in Grönfors (ed.) *Damage from Good* (1978), 69, 77; Bulow, "'Dangerous' Cargoes: the Responsibilities and Liabilities of the Various Parties" [1989] *LMCLQ* 342, 344.

[97] Abdul Hamid, *Loss or Damage from the Shipment of Goods, Rights and Liabilities of the Parties to the Maritime Adventure* (Diss. Southampton 1996), 53.

[98] *Ibid.*

[99] *The Athanasia Comminos* [1990] 1 Lloyd's Rep. 277, 282.

[100] Jackson, "Dangerous Cargo- A Legal Overview" in *Maritime Movement of Dangerous Cargoes- Public Regulations Private Liability* Papers of One Day Seminar (1981), A10.

whether the cargo is dangerous or not. In other words, a cargo is not dangerous merely that its nature is unknown by the carrier, but liability of the shipper is likely to arise if the nature is unknown by the carrier.

dd) Dangerous as particular characteristics undisclosed

Under certain circumstances a cargo that could have been carried safely had the master been made aware of its particular characteristics may be dangerous if he is not given that information.

A cargo of iron ore concentrate was loaded in *Micada v. Texim*,[101] but the master was misled as to its moisture content and was not informed that the moisture content was such that shifting boards should have been fitted. Shifting boards were not fitted and, as a result of the cargo shifting, the ship developed a severe list and had to put into a port of refuge. It was held that these goods must be considered as being *dangerous*. The danger consisted in the fact that the cargo was not what it seemed to be.[102] It is worth noting that iron ore is listed in the BC Code. However, it could be said that it is not correct to say that cargo is dangerous when it is not what it seems to be, but rather that it might be dangerous if the moisture content exceeds a certain level. In other words, the danger is in the moisture nature of the cargo rather than in the non-disclosure of its moisture content.

ee) Dangerous as cargo has different characteristics than usual

There may be situations where the cargo is accurately and precisely described in the contract of carriage. From such description the carrier will know or is supposed to know of the normal hazards accompanying the cargo. However, there may be cases in which a cargo looks safe and, according to its description, would normally be safe, but, owing to some special and not obvious feature, it is dangerous.

In the *Athanasia Comminos*,[103] a cargo of coal had emitted methane gas, which mixed with air and caused an explosion on board. The ship was damaged and four seamen suffered serious personal injuries.[104] It was said that the character of goods play an important part but it is not the only factor. Equally important are the knowledge of the shipowner as to the characteristics of the goods and the care with which he carries them in the light of that knowledge. The court drew a dis-

---

[101] [1968] 2 Lloyd's Rep. 57.

[102] *Ibid.* at 62. In *Health Steel Mines, Ltd. v. The Erwin Schroeder* [1969] 1 Lloyd's Rep. 370, a cargo of copper concentrate was shipped. However, a certificate of analysis was obtained and, because of the concentrate's nature, shifting boards were fitted, the ship listed. It was found that a cargo of this nature with these characteristics is a dangerous cargo for a vessel to have on board.

[103] [1991] 2 Lloyd's Rep. 277

[104] In particular, by reason of the opinion that no special precautions would have been considered necessary by a prudent owner at the relevant time.

tinction between "extra hazardous goods"[105] and "goods which are neither dangerous nor safe". The court continued that the carriage of coal involves hazards greater than those associated with inert goods; but they are hazards which could be overcome if the shipowner had the necessary knowledge, skill and equipment; and this is so even if the particular cargo brings with it a risk greater than that which is usually associated with the carriage of coal. In such a case, it is not correct to start with an implied warranty as to the shipment of dangerous goods and try to force the facts within it; but rather to read the contract and the facts together, and ask whether, on the true interpretation of the contract, the risks involved in this particular shipment were risks which the plaintiffs contracted to bear. It was found that coal cargo had no special characteristics at the time of shipment.[106] Consequently, it can be said that if the particular characteristics are not wholly different than usual, cargo may not be deemed dangerous. In other words, if the particular characteristics are completely different than usual, cargo may be deemed dangerous.

ff) Innocuous goods causing damage may be dangerous

Although it is obvious that the subject of damage in transit to a large extent arises from inherently dangerous cargo or cargo becoming dangerous under special circumstances, there are cases in which damage arises from totally innocuous cargo or from the defects in the packages themselves. Cases dealing with defective packing often involve dangerous or hazardous cargo that leaks from a defective container, drum or otherwise, but there are also cases involving a situation where an innocuous cargo causes damage. In common law, a cargo may be dangerous if it is dangerously packed[107] or necessitates cleaning of the ship.

For instance, in *Ministry of Food v. Lamport & Holt Line*,[108] where cargo of tallow contaminated another cargo of maize, the owner argued that it was a startling proposition that anything so placed and innocuous as tallow should be dangerous. It seemed that goods are dangerous in the sense that they can do damage to other goods. Therefore, it was said that there is not much difference between the dam-

---

[105] It was asserted that the term "extraordinary hazards" should not be equated with hazards which are intrinsic in the goods. It actually means hazards arising from goods which are extraordinary to the knowledge of the carrier. Abdul Hamid, *Loss or Damage from the Shipment of Goods, Rights and Liabilities of the Parties to the Maritime Adventure* (Diss. Southampton 1996), 56.

[106] See also *The Atlantic Duchess* [1957] 2 Lloyd's Rep. 55, where butanized crude oil caused fire and explosion. It was held that the shipowners had failed to prove that the butanized crude-oil cargo shipped by the charterers was outside the contractual description, or that butanized crude oil involved any special hazards. In *Mediterrenean Freight Services Ltd. v. BP Oil Internatinal Ltd. (The Fiona)* [1993] 1 Lloyd's Rep. 257, 258 the fuel-oil cargo had dangerous characteristics which were wholly different from those commonly associated with fuel-oil cargoes.

[107] Boyd/Burrows/Foxton, *Scrutton on Charterparties* (1996), 105.

[108] [1952] 2 Lloyd's Rep. 371.

age done, for instance, by corrosives or by a commodity which leaks something less dangerous. .[109]

Moreover, in the *SIG Bergesen DY and Co. v. Mobil Shipping (The Berge Sund)*,[110] the owners argued that a cargo is dangerous if it or its residue contaminates another cargo on the next voyage. That would mean any cargo would be dangerous if it contaminated in the absence of effective cleaning. The court accepted in principle that the shipment of a cargo which, unknown to the carrier, would necessitate extensive decontamination of the ship before she was fit to load her next cargo might amount to the shipment of dangerous cargo. However, this proposition was deemed to be too wide, since it would mean that any cargo would be dangerous if it contaminated in the absence of effective cleaning.[111] It is said that this is a matter of degree and the mere necessity to clean after the voyage cannot render the cargo dangerous.[112]

gg) Legally dangerous goods

The notion of "harm" in the definition of "danger" connotes physical danger in the form of personal injury, or damage to ship or property on board.[113] Nevertheless, in English common law the concept of dangerous goods is potentially wider, because it embraces not only goods which are physically dangerous but also those which are likely to subject the ship to legal or political risks, causing detention or confiscation, or delay.[114] That is to say, the application of the dangerous goods provision was also extended to *unlawful goods*.

In *Mitchell Cotts v. Steel*,[115] a ship was chartered on a voyage from Basrah to Alexandria with a cargo of rice. After the voyage had commenced, the charterers

---

[109]   *Ibid.* at 382. As the carrier was aware of the nature of the goods, there was no breach of any duty. However, in *Goodwin, Ferreira & Co. Ltd. and Others v. Lamport & Holt, Ltd.* (1929) 34 Ll. L. Rep. 192, a cargo of machinery came out of its case, dropped to the bottom of the lighter, made a hole in it, let in seawater and damaged the other cargo of cotton yarn. The case was decided as being one general cargo case.

[110]   [1993] 2 Lloyd's Rep. 453.

[111]   *Ibid.* at 463.

[112]   Cooke/Young/Taylor, *Voyage Charters* (2001), 150. The authorities offer little guidance on where the line is to be drawn. In the *Berge Sund* case, the need to clean the ship, which took about 10 days, was not held to render the cargo dangerous. In the *Bela Krajina* [1975] 1 Lloyd's Rep. 139 case, the fact that the vessel required a week to clean was held not to raise the inference that more than normal cleaning and washing was required, and in *The Giannis NK* [1998] 1 Lloyd's Rep.337 case, the need to fumigate the vessel after carriage of the infested cargo was not held to render the cargo dangerous as regards the vessel. By contrast, in *The Orjula* [1995] 2 Lloyd's Rep. 395 case, the contamination of the vessel by hydrochloric acid, which required cleaning by specialist contractors, was held to amount to physical damage.

[113]   Abdul Hamid, *Loss or Damage from the Shipment of Goods, Rights and Liabilities of the Parties to the Maritime Adventure* (Diss. Southampton 1996), 61.

[114]   Boyd/Burrows/Foxton, *Scrutton on Charterparties* (1996), 105; Cooke/Young/Taylor, *Voyage Charters* (2001), 153.

[115]   [1916] 2 K.B 610.

asked the owners to agree to a change of destination to Piraeus, and the owners agreed. The charterers, but not the owner, were aware that rice could not be discharged at Piraeus without the permission of the British Government. The ship was detained at Piraeus for 22 days while attempts were made to obtain permission, which was eventually refused. The shipowners claimed damages for detention. The claim succeeded and it was held that the shipper undertakes that he will not ship goods likely to involve unusual danger or delay to the ship without communicating to the owner facts which are within his knowledge indicating that there is such risk, if the owner does not and could not reasonably know those facts.[116]

However, this principle extends the duty of the shipper and goes one step further beyond the dangerous goods principle.[117] As the word "dangerous" connotes physical damage, it becomes manifestly inappropriate to apply it to cases where the loss to the carrier takes the form of delay or detention. But the common law undertaking is to be understood as a general approach to liability which should include both physical loss and economic loss arising from delay or detention of the ship.[118]

It is said that the basis of this general principle has been traced to a passage in Abbott's treatise, where the author states that "The merchant must lade no prohibited or uncustomed goods, by which the ship may be subjected to detention or forfeiture".[119] It should be noted, however, that there the author was discussing the situation where a shipment caused detention or forfeiture of the vessel, and not

---

[116] By contrast in *Owners of Spanish S.S. Sebastian v. De Vizcaya* [1920] 1 K.B. 332, a cargo of coal was contracted to be loaded and delivered to a port in Spain. Subsequently, a proclamation prohibited the export of coal to Spain except with a licence. At the loading port the vessel was detained while waiting for the licence, which, without default on the part of the charterers, was not obtained until 2 weeks later. The shipowners relied on the implied common law obligation as propounded in the *Mitchell Cotts* case. It was, however, held that as the owners had knowledge at the time the cargo was loaded that it was necessary to obtain an export licence, which might involve delay, the charter was not liable for the detention. See also *Chandris v. Isbrandtsen-Moller Co. Inc.* [1951] 1 K.B. 240.

[117] The principle in *Mitchell Cotts v. Steel* is not extended so far to entitle the shipowner to recover damages for *any* expense or delay caused by the nature or condition of the goods. In *Transoceanica v. Shipton* (1923) 1 K.B. 31, the presence of stones in a cargo of barley prevented the cargo from being discharged by spout and delayed the vessel for a day and a half. It was held that the cargo was such that it could be discharged by the appliances normally in use at the discharged port. In *Rederi Aktiebolaget Transatlantic v. Board of Trade* (1924) 20 L.l.L.Rep. 241, it was held that the charterer was in breach of the implied warranty by loading a cargo of heavy locomotives weighing up to 16 tons when they had been informed before the charterparty was made that the ship's tackle could only lift 5 tons and at the discharging port there was no equipment capable of handling the goods. The distinction between these cases is that in the former case the delay and inconvenience were not of major significance, whereas in the latter case the condition of the cargo caused serious consequences.

[118] *Giannis NK* [1994] 2 Lloyd's Rep. 171, 179.

[119] Abdul Hamid, *Loss or Damage from the Shipment of Goods, Rights and Liabilities of the Parties to the Maritime Adventure* (Diss. Southampton 1996), 64.

physical damage to the ship or its cargo. Therefore, it can be contended that the application of the common law undertaking to loss in the form of delay or detention does not involve a purported extension of the dangerous goods principle.[120] If there were any extension, it might be the opposite.

On the other hand, in *the Giannis NK* case, the House of Lords declined to express a view on whether goods may be of a dangerous nature within the meaning of Art. IV.6 of the Hague/Hague-Visby Rules if they are liable to cause delay or seizure of the ship and cargo through the operation of some local law.[121] Clearly, if the risk is one of delay only as opposed to seizure, the case is less strong.[122] On the other hand, it is said that if the Hague Rules do not deal with non-physically dangerous cargo, one cannot determine the rights of the parties in relation to such cargo by reference to such rules.[123]

hh) What is meant by "goods of inflammable, explosive or dangerous nature" in Art. IV.6 of the Hague/Hague-Visby Rules

(1) Dangerous not restricted to preceding words

Art. IV.6 of the Hague/Hague-Visby Rules provides that "Goods of an inflammable, explosive or dangerous nature to the shipment…". The first two designated types of danger both indicate combustion of some sort. The provision in the second paragraph indicates merely that a "danger to the ship or cargo" is contemplated and that is the governing criterion. However, the word "dangerous" is given a broad meaning, not limited to the inherent properties of the goods themselves nor the sort of danger posed. It is contended that the danger may not simply be to the vessel but may extend to other goods on the vessel and it may be affected by the laws at the ports of call or discharge.[124]

In The *Giannis NK*,[125] a cargo of groundnut and a cargo of wheat pellets were loaded in the Caribbean, where there were strict quarantine and phytosanitary regulations. At the second port of discharge, the groundnut pellets were found to be infested with live khapra beetles. This infestation was inherent in the cargo upon shipment despite fumigation and without the knowledge of the shipowners, the charterers or the shippers. Because the Khapra beetle was a voracious consumer of foodstuffs and thus undesirable, the vessel was ordered by the local agricultural authorities to dump the entire remaining cargo at sea or to return it to the country of loading. The remaining wheat pellet cargo was not itself threatened by the infestation spreading to it, but it was likely that the consequence of the infestation of the groundnut cargo was that it would have to be destroyed too, as indeed happened when both cargoes were dumped at sea. The vessel was then extensively fumigated so that she was fit for further trading.

---

[120] *Ibid.*
[121] *The Giannis NK* [1998] 1 Lloyd's Rep. 337, 341, 346.
[122] Treitel/Reynolds, *Carver on Bills of Lading* (2005), 645.
[123] *The Giannis NK* [1998] 1 Lloyd's Rep. 337, 338.
[124] Cooke/Young/Taylor, *Voyage Charters* (2001), 1006.
[125] *The Giannis NK* [1998] 1 Lloyd's Rep. 337.

The shipowner sued both shippers and charterers. The claim, based on Art. IV.6 of the Hague-Visby Rules, was that the groundnut pellets cargo was dangerous cargo, presenting a physical danger to the vessel, which required fumigation, and to the other wheat pellet cargo, which also had to be dumped at sea. The House of Lords held that the cargo was dangerous within the meaning of Article IV Rule 6. It was of a "dangerous" nature on shipment because it was liable to give rise to the loss of other cargo by dumping at sea and, presumably, to the quarantining of the vessel until after she had been fumigated. What made the cargo dangerous was the fact that the shipment and the voyage were to countries where the imposition of quarantine and an order for the destruction of the entire cargo was to be expected or at least a natural and not-unlikely consequence of the presence of the infested goods. Accordingly, the word "dangerous" was given a broad meaning, not confined to goods of inflammable or explosive nature or their like.[126] It was said that it would be wrong to apply the *euisdem generis* rule to the words "goods of inflammable, explosive or dangerous nature" as these are disparate categories of goods.[127]

The idea that "dangerous cargoes" cover those whose danger is that of quarantine blight in a world where dangerous cargoes are categorized in the IMDG Code and other codes is surprising. The words "inflammable" and "explosive" do not suggest this, and it is not entirely satisfactory to extend the meaning of the word "dangerous" in this way.[128] Art. IV.6 aims at regulating extraordinarily dangerous things such as chemicals and chemical effects that may pose unreasonable dangers to ship or cargo; therefore, infestation should not fall under Art. IV.6.[129]

(a) Eiusdem generis rule

The so-called *eiusdem generis* rule lies down that where a list of specific items is followed by general words such as "any other cause", the general words should be interpreted as being restricted to things of *the same kind* as the specific examples.

---

[126] *Ibid.* at 338.

[127] *Ibid.* at 346.

[128] Treitel/Reynolds, *Carver on Bills of Lading* (2005), 645. See also *Acatos v. Burns* (1878) 3 Ex.D. 282. By contrast, it is asserted that the court's approach is a realistic analysis of the risks accepted by a shipowner in carrying a shipper's cargo with unknown characteristics. Thus, the use of word "danger" in this context should be reconsidered. Steel, "Dangerous Beetles? The Hague Rules and The Common Law" [1996] *IJOSL* 229, 230.

[129] In a U.S case, the *Stevenson & Co. Inc. v. Bags of Flour* (629 F.2d. 338), whose facts are very similar to that of the *Giannis NK,* an in rem action, stemming from insect infestation of wheat flour, was brought by the carrier, which claimed a lien on the cargo of one of the vessel for freight, detention and expenses. It was held that infestation began either on the rail cars or at three independent mills supplying substantial amounts of the flour. There was substantial evidence that neither warehouse nor ships were the source of infestation. Flour from previous shipments had been recently stored in the warehouses without any problem and there was evidence that ships and warehouses were relatively clean. It was found that infestation occurred prior to loading. The carrier recovered for damages and detention.

In charterparty cases, its application has been sporadic and unattended by any generally accepted principles. For the *euisdem generis* rule to apply, there must be sufficient indication of a category that can properly be described as a *class* or *genus.*[130] The nature of the genus is gathered by implication from the express words which suggest it. Usually these consist of a list or string of substantives or adjectives.[131]

In *Chandris v. Isbrandsen,*[132] the court refused to apply it to a provision in a charterparty which prohibited the shipment of "acids, explosives, arms, ammunition or other dangerous cargo" on the ground that there was no presumption to the effect that it should be applied and nothing in the context to show that it was intended to apply.[133] In this case, the charterparty incorporated the provisions s.4 sub-s. 6 of the U.S. COGSA, which is identical with Art. IV of the Hague/Hague-Visby Rules, and with the knowledge of and consent of the master, but not of the owner, the charterer shipped turpentine, which is highly flammable. The ship was delayed 16 days beyond the lay-days. It was held that the charterers were in breach of contract in shipping the turpentine, since this was a dangerous cargo within the ordinary meaning of those words, and the meaning of the words was not restricted to cargoes like acids, explosives or ammunition.

On the question of whether turpentine was "like" any of the goods specified, it was found that turpentine was a volatile inflammable liquid. Its flash point was $90°$-$95°F$, at which it gave off a vapour, which, with air in certain proportions, formed an inflammable and explosive mixture. Generally, the risk in the carriage of turpentine was inflammability rather than explosion, but in a confined place, if the temperature was sufficiently high, an explosion involving considerable concussion might result. It is in Class 3 of the IMDG Code. The chief danger in the carriage of acids was corrosion, though tainting of certain types of other cargo might also result. Turpentine did not cause corrosion, but could taint certain types of cargo. It was held that general words prima facie be considered as having their natural and larger meaning and not to be restricted to things *euisdem generis* with

---

[130] The *euisdem generis* rule has not been applied to the "q" clause exception of the Hague/Hague-Visby Rules Art. IV.2 in *Potts & Co. v. Union SS. Co. of New Zealand* [1946] N.Z.L.R 276, 286, where it was held that there was no genus or class encompassing the (a) to (p) exceptions in Art. IV.2. Consequently, it was held that the *euisdem generis* rule could not apply and that the words "any other cause" must be given their ordinary and natural meaning. Most authorities have also taken the view that the q-clause exception is to be construed broadly. Generally speaking, however, the Hague/Hague-Visby Rules exceptions must be interpreted restrictively. Tetley, *Marine Cargo Claims* (1988), 515 f.

[131] Bennion, *Statutory Interpretation* (1997), 956.

[132] [1951] 1 K.B. 240.

[133] *Ibid.* at 246. Other charterparty cases, however, in the context of liberty clauses and laytime exception clauses, have adopted the approach that the *euisdem generis* rule is generally applicable unless there is something in the language or the context to rebut this application. Cooke/Young/Taylor, *Voyage Charters,* (2001) 35. See *Knutsford v. Tillmans* [1908] A.C. aff'd [1908] 2 K.B. 385; *Thorman v. Dowgate* [1910] 1 K.B. 410.

those previously enumerated, unless there was something to show an intention so to restrict them.[134]

The *euisdem generis* rule might be explained by reasoning that the drafter, on working down the list, would keep this in mind when writing the general term at the end. It can be deduced that the final term was probably intended by the legislature to be limited to the terms in the list. In interpreting the rule, it is asserted that the drafter must be taken to have inserted the general words in case something which ought to have been included among the specifically enumerated items had been omitted or it is assumed that the general words were only intended to guard against some accidental omission in the objects of the kind mentioned and were not intended to extend to objects of a wholly different kind.[135] The validity of this rule depends on the presumed competence of the drafter.

In *Chandris v. Isbrandtsen-Moller,* it was said of commercial contracts, which are known to be frequently ill-drawn: "The presumption against surplusage is of little value in ascertaining the intention of the parties to commercial documents as many great commercial judges have recognized."[136] The same applies to enactments which are the subject of disorganized composition. However, where there is a "genus", words should be given a relatively broad meaning but one that is limited to "genus".

On the other hand, despite *Chandris,* it was suggested that 'any other goods of a dangerous nature' is probably to be construed *euisdem generis* with the goods specified before."[137]

(b) Euisdem generis rule and Art. IV. 6 of Hague/Hague-Visby Rules

Although not applying *eiusdem generis* might be justified on the basis that turpentine is included in the IMDG Code, the same cannot be said with regard to the purpose of the Hague Rules Art. IV.6 for kaphra beetles, whose mere danger was the fact that the shipment and voyage were to countries which imposed quarantine and an order for the destruction of the entire cargo.

The drafters of the Hague Rules were aware that that any cargo could be dangerous under certain circumstances. During the discussions, it was pointed out that very ordinary cargo may at some time be ignitable.[138] However, Art. IV.6 was aimed at regulating not *any* dangerous cargo but *exceptionally* dangerous cargoes, particularly *chemicals*.[139] Therefore, it is questionable whether one should extend the meaning of dangerous cargo under Art. IV.6 to any cargo causing damage, since the Article was originally intended to govern only exceptional dangers.

Art. IV.6 of the Hague/Hague-Visby Rules states "Inflammable, explosive or dangerous nature…", e.g. chemical hazards. Indeed, goods in Classes 1 to 5 of the

---

[134] *Chandris v. Isbrantsen-Moller Co. Inc.* [1951] 1 K.B. 240, 244 ff.
[135] Bennion, *Statutory Interpretation* (1997), 955.
[136] Bennion, *ibid.; Chandris v. Isbrantsen-Moller Co. Inc.* [1951] 1 K.B. 240, 244.
[137] Colinvaux, *Carver's Carriage by Sea* (1982) 846 fn. 84.
[138] Sturley, *Legislative History of the Carriage of Goods by Sea Act* (1990) Vol. 1, 272 f.
[139] *Ibid.*

IMDG Code are explosion and/or fire hazards.[140] Therefore, in this regard, Art. IV.6 is in conformity with the danger classification of the IMGD Code. Although the hazard classification of the IMDG Code is not limited to inflammability and explosivity, it would not be problematic to extend hazards to toxicity and corrosion, as they certainly fall under chemical hazards. However, what would be controversial is that the IMDG Code covers not only chemicals but some very ordinary cargoes which are not chemicals but may pose chemical hazards, such as coal or cotton. Moreover, it is highly difficult to justify under Art. IV.6 those bulk cargoes whose mere danger is to affect the stability of a vessel. However, it is at least clear that Art. IV.6 was intended to regulate *exceptional* dangers.

(2) Dangerous goods and inherent vice

(a) Inherent vice or defect defence

The inherent vice or defect of the cargo is one of the exculpatory exceptions most often relied on by carriers. The generally accepted definition of "inherent vice" in carriage of goods cases is the unfitness of the goods to "withstand the ordinary incidents of the voyage, given the degree of care which the shipowner is required by the contract to exercise in relation to the goods."[141] That means the risk of deterioration of the goods shipped as a result of their natural behaviour in the ordinary course of the contemplated voyage without the intervention of any fortuitous external event or casualty.

Hague-Visby Rules Art. 2(m) absolves the carrier from damages caused by "wastage in bulk or weight or any other loss or damage arising from inherent defect, quality or vice of the goods". The carrier is not liable for damage caused by the nature of the cargo itself; however, he is not excused from exercising due care to preclude or minimize loss or damage resulting from this exception.[142]

It should be pointed out that the official French version of the Hague Rules uses the term "vice caché" which means "hidden defect" and "vice propre" which means "inherent vice".[143] Accordingly, an inherent vice is one which is an innate or natural or normal quality of the goods. For instance, it is an inherent vice of flour that it shrinks and loses weight with the elapse of time. On the other hand, a hidden defect is something that is hidden and defective in the cargo and not nor-

---

[140] Watt/Burgoyne, "Know Your Cargo" [1999] 13(5) *P&I Int'l* 102.

[141] Boyd/Burrows/Foxton, *Scrutton on Charterparties* (1996), 224; Tetley, *Marine Cargo Claims* (1988), 479 ff.

[142] Thus, where the carrier is cognizant of the perishable nature of the cargo, it may be required to exercise greater care than that required for non-perishable cargo. Each case must be decided on its particular facts.

[143] Tetley, *Marine Cargo Claims* (1988). 479. The official French text is: "4(2)(m) de la freinte en volume ou en poids ou de toute autre perte ou dommage résultant de vicé caché, nature spéciale ou vice propre de la merchandise." The provision makes it clear that the exception refers to both hidden defect and inherent vice.

mally expected to be found there, although the cargo may have the propensity to attract such defect such as infestation of flour by tiny larvae.[144]

This exemption embodies the equitable principle that the carrier should not be liable in the absence of fault. The theory and policy of the exemption is that the shipper rather than the carrier should know the inherent characteristics of the goods shipped and should have the responsibility of guarding against loss or damage.[145]

(b) The overlap of dangerous goods obligation and inherent vice defence

In some situations, the facts which allow for the consideration of a cargo as dangerous goods may also be analyzed as an application of the inherent vice defence. A cargo of coal which spontaneously combusts may damage the ship and other cargo, so as to give the carrier a cause for action under the dangerous goods obligation. At the same time, if the suit is brought by the cargo-owner for loss or damage to the coal, the carrier may plead in his defence the exception of inherent vice.[146] It appears that the obligations implied by common law regarding dangerous goods and the common law defence of inherent vice originate from the same source; thus, in some cases, the two doctrines may overlap.[147]

(c) Distinction between dangerous goods and inherent vice or defect

The distinction between goods which are of a dangerous character and goods which merely suffer from inherent vice is not always easy to draw, yet it is important. It is asserted that inherent vice constitutes a defence to a claim for damages arising out of loss or deterioration in the course of carriage but it does not itself involve any breach of duty by the shipper.[148] Accordingly, where the condition of the goods on shipment is such that they are liable to cause injury to persons or damage to the ship or other goods, or even serious delay to the voyage, they fall within the category of dangerous goods.[149]

This assertion, however, is doubtful, because whether a particular commodity suffers from an inherent vice is a mixed question of law and fact.[150] The determination of the applicability of inherent vice or defect exception often turns

---

[144] The composite inherent defect as erroneously enunciated in the English translation of the Hague Rules has become commonplace in jurisprudence. It is, nevertheless, suggested that "inherent defect" covers both "hidden defect" and "inherent vice". Tetley, *Marine Cargo Claims* (1988) 480.

[145] Schoenbaum, *Admiralty and Maritime Law* (2004) Vol. 1, 704 f.

[146] *Greenshields, Cowie v. Stephens* [1908] A.C. 431, where maize sprouted; *The Amphion* [1991] 2 Lloyd's Rep.101, where fishmeal heated; *Accinanto v. Ludwig* [1953] 1 Lloyd's Rep. 38, where ammonium nitrate fertilizer spontaneously combusted.

[147] Abdul Hamid, *Loss or Damage from the Shipment of Goods, Rights and Liabilities of the Parties to the Maritime Adventure* (Diss. Southampton 1996), 40.

[148] Cooke/Young/Tailor, *Voyage Charters* (2001), 152.

[149] *Ibid.* at 153.

[150] Schoenbaum, *Admiralty and Maritime Law* (2004) Vol. 1, 704.

on the burden of proof. The carrier cannot rely on the excuse that the loss has been due to an inherent weakness of the goods unless he has used such care with them as their nature demanded.[151] Moreover, where a loss which may be traced to an inherent quality or defect of the goods has arisen not from the ordinary development of that quality or defect, but from adventitious causes introduced by the carrier, the inherent vice defence does not apply.

Furthermore, the inherent vice defence relates to insect infestation of food cargo; particularly grain is one of the most difficult applications of inherent vice. The issue is generally determined on a case-by-case basis, depending on whether the insect infestation occurred or worsened as a result of the carrier's negligence or before the shipment.[152]

Fishmeal is a good example of cargo which may fall under both inherent vice and dangerous cargo. Fishmeal is listed in the IMDG Code. When fishmeal causes damage to a shipment, it is not correct to say from the outset that it was due to the dangerous nature of fishmeal. As explained in Part 4, the shipper is not liable for "shipping dangerous goods" but "shipping dangerous goods without informing the carrier of the dangerous nature of the cargo." In most cases, the carrier will know that he is shipping fishmeal and so liability for damages arising from the shipment of fishmeal will depend on the facts of the case, i.e. whether it was properly prepared for the shipment or not. If improperly handled, that will be the fault of the shipper. In the same way, improper and careless handling by the carrier will be the fault of the carrier.

In principle, there might seem to be no good reason in drawing a distinction between damage such as that arising in inherent defect cases and in the shipment of dangerous cargo.[153] However, as a different standard of liability is applied, there should be a difference between dangerous goods and ordinary goods which may cause damage.

ii) Basis of liability for delay

With regard to cargo damage, cargo which is delayed although not physically damaged may suffer loss in value because of a fall in the market price. Histori-

---

[151] Colinvaux, *Carver Carriage by Sea* (1982) Vol. 1, 15.

[152] The infestations which occur in ships originate either in the commodity before loading or from cross-infestation from residues of previous cargoes or other infested cargoes being carried. The condition of goods prior to loading is a consideration in proving inherent vice or defect. It is often difficult, however, to determine the precise origin of particular infestations and the issue is generally determined on a case- specific basis. Little or no idea can be obtained about the propensity of a cargo to infest, sprout etc. by watching it being loaded onto the ship, and the likelihood of inherent defect arising can only be assessed after obtaining information as to the origin of the grain, the conditions under which it has been handled and stored, and its heat, moisture etc. content, all of which are usually more accessible to the shipper than to the carrier. The reason is that this procedure is more difficult to prove rather than holding the shipper strictly liable by applying Art. IV.6.

[153] Cooke/Young/Tailor, *Voyage Charters* (2001), 153.

cally neither maritime law nor marine insurance recognizes claims for delay because of the inherently dangerous unpredictable nature of maritime commerce.[154] Likewise, neither the Hague nor the Hague-Visby Rules refer specifically to delay. The question was left to national laws.[155] The word "damage" appears in different articles: damage Art. 3 (6); Art. 4 (1), (2),(3),(4), damage in connection with goods (Art. 3(8)), damage or in connection with goods (Art. 4(5) and(6)). In common law, "damage or in connection with goods" was interpreted as covering not only cases of physical damage but also delay,[156] whereas in Continental law, contracting states incorporated the expression "loss or damage" in Art. 4(1),(2), and (4) into their domestic legislation as "loss of or damage to goods" and interpreted it as "physical damage."[157]

Today the oceans of the world are not considered as dangerous and damages for delay in delivery are awarded on occasion. In English law, in general, even when there is no physical damage or deterioration, damages for loss of market caused by delay in the delivery of the cargo would be recoverable if the carrier knew or could be expected to have known the peculiarities of the cargo at the time of contracting or if special circumstances were communicated to the carrier.[158]

Under the Hamburg Rules Art. 5.1, the carrier can be liable for delay in delivery. Delay in delivery is defined in Art. 5.2 as occurring when the goods have not been delivered at the port of discharge within the time agreed in the contract of carriage. Art. 5.2 continues by reference to the time which it would be reasonable to require of a diligent carrier, having regard to the circumstances of the case. The provisions on delay were rather controversial and carriers feared open-ended liability. Furthermore, Art. 6.1 (b) provides a special limit of liability that is not more than two and half times the freight payable on the delayed cargo.

Under these circumstances, it is hard to justify extending the meaning of dangerous goods which only cause delay or detention on the basis of strict liability, whereas the carrier's liability for delay applies under certain circumstances and is subject to limitation and, more importantly, on the basis of fault liability.[159] In the shipping business, time is money. Due to new security regulations,[160] delays are a fairly new area of disputes. Under Hague/Hague-Visby Rules there is still scope

---

[154] Tetley, *Marine Cargo Claims* (1988), 309.
[155] Sturley, *Legislative History of the Carriage of Goods by Sea Act* (1990) Vol. 1, 498, 514.
[156] Sturley, *Legislative History of the Carriage of Goods by Sea Act* (1990) Vol. 3, 205; Schoenbaum, *Admiralty and Maritime Law* (2004) Vol. 1, 734; Colinvaux, *Carver Carriage by Sea* (1982) Vol. I, 311 ff.
[157] Karan, *The Carriers Liability under the International Maritime Conventions the Hague, Hague-Visby, and Hamburg Rules* (2004), 217 f.
[158] *Hadley v. Baxendale* (1854) 9 Ex.C.341, 156 E.R. 145; Gaskell/Asariotis/Baatz, *Bills of Lading: Law and Contracts* (2000), 342; Tetley, *Marine Cargo Claims* (1988), 309.
[159] See *infra* Part 4 and 5.
[160] See *infra* Part 3.

for liability for delay outside the ambit of Art. IV.6 under Art. IV.3.[161] Why should the carrier be in more favourable position?[162]

b) Meaning of dangerous in American law

Although the definition of what constitutes a dangerous cargo is somewhat elusive, American law on dangerous goods deals more with cargoes which may cause physical damage rather than legal obstacles. Cargoes listed in the 49 CFR 171-180 or 46 CFR 146-154, 173 and the IMDG Code and other codes are deemed to be dangerous in general. However, equally important are the facts of the cases. The carrier's knowledge of the dangerous nature of the cargo is of great importance in determining liability. As in English law, bulk cargoes creating stability dangers are often considered in dangerous cargo cases. From the extensive list of regulations, particularly coal, fishmeal, turnings and borings have been the subject matter of dangerous cargo cases in which they have caused damage.[163]

aa) Dangerous because different from contractual description

As in English law, cargo may be dangerous if it is different from the one described. In the *Santa Clara*[164] case, copper concentrates were loaded instead of

---

[161]   Steel, "Dangerous Beetles in the House of Lords – Shippers Absolutely Liable" [1998] (2) *IJOSL* 119, 120.

[162]   The CMI/UNCITRAL draft instrument Art. 27 obliged the shipper to provide the carrier with the necessary information, instructions and documents. According to Art. 28, this information, instructions and documents must be accurate, complete and timely. The shipper's liability for failure to comply with these obligations is strict. However, why the shipper should be liable, irrespective of his own fault, for failure to provide any such information, instructions or documents or for delay is not clear. Asariotis, "Main Obligations and Liabilities of the Shipper" [2004] *TranspR* 284, 290. Later on, Art. 27 became 30 and Art. 31 was drafted, including in parentheses a reference to delay as a basis of liability of the shipper. A/CN.9/WG.III/WP.56. However, the issue of delay was particularly problematic as a basis for the shipper's liability, since it could expose the shipper to enormous and potentially uninsurable liability. A/CN.9/591 parag.143 ff. See *infra* p. 210 f.

[163]   *Pitria Star Navigation Co. v. Monsanto Co.* 1984 WL 3636 (E.D.La 1984), *The Stylianos Restis* 1974 AMC 2343 (S.D.N.Y 1972), *Poliskie Line Oceanic v. Hooker Chemical Corp.* 499 F.Supp. 94, *International Marine Development Corp. v. Lakes Shipping and Trading Co. (The M/V Gilia)* (1975) S.M.A No: 931, *Conti Shipping v. Bomar Resources Inc. (The Continental Trader)* 1986 S.M.A No. 2211, *Tramp Shipping Co., Julianne Shipping Corp. v. Amalgament Inc. (The M.V. Kapetan Antonis)* 1988 S.M.A. 2516, *Ente Nazionale per l'Energia Elettrica v. Baliwag Navigation Inc.*605 F.Supp. 355, *Colormaster Printing Ink Co. v. S.S.Asiafreighter* 1991 WL 60413 (S.D.N.Y), *Ionmar Compania Naviera S.A v. Olin Corp,* 666 F.2d 897, *Borgships Inc. v. Olin Chemicals Group* 1997 WL 1241127 (S.D.N.Y), *United States v. M/V Santa Clara I,* 887 F.Supp. 825. *A/S Gylfe v. Hyman-Michaels Co. (The Gyda)* 304 F.Supp. 1204, 1971 AMC 2041, *Pt. Karana Line v. Eddie Steamship Co. Ltd. (The Kartini,)* (1984) AMC No. 1958, *Boykin v. China Steel Corporation,* 73 F.3d 539, 1996 AMC 920 (4th Cir. 1996) .

[164]   281 F.725 (2d Cir. 1922).

copper ore. The vessel became unseaworthy because the stowage and securing of the cargo was insufficient. The cargo was dangerous, as it was different from the one described. Similarly, in *Boykin v. China Steel Corporation*,[165] the cargo of coal was incorrectly described as being Category A (the most common type of coal, which is not dangerous) when, in fact, it was Category B, a highly volatile coal. The cargo was found to be dangerous.

bb) Goods which become dangerous under special circumstances

A cargo which is not dangerous *per se* can become dangerous due to a special set of circumstances. For instance, in *Narcissus Shipping Corp. v. Armada Refeers, Ltd*[166] the cargo carried was orange juice which was packed in plastic bags and then loaded into plastic drums. The drums were shipped as break bulk cargo secured by stanchions. The vessel experienced a severe list. It was held that a voyage charterer was liable because of his failure to inform the owner, master or time charterer of the dangerous properties of the cargo.[167] However, generally solid bulk cargoes those affecting the stability of the ship fall under this heading.

In *Sucrest Corp. v. M/V Jennifer*,[168] the vessel had a cargo of sugar which shifted during the voyage because of a biological degradation of the cargo. The vessel developed a severe list and the master intentionally went aground in order to save the vessel and her cargo. The casualty was the first instance in which the maritime or scientific community learned of the thixotropic properties of raw sugar.[169] Similarly, in *P. Brown Jun. & Co. v. Minas de Matahambre (The Nord Amerika)*,[170] the cargo of copper concentrate became colloidal when an excess amount of moisture and oil separated. The cargo, therefore, shifted and caused the vessel to list, forcing it to seek refuge. Arbitrators concluded that "although copper concentrate *per se* is not dangerous cargo, such colloidal state is both dangerous and injurious within the terms of the NYPE Time Charterparty."[171]

By contrast, in *Westchester Fire Ins. Co. v. Buffalo Housewrecking & Salvage Co.*,[172] a barge was destroyed by fire when its cargo of turnings and borings overheated. The bill of lading contained the provision that the shipper should be liable in the event that the cargo was dangerous, unless full written disclosure of its character was made to the barge owner. In dealing with the threshold issue of whether the cargo was dangerous, the court stated that, although listed in Class 4 of the IMDG Code, turnings are not considered to be explosives or dangerous, but are well-known articles of commerce. The findings of the Bureau of Explosives read that "iron turnings, borings, filings, when in large bulk, have a fire hazard as

---

[165] 73 F.3d 539.
[166] 950 F.Supp. 1129.
[167] *Ibid.* at 1143.
[168] 455 F.Supp. 371.
[169] Likewise the charterer had no actual or constructive knowledge of the inherent dangers of the cargo as the cargo of bulk raw sugar. *Ibid.* at 385.
[170] 1931 A.M.C 1637.
[171] *Ibid.* at 1642 f.
[172] 40 F.Supp. 378.

they oxidize spontaneously if wet and the oxidation may produce enough heat for ignition. This risk is not sufficient to cause material to be classed as inflammable by ICC (Interstate Commerce Commission) Regulations and material is accepted by steamship companies." The court noted that turnings may become dangerous if they contain an excessive amount of moisture or waste material. As the turnings were dry and free of waste materials when loaded, they were not dangerous.[173]

On the other hand U.S. courts did not have any occasion to consider the aspect of the *Giannis NK*, i.e. ground-nut extraction meal pellets could constitute dangerous goods under Art. IV.6 of Hague/Hague-Visby Rules.[174] It will be interesting to see U.S. courts attitude on the similar case.

cc) Goods may be dangerous even if not listed in the regulations

A cargo may be deemed dangerous although it is not listed in the regulations. In *Borgships Inc. v. Olin Chemicals Group*,[175] a cargo of "dichloroisocyanuric acid salts dehydrate ("SDIC")" was shipped. SDIC was specifically excluded from the list of hazardous materials subject to Hazardous Materials Regulations (HMR).[176] Moreover, the shipper affirmatively identified the SDIC material as "non-hazardous" on the bill of lading information it gave to the carrier. The vessel encountered heavy weather while at sea. A number of containers were lost overboard and some of the remaining containers, including the one with the SDIC, were welded together to prevent loss. On arrival, the ship's crew used a blow torch to cut the containers free and allow their offloading. During the cutting operation, the SDIC in the container ignited, causing a fire and a large amount of smoke. The port and local businesses recovered on nuisance claims against the carrier. The carrier sought recovery from the shipper. The court concluded that "compliance with DOT regulations does not satisfy, as a matter of law, a shipper's duty to warn"; and the cargo was found to be dangerous.[177]

By contrast, in the *Internav Ltd. v. Scanbulk Ltd. (The Wismar)*[178] case, the owner instructed the master to refuse the loading of direct reduced iron-ore pellets (DRIP). It was concluded that in determining whether DRIP was in fact dangerous cargo, IMO and U.S. Coast Guards regulations were especially relevant since those were regulatory bodies expressly referred to in the charter. Because DRIP

---

[173]   *Ibid.* at 382.

[174]   Robertson/Sturley, "Recent Developments in Admiralty and Maritime Law at the National Level and in the Fifth and Elevent Circuits" [2003] 27 *Tul. Mar. L.J* 495, 520 f.

[175]   1997 WL 124127 (S.D.N.Y).

[176]   Although 49 C.F.R §§ 171.1 (that was in force at that time) list "dichloroisocyanuric acid salts" as hazardous, the regulations specifically excluded SDIC from the list of hazardous materials: "The dehydrated sodium salt of dichloroisocyanuric acid is not subject to the requirements of this subchapter". 49 C.F.R § 172.102. The IMDG Code also excluded SDIC from its requirements. 1997 WL 124127 at 1.

[177]   1997 WL 124127 at 4.

[178]   SMA 1454 (Arb. at N.Y 1980).

was neither listed in the IMDG Code nor mentioned by U.S. Coast Guard regulations, it was held that DRIP was not dangerous.

However, in the recent *Senator Linie GmbH. & Co. KG v Sunway Line, Inc.*[179] case, TDO, a white, odorless powder used as a reducing agent and in the bleaching of protein fibers, was carried on the *Tokyo Senator*. At the time of shipment, TDO was not named as a hazardous or dangerous cargo in the IMDG Code or in the HMR. The TDO shipment was not listed on the *Tokyo Senator*'s hazardous cargo manifest. It was not until 1998 that TDO was specifically listed as a dangerous material in the IMDG Code and not until 1999 in the Hazardous Materials Regulations.[180] TDO, however, was held to be dangerous.[181]

c) Meaning of dangerous in German law

aa) Definition of dangerous goods in GGBefG

HGB § 564b reflects Art. IV.6 of the Hague/Hague-Visby Rules regulating the liability of the shipper when shipping dangerous goods. As in the Hague/Hague-Visby Rules, § 564b provides only for "...inflammable, explosive, or other dangerous goods..." and does not define dangerous goods either.

It is, therefore, unclear which goods, besides explosives and inflammables, fall under § 564b. GGBefG defines dangerous goods as "goods which due to their nature, character or states associated with transport are considered dangerous for public safety or order, particularly for general public, important public properties, life and heath of human as well as animals and objects".[182] It is, however, controversial whether this broad definition of dangerous goods would be applicable in § 564b, because such a definition covers any type of cargo which, although not intrinsically dangerous, may cause danger under certain circumstances. While some authors are of the opinion that the definition of GGBefG is applicable for the purpose of § 564b,[183] general opinion rejects the adoption of the definition in GGBefG for the purpose of § 564b due to its broad extent.[184]

---

[179] 291 F.3d 145, *2002* A.M.C. 1217.

[180] In *Colormaster Printing Ink v. S.S. Asiafreighter,* 1991 WL 60413 (S.D.N.Y), there was a cargo of "arsine" gas, which is not specifically listed in the HMR, but comes under 'N.O.I.' 'N.O.I' denotes 'not otherwise indicated' and is used to label explosive or otherwise dangerous material not specified in the HMR. As a poisonous and dangerous commodity, it is recognized as falling within the category of Class A poisons. A chart in the HMR indicates that packages containing Class A poisons must be labelled with a "poison gas" label and be stowed "on deck under cover". Therefore, it was negligence to fail to label the container properly and to fail to state on the transfer receipt that the container held dangerous cargo.

[181] *Senator Linie GMBH & Co. v. Sunway Linie, Inc.* 291 F.3d 145, 152 f.

[182] GGBefgG § 2(1)

[183] Trappe, "Haftung beim Transport gefährlicher Güter im Seeverkehr" [1986] *VersR* 942, 944.

[184] Rabe, *Seehandelsrecht* (2000), 444; Gündisch, *Die Absenderhaftung im Land- und Seetransportrecht* (1999), 193.

bb) General or concrete dangerous nature of the goods

In general, in the sense of HGB, dangerous goods are deemed goods which due to their *general* physical or chemical properties may damage ship or other property.[185] This view employs *general/concrete* criterion to distinguish between dangerous and non-dangerous goods.[186] Accordingly, a danger arising from a particular state of non-dangerous goods is not sufficient for them to be deemed dangerous goods. To be dangerous, danger should be inherent in the nature of the good.

On the other hand, a contrary view approaches the "general/concrete" criterion as a "general/concrete dangerous" criterion, e.g. "a concrete danger" can be created by cargo which is not dangerous.[187] Wholly harmless cargo, such as straw or cotton, if carried at a high temperature and in a moist state, can be dangerous due to their tendency to inflammability. For instance, sulphur chips which corroded the equipment of the ship;[188] bisulfate which was declared as rock salt on the bill of lading and damaged the ship;[189] and raw anthracene which in hot weather released oily, penetrating fluid and thus polluted ship's holds were deemed to be dangerous cargoes.[190] Moreover, in many cases coal and briquettes which self-heated[191] and caused damage were also found dangerous within the meaning of § 564b. However, it is thought it is doubtful if the same applies in cases where an organic substance such as cotton or tapioca spontaneously ignites at high temperature in combination with moisture. It is contended that the application of § 564b in such a case is hardly justified.[192]

On the other hand, the Court of Appeal held that highly moist zinc residue which became colloidal and caused the danger of capsizing the ship "*Neuwarder Sand*"[193] was not dangerous on the ground that it did not have general properties to exceptionally damage the ship or other cargo; but only particular loading endangered the safety of the ship due to its high moisture content.[194] Similarly, in "*Viking Bank*", excessively moist calcium fluoride concentrate was held not to be

---

[185] *Ibid.*

[186] The "general/concrete" criterion was first used by Gramm in 1938. Gramm, *Das neue deutsche Seefrachtrecht nach den Haager Regeln* (1938), 103.

[187] Trappe, "Transport gefährlicher Güter, Unfallursachen und beteiligte Ladungen" [1988] *TranspR* 396 399.

[188] RG, HansGZ 1916, No. 9. Although sulphur chips were dangerous, the shipper was not liable as he declared the cargo and its corrosive properties. However, it is asserted that according to the general/concrete criterion, disulfide should not have been classified as dangerous, because bisulfide in a dried state is not dangerous. Gündisch, *Die Absenderhaftung im Land- und Seetransportrecht* (1999), 195.

[189] RGZ 93, 163.

[190] BGH 75 *VersR* 824.

[191] RG 170, 133.

[192] Rabe, *Seehandelsrecht* (2000), 445. This is not to say that shippers of such goods are not liable, but rather that liability is assessed according to §564.

[193] BGH 11.3.1974 (II ZR 45/73, Düsseldorf); 74 *VersR* 771.

[194] Namely it must be proved whether the high moisture content of the cargo was known or ought to have been known by the shipper.

dangerous cargo in the sense of § 564b. It must be noted that this is not to say that the shipper was not liable.[195] In these cases, the shipper's liability was established on the ground of fault under § 564 instead of application of the special dangerous cargo provision § 564b, which provides for strict liability.[196]

It is also to be noted that by contrast to English law cargoes which damage other goods due to contaminated or infested state or drenching the other goods due to damaged container is not sufficient to be regarded dangerous.[197] It is because in such cases danger arises from concrete state of individual case rather than general.

cc) Goods listed in the regulations on the transport of dangerous goods

Although the inconsistency of jurisprudence creates uncertainty, it is at least clear that goods listed in GGVSee which implements the IMDG Code and other codes, are considered dangerous and likewise fall under § 564b.[198] Therefore, although seemingly innocuous, cargoes such as straw, hay, dry vegetable fiber cotton flax, hemp, jute, sisal, matches[199] or coal, oily cotton scrap, wet cotton, moisture vegetable fibers and fishmeal[200] are classified as dangerous due to their tendency to inflammability or self-heating properties in the IMDG Code. Furthermore, GGVSee is deemed an important indicator for § 564b. In general, therefore, it cannot be concluded that goods which the aforementioned Regulations do not deal with also fall under § 564b.[201]

However, what is not clear is whether the cargoes whose mere danger is the effect on stability are to be considered dangerous or not. GGVSee § 2(2) set out which cargoes are considered to be dangerous for the purpose of the Regulations. Accordingly, dangerous bulk substances are those which are classified as dangerous in the BC Code.[202] Cargoes in solid form in bulk are considered dangerous which are covered by the IMDG Code,[203] namely cargoes listed in Appendix B of the BC Code, which pose chemical hazards, but not those in Appendix A and C, which affect the stability of the ship. Thus, this is an aspect to be considered.

---

[195] German law distinguishes between ordinary cargo and dangerous cargo, but requires the shipper of both cargoes to give notice of the characteristics of the cargo. The difference is the standard of liability, i.e. the former is subject to fault-based liability, the latter to strict liability. See *infra* Part 4.

[196] Hamburger Schiedspruch v. 25.2.1986, Trappe, "Haftung beim Transport gefährlicher Güter im Seeverkehr" [1986] *VersR* 942, fn. 9.

[197] Rabe, *Seehandelsrecht* (2000), 445.

[198] *Ibid.* at 444; Gündisch, *Die Absenderhaftung im Land- und Seetransportrecht* (1999), 195 f.

[199] Class 4.1 of the IMDG Code.

[200] Class 4.2 of the IMDG Code.

[201] Abraham, *Das Seerecht in der Bundesrepublik Deutschland* (1978) 1. Teil, 526.

[202] GGVSee Art. § 2(2).2.

[203] SOLAS Chap. VII.A-1.7.

d) Meaning of "dangerous" in the CMI/UNCITRAL draft instrument on the carriage of goods

aa) No distinction between dangerous and non-dangerous goods

In contrast to the Hague-Visby Rules, the original draft instrument contained no specific reference to dangerous goods. The draft instrument lifted the distinction that current conventions make between dangerous cargo and ordinary cargo. Such a distinction was considered out of date, because the notion "danger" has acquired a more relative character nowadays in the light of existing case law.[204] Instead, the shipper is obliged to ensure that the goods "will withstand the intended carriage ... and will not cause injury or damage".[205]

bb) Drafting specific provision dealing with dangerous goods

However, during the later discussions a proposal was made for the replacement of Draft Articles 29 and 30, that govern the right and liability of carrier and shipper towards each other by a provision which also mentions specifically dangerous goods.[206] As to the substance of the proposal under which the shipper should inform the carrier of the dangerous nature of the goods and of the necessary safety measures, a concern was expressed that the proposed rule might be unnecessary and its effect uncertain, unpredictable and excessively onerous for the shipper, particularly in view of existing case law in a number of countries, under which goods, although not identifiable as dangerous before carriage, could later be declared dangerous by courts adjudicating the claim, for the sole reason that they had caused the damage. The view was expressed that the issue of dangerous goods was sufficiently covered in the draft instrument, for example in Draft Art. 12[207] and 27[208] which appropriately avoided using the notion of "dangerous goods" itself.

---

[204] Van der Ziel, "The UNCITRAL/CMI Draft for a New Convention Relating to the Contract of Carriage by Sea" [2002] *TranspR* 272.

[205] Draft instrument Art. 25. Draft instrument available at <www.uncitral.org>.

[206] Proposal provided that ".... 2. If the shipper has delivered dangerous goods to the carrier or the sub-carrier without informing the carrier or sub-carrier of the dangerous nature of the goods and of necessary safety measures, and if the carrier did not otherwise have knowledge of the dangerous nature of the goods and the necessary safety measures to be taken, the shipper is responsible for the damage or loss sustained by the carrier." A/CN.9/552 para. 139.

[207] Draft instrument contains 2 two different variations of Art. 12: Variant A. "Notwithstanding articles 10, 11, and 13(1), the carrier may decline to load, or may unload, destroy, or render goods harmless or take such other measures as are reasonable if goods are, or reasonably appear likely during its period or responsibility to become a danger to persons or property or an illegal or unacceptable danger to the environment." Variant B "Notwithstanding articles 10, 11, and 13(1), the carrier may unload, destroy or render dangerous goods harmless if they become an actual danger to life or property".

[208] Art. 27 provides "The shipper shall provide to the carrier the information, instructions, and documents that are reasonably necessary for: (a) the handling and carriage of the goods, including precautions to be taken by the carrier or performing party (b)

The discussion focused on the definition of dangerous goods. It was generally felt that, should a provision expressly referring to the notion of dangerous goods be retained, a definition should be provided in the draft instrument.[209] The only possible reference was said to be the definition in the international Convention on Liability and Compensation for Damage in Connection with the Carriage of Hazardous and Noxious substances (HNS) by Sea, but considerable doubts were expressed regarding the appropriateness of introducing such a definition in an international trade law instrument. Support was also expressed for addressing in the definition the issue of goods that became dangerous during carriage.

In conclusion, it was decided that a specific provision should be inserted at an appropriate place in the draft instrument to deal with the issue of dangerous goods, based on the principle of strict liability of the shipper for insufficient or defective information regarding the nature of the goods.[210] Upon this decision a provisional Draft article on dangerous goods was drafted, based on the definition provided in the HNS Convention.[211]

With regard to the provisional Draft article based on the definition in the HNS Convention, a majority of the delegations stated that they preferred either no definition at all or a more general and simplified definition than that proposed.[212] The reason for not using the definition in the HNS Convention was that the HNS Convention fulfils a public interest, i.e. protecting the environment and third parties, rather than a private one, and that a technical definition like this one always runs the risk of soon being out of date.[213] Those who preferred a general definition indicated that a general definition might inhibit the courts from applying varying interpretations of the notion of dangerous goods and so promote uniformity.[214] It was also suggested that a definition of dangerous goods should also clarify whether illegal cargo, such as contraband, would fall under this category.[215] The Secretariat proposed a more general definition in Article 33.1, which provides that:[216]

"Dangerous goods" means goods which by their nature or character are, or reasonably appear likely to become, a danger to persons or property or an illegal or unacceptable danger to the environment.[217]

---

compliance with rules, regulations and other requirements of authorities in connection with the intended carriage, including filings, applications, and licenses relating to the goods. ..." A/CN.9/WG.III/WP.32.
[209] A/CN.9/552 p. 33 f.
[210] A/CN.9/552 pp.33-34
[211] A/CN.9/WG.III/WP.39, para.19. See HNS Convention Art.1.5 in appendices.
[212] A/CN.9/WG.III/WP.55 para.32.
[213] *Ibid.*
[214] *Ibid.*
[215] A/CN.9/591 para.158.
[216] A/CN.9/WG.III/WP.56 p. 31.
[217] After discussions it was decided that the words "or an illegal or unacceptable danger" should be deleted since they failed to add meaning to the term "danger to the environment". A/CN.9/591 para.160 f.

F. What should be considered dangerous?

I. Nature of the goods

The foregoing examination of three legal systems shows that the concept of dangerous cargo is unclear, controversial and vague. Goods can cause various sort of damage in many different situations in sea carriage. Goods may physically damage the ship and other goods on board or may cause non-physical damage and other extra costs. In both situations, the incurred damage can be quite substantial and the ensuing legal issues highly complex.

From the point of view of the carrier, dangerous goods are goods which pose unforeseeable hazards to the ship and cargo. It follows from case law that whether the goods are dangerous or not is subject to an individual evaluation of the occurrence. The court looks at many different factors and it is not sufficient that the goods in question are generally considered dangerous. Virtually each legal system is inconsistent in its jurisprudence and this creates uncertainty.

The problem of "what a dangerous good is" can be approached either as one of category or as an integral part of the larger question of sharing risk.[218] In a sense, it is always an aspect of that larger question, but to isolate it from the rest carries the danger that it will be the goods which are focused on. Although the knowledge of the carrier and the taking of necessary precautions regarding the dangerous nature of the goods are important factors in allocating liability, using this method to determine whether a good is dangerous or not would not be proper. That is because this approach covers all types of goods, whatever dangers they may create, and treats the category of "dangerous goods" as at most non-existent and at least immaterial.[219]

It is obvious that any goods may be *dangerous* or may be regarded as *dangerous* under certain circumstances, whether or not they are within the codes or may qualify to be within the codes. Any good likewise requires special attention and proper packing; if that is not done, a "dangerous situation" may be created. Moreover, any goods may cause delay or detention due to stringent safety and security regulations. However, it seems that the purpose of the specific provision is to govern "dangerous cargoes" rather than "dangerous situations" and for the latter, general provision is thought to be sufficient protection.

II. Public regulations private liability

Public regulations and private liability might seem odd at first glance, but the main reason for regulating dangerous goods specifically for the purpose of carriage

---

[218] Jackson, "Dangerous Cargo – A Legal Overview" in *Maritime Movement of Dangerous Cargoes – Public Regulations Private Liability*, Papers of One Day Seminar (1981), A8.

[219] *Ibid.* at A10.

contracts are the safety concerns with the ship, crew and cargo onboard. Therefore one cannot ignore the role of public regulations in this field and the link between regulations and liability in contracts. If there were no safety concerns, there would probably not be any specific rules on dangerous goods. A general liability scheme would be sufficient to allocate damages and expenses.

III. Goods listed in the IMDG Code and other codes

Without a certain category of dangerous goods, uncertain and unpredictable outcomes are unavoidable. While protecting public safety, the IMDG Code also serves this aim. Despite the fact that legal systems have inconsistency with regard to what constitutes dangerous cargo, cargoes listed in the IMDG Code, or in domestic laws implementing the IMDG Code and other codes, are deemed to be dangerous. However, under the set of circumstances when the shipper is not liable, a false impression is created that despite the fact that cargo is listed in the codes, it is not dangerous. Commonly transported dangerous goods are listed in the DGL. The DGL provided by the IMDG Code and other codes are not exhaustive. This is also expressly mentioned in the codes. Therefore, DGL is not taken to be exclusive in the sense of relieving the shipper of liability if he has come across some new kind of cargo, not listed in the IMDG Code, which has manifestly dangerous characteristics[220] or insufficient or improper instructions, such as stowage.[221] It is the onus of the shipper to identify the nature of the cargo and take the necessary precautions accordingly.

Many cargoes listed in the IMDG Code or other codes are inherently dangerous, although damage arising from dangerous cargoes is not always a result of the intrinsic nature of the cargo. The damage is often a combination of outside influences and some inherent nature of the cargo itself.[222] Likewise, the IMDG Code includes wholly innocuous cargoes, such as cotton, which under certain circumstances can become highly dangerous. However, nobody suggests that the shipper is strictly liable for the shipment of cotton cargo in the case of damage thereto.

IV. Dangerous goods in bulk

Traditionally maritime transport of dangerous cargoes was limited mainly to packaged consignments. However, limiting dangerous cargoes only to packaged cargo ignores the fact that gases and liquid chemicals are carried in bulk, which is obviously dangerous. Considering the IMDG Code as the sole indicator of dangerous cargoes would lead to confusion as other codes govern bulk dangerous cargoes.

---

[220] Tiberg, "Legal Survey", in Grönfors (ed.), *Damage from Goods* (1978), 9, 17.
[221] *In re M/V Harmony and Consolidated Cases* 394 F.Supp.2d 649.
[222] For instance, improper stowage when two incompatible goods are stowed next to each other, which creates danger. The damage would not have occurred if it was not for this improper stowage.

National legislations, likewise, generally cover both packaged and bulk as dangerous. Thus, substances included in the IBC, IGC and BC codes posing chemical hazards, with the exception of Appendix A and C of the BC Code, should be deemed dangerous for the purpose of carriage contracts. The differences between packaged and bulk dangerous goods is that while there are thousands of packaged dangerous goods and mostly their dangerous character is unknown, the number of bulk dangerous goods is rather limited and their dangerous character generally well known in the trade.

V. Is the effect on stability a danger?

Grain cargoes and cargoes in Appendixes A and C of the BC do not pose a hazard in the classification of the IMDG Code, but due to biological degradation they may affect the stability of a vessel. It is obvious that Art. IV.6 of the Hague/ Hague-Visby Rules was aimed at regulating *exceptionally* dangerous goods, particularly *chemicals*. Therefore, a danger to the stability of the vessel, although exceptional, is hardly justified as falling under Art. IV.6.[223] This is an aspect to be considered. It is to be noted that not deeming such cargoes to be dangerous does not mean that there would be no liability for shipping such cargoes.[224]

VI. Exceptionally dangerous goods

There may seem to be no good reason in drawing a distinction between exceptionally dangerous cargoes, such as explosives and inflammables, and seemingly innocuous cargo, such as apples or groundnuts. Damage is still damage, even when caused by an innocuous cargo. However, there are two reasons that require a distinction to be drawn between such goods: standard of liability and significance of risk and potential damage thereof.[225]

---

[223] The reason why the issue did not come out at the time the Hague Rules were being considered may be that at that time there was no concern with regard to such cargoes. Indeed, before the Second Word War there was no real demand for special bulk carriers. Seaborne trade of all mineral ores only amounted to 25 million tons in 1937 and this could be carried in conventional tramp ships (freight vessels). By the 1950s, however, movements of bulk cargoes were increasing. Very often ores and other commodities were found far away and the most convenient and cheapest way of shipping them was by sea. Companies in the United States, Europe and increasingly in Japan began to build ships designed exclusively for the carriage of bulk cargoes. As demand increased and shipbuilding technology advanced, these ships tended to become bigger in size and carrying capacity. So the dangers of bulk carriage appeared after the Hague Rules were promulgated.

[224] *Infra* Part 4.

[225] Falkanger/Bull/Brautaset, *Introduction to Maritime Law* (1998), 297.

VII. Standard of liability, significance of risk and potential damage

If the basic liability rule is one of strict liability for one or more parties as against others for all goods, there is no problem of classification. Assuming that the basic rule requires fault, all goods may be treated in a standard way with the duty of care related to the risk. But a problem arises if a different liability is attached because of a degree of danger thought to follow from the carriage of the goods arising from the "nature" of the goods. This type of approach makes it possible to apply special rules to a category, but such an approach is only possible if the category is limited and only necessary if distinct liability rules are to apply. These rules may differ from what is standard because of their basis, either strict or fault, or because of the person responsible or both. Moreover, the category must be based on the likelihood of serious damage rather than any inherent danger of the goods. Although a distinction has been clearly recognized by those whose function is to regulate and supervise the carriage of potentially dangerous goods, the distinction has not been so clearly drawn in many cases.

It has been asserted that the correct approach is that there should be a general approach to liability in relation to the shipment of goods which cause loss to a carrier without prior qualification of the character of such goods.[226] However, as examined in Part 4, in the current scheme, the basic rule for liability in sea carriage is fault and the application of strict liability is limited. It would not be just and proper to provide a more advantageous position to the carrier compared with the shipper when the basis of the carrier's liability is fault.

Accordingly, it is submitted that the meaning of dangerous goods should be understood in the sense of the IMDG Code and other codes in relation to dangerous cargoes, but also should be flexible to comprise new, synthetic goods or goods whose dangers have been recently identified. Given the harsh nature of the strict liability rules, the term "dangerous" should be restricted to cargoes constituting significant risks and causing damage therefrom.

---

[226] Abdul Hamid, *Loss or Damage from the Shipment of Goods, Rights and Liabilities of the Parties to the Maritime Adventure* (Diss. Southampton 1996), 50.

Part 3: Duties of the parties in relation to dangerous goods

A. In general

It is clear that a shipper is not restricted to shipping only goods which carry no risk at all and also true that all goods present some risk.[1] Responsibility for such risks or for avoiding or minimizing their consequences may of course fall on the shipper, who may be liable for losses suffered by the carrier or third parties as a result of the shipment of such goods. Otherwise it is the carrier's duty to provide a seaworthy ship which is fit for the contracted cargo. Furthermore, it is a normal part of a carrier's duties to take appropriate measures to avoid loss resulting from risks of which he is or should be aware. He assumes all risks of accidents attributable to a failure to carry in that manner.

Under a charter contract, the owner and the charterer have various duties. For instance, the owner must provide a seaworthy ship, exercise reasonable care for the cargo and provide for carriage without unnecessary deviation. Depending on the type of charter, the charterer has different duties, but in general he should have the cargo ready for loading, nominate the loading port and pay freight. Loading and discharging are the responsibility of both carrier and charterer and, in the absence of a provision to the contrary, their duties are determined by the custom of the port.[2]

The Hague/Hague-Visby Rules contain very few rules on shipper's duties and liabilities. This is because the main purpose of the Hague/Hague-Visby Rules was the establishment of mandatory minimum standards of carrier liability in the context of contracts of carriage on the carrier's standard terms.[3] Accordingly Art. IV.3 sets out a general rule to the effect that the shipper shall only be liable to the carrier in cases of fault on the part of the shipper or his servants or agents. This general rule is supplemented by two more specific rules contained in Art. III.5 and Art. IV.6. Furthermore, under the Rules the carrier has two main duties: to provide a seaworthy ship and to care for cargo in Art. III.1 and III.2 respectively. However, there are duties customary in trade or arising from contract law.

---

[1] Rose, "Cargo Risks: 'Dangerous' Goods" [1996] 55 *Cam. L.J.* 601, 604.
[2] In the absence of express agreement of custom, it is the duty of the charterer at his own risk and expense to bring the cargo alongside and lift it to the ship's rail; it is then the duty of the carrier, through his master, to receive and stow the cargo properly.
[3] Asariotis, "Main Obligations and Liabilities of the Shipper" [2004] *TranspR* 284, 285.

The transportation of dangerous goods is by nature multi-modal, which consequently introduces intrinsic and potential dangers by virtue of the involvement of the "links" in the transport chain.[4] Each member of this chain has certain responsibilities, which inevitably leads to confusion over which people and at what stage the requirements of the IMDG Code and other codes are to be met and verified. There are three key participants in the transport chain prior to the cargo transport unit being loaded onto the ship: the shipper, who initiates or consigns the transport movement of dangerous goods; the packer, who is responsible for packing the goods; and the marine carrier. Each person in the transport chain has a responsibility for the safe movement of dangerous cargo and each is governed by comprehensive regulations. In practice, there are regulations for almost all types of cargoes, regardless of whether they are considered dangerous or not.

While discussions continue on the relevancy of the IMDG Code to the contract of carriage, dangerous cargoes are to be carried in accordance with the requirements of the Code and national enactments of it. Dangerous Cargo clauses in bills of lading require the shipper to observe and comply with laws, regulations or requirements.[5] The IMDG Code prescribes the necessary method of handling goods that require special care. The IMDG Code prescribes particular actions in certain parts but the responsibility for carrying out the action is not specifically assigned to any particular person. Such responsibility may vary according to the laws and customs of different countries and the international conventions into which these countries have entered. For the purpose of the Code, it is not necessary to make such an assignment but only to identify the action itself. The fault of a particular party is assessed in accordance with the requirements of the IMDG Code.

B. Duties of the shipper

In general, a shipper will be liable for the payment of freight charges, as well as demurrage in some cases. In addition, certain clauses deal with the allocation of responsibility for those parts of the contractual performance which require joint actions from both carrier and shipper, such as loading, stowage and delivery of the goods. Apart from these responsibilities, the shipper is allocated certain obligations which are either in his exclusive sphere of influence and necessary prerequisites to the contractual performance by the carrier or are based on the special knowledge of the shipper.

Although the Hague/Hague-Visby Rules contain only two provisions concerning shipper's duties, this does not mean that the shipper has no other duties. Duties arise from general law, statutes or are naturally implicit in the contract.[6] Whatever

---

4   Williams "The Implications of the ISM Code for the Transport of Packaged Dangerous Sea", in *International Symposium on the Transport of Dangerous Goods by land Waterways* (1998), 117, 120.

ast Africa/Mauritius Service Bill cl.19; P&O Nedloyd Bill cl.19.

Carver *Carriage by Sea* (1982) Vol. 1, 17.

the type of contract of carriage, the shipper must deliver the goods properly, i.e. goods must be ready for intended carriage in such a condition that they will withstand the intended carriage. Dangerous goods must be properly classified, packed, marked and labelled and handled in accordance with the regulations. It is to be noted that Hamburg Rules requires the shipper must mark or label in a suitable manner dangerous goods dangerous.[7] However, notwithstanding the requirements for notification and packaging, it is a fact that many consignments of dangerous chemicals are shipped, particularly in freight containers, without the appropriate labelling, packaging and notification.[8]

Here a distinction is necessary between bulk and packaged dangerous cargoes. The physicals characteristics of bulk cargoes are unique and differ from that of packaged goods. In terms of quantity dangerous goods in bulk have greater potential to cause more damage than those which are packed. However as they are not packaged, there is no element of concealment. Moreover, there is no need, nor it is practical, for bulk dangerous goods to be marked and labelled. Therefore, below explanations will be mostly regarding packaged dangerous goods which in practice cause more troubles.

I. Classification of dangerous goods

When it is agreed that dangerous goods are to be carried, the first step is to determine the class of the goods, so that all necessary measures can be taken as required.[9] The classification must principally be made by the shipper[10] in accordance with the IMDG Code. As mentioned previously, if the substance is not specifically listed in the extensive Dangerous Goods List (DGL), this does not mean that it is not dangerous. It is the onus of the shipper to determine the class of the substances according to provided criteria.

II. Packing of dangerous goods

1. Purpose of packing

Traditionally dangerous cargoes are carried in packaged form for easy identification, durability and easy and safe recovery in case of incidents. Packing is the assembly of items into a unit, intermediate or exterior pack with necessary blocking, bracing, cushioning, weatherproofing, reinforcement and marking. The transport of dangerous goods is associated with special risks because of the property of the transported substances, i.e. that of being dangerous. The packing has an ex-

---

[7] Art. 13.1.
[8] Williams, "The Implications of Shipping Dangerous Cargo", in *Pursuit and Defence of Cargo Claims,* 10th & 11th May 1991, 2.
[9] See *supra* Part 2.
[10] Where specified in the IMDG Code, the classification shall be made by a competent authority.

tremely important role with respect to minimizing such risks.[11] Packing emphasizes two main purposes: protection and handling.

Packing should protect the contents in such a way that neither their performance nor their reliability is affected by:

– outside mechanical forces such as impact or vibration
– contamination of undesirable substances, e.g. water or air
– climatic conditions, e.g. heat or cold.

and should facilitate

– transport and storage of the product
– opening and remaining opened when needed
– closing and remaining closed when needed
– removal of content from the packing
– complete emptying of the packing

2. Packing obligation

The Hague/Hague-Visby Rules contain no provision imposing a duty on the shipper to pack sufficiently, but instead relieves the carrier from liability. It is historically one of the earliest contractual defences available to carriers and is still among the most important exculpatory exceptions to be found in the Hague/Hague-Visby Rules which provides that "Neither the carrier nor the ship shall be responsible for loss or damage arising or resulting from insufficiency of packing."[12] Insufficiency of packing is a development of the "inherent vice" exception to packing, which of course is not inherent. The exception is closely connected with several other major defences of the carrier.[13] The provision does not mean expressly that the shipper has an obligation to pack. However, the duty of the shipper to pack properly is a natural part of his duties and the specific requirements as to packing can be found as part of the statutory or contractual obligations of the shipper.[14] In practice, unless otherwise agreed by parties in the contract of carriage, goods are packed by shippers.[15]

Potential liabilities of the shipper often involve defects in the packages. Although in most cases the shipper does not pack the goods himself, it is his duty to

---

[11]   Willinger, "Rechtliche Grundlagen für die Verpackungen gefährlicher Güter" [1981] *TranspR* 81.

[12]   Art. IV.2.(n).

[13]   Act of omission of the shipper or owner of the goods Art. 4.2.m; inherent vice and hidden defect of the goods Art. IV.2.(o); insufficiency or inadequacy of marks Art. 4.2.(p); latent defects of the ship and any other cause arising without the actual fault or privity of the carrier Art. IV.2.(q).

[14]   Gündisch, *Die Absenderhaftung im Land- und Seetransportrecht* (1999), 158 f; Wong K.K, "Packing Dangerous Goods" [1976-77] 8 *J. Mar L. & Com.* 387, 387 ff.

[15]   Karan, *The Carriers Liability under the International Maritime Conventions the Hague, Hague-Visby, and Hamburg Rules* (2004), 308.

ensure sufficiency of packing under the carriage contract. Failure to do so amounts to his fault and he is responsible for the consequences.[16]

3. Sufficiency of packing

Sufficient packing is normal or customary packing in the trade. Such packing should permit the goods to withstand the normal hazards likely to be encountered on the specific voyage contemplated and to prevent all but the most minor damage under normal conditions of care and carriage.[17] There is no single criterion for insufficiency of packing. Each case must be considered in its own facts. The degree of packing required and the care to be taken relate to the cargo to be carried. Some cargoes require a higher degree of care and packing in a certain way, as in the case of dangerous goods in packaged form. In each instance, the degree of care that can reasonably be expected of the shipper and the degree of care that can reasonably be required from the carrier must be well balanced.

Dangerous cargoes are packed either by the shipper himself or by a third-party packer. Whoever performs it, packing is very important task. He is responsible for following the comprehensive and relatively complex legislation which must be observed in preparing dangerous goods for transportation.[18] The importance of the packer's role is apparent when his key responsibilities are examined: packages must be suitable; containers and other packaging must be in a good condition and appropriate for the goods; packing must be in accordance with regulations, and with new technology.[19] Packing is the most important aspect of the handling of dangerous goods. It is generally believed that if the packing is suitable, the risk of a serious incident occurring is greatly reduced. The packing of dangerous goods plays a significant role in protecting other cargo, the ship and those onboard. The

---

[16] Hague-Visby Rules Art. IV.3. Problems can arise where insufficiency of packing causes damage to other cargo. In *Goodwin, Ferreira & Co. v. Lamport & Holt Line Ltd.* (1929) 34 Lloyd's Rep. 192, it has been suggested that although the exception is primarily directed at the packing of goods damaged, there is nothing to prevent its being used in such a situation. However, this case was decided on the basis of a "catch-all" exception, i.e. Art. IV.2.(q). On the other hand, this may also fall within the question of care under Art. III.2 in respect of the "other" cargo, depending on the circumstances. If the carrier is held responsible for the other cargo that was damaged by insufficiently packed cargo, he may nevertheless seek indemnification from the shipper of the first cargo after compensation. Tetley, Marine *Cargo Claims* (1988), 505 f.

[17] Tetley, *Marine Cargo Claims* (1988), 491. In *"Bamfield v. Goole etc.,* [1910] 2 K.B. 94, unknown to both the shipper and the carrier, the substance underwent a change during the transit: the packing proved insufficient and poisonous gases were given off.

[18] Wong K.K., "Packing Dangerous Goods" [1976-7] 8 *J. Mar L. & Com.* 387, 395 ff. Training in this area is, therefore, essential and management arrangements in the organization carrying out the packing must ensure that this training is provided.

[19] Williams, "The Implications of the ISM Code for the Transport of Packaged Dangerous Goods by Sea", in *International Symposium on the Transport of Dangerous Goods by Sea and Inland Waterways* (1998) 117, 120.

case law dealing with packing defects often involves dangerous goods that leak or escape from a defective container, drum or otherwise.[20]

Dangerous goods must be packed in good quality packagings which shall be strong enough to withstand the shocks and loadings normally encountered during transport. Packagings must be contructed and closed so as to prevent any loss of contents when prepared for transport which may be caused under normal conditions of transport, by vibration, or by changes in temperature, humidity or pressure and must be closed in accordance with the information provided by the manufacturer. For packing purposes, substances other than those of classes 1, 2, 5.2, 6.2, 7 and self reactive substances of class 4.1, are assigned to three packaging groups in accordance with the degree of danger they present:[21]

- Packing group I: substances presenting higher danger
- Packing group II: substances presenting medium danger
- Packing group III substances presenting low danger.

III. Stowage and segregation of dangerous goods

Stowing and lashing the cargo are part of the operation of loading. It refers to the placing of the goods in a ship's spot or a container.[22] In the absence of an express contract or custom, stowage is the duty of the carrier through the master.[23] In other words, if there is an express contract or custom, stowage is done by the shipper. The final responsibility for proper stowage, however, remains in all circumstances with the carrier. Questions of stowage are under the absolute control of the master of the vessel and as such he has the final say as to how stowage is to be effected.[24]

Goods on board ships are subject to all sorts of stresses. Of particular significance are those imposed during handling, transfer between ship and shore and bad weather. Significant stresses also arise on shore.[25] This means that poor stowage or securing of packages within the cargo transport unit is liable to cause serious problems both at sea and on land.[26] The relative complexity of this transportation process manifests itself in the numerous guidance documents and regulations

---

[20]   Maloof/Krauzlis, "Shipper's Potential Liabilities in Transit" [1980] 5 *Mar. Law* 175. See also *Hoey v. Hardie* (1912) 29 30 N.S.W; *The Zhulia* 235 F.433, 434, (E.D.N.Y 1916).

[21]   IMDG Code 2.0.1.3.

[22]   Karan, *The Carriers Liability under the International Maritime Conventions the Hague, Hague-Visby, and Hamburg Rules* (2004), 202.

[23]   Colinvaux, *Carver Carriage by Sea* (1982) Vol. 2, 831.

[24]   This is so not only because of the carrier's responsibility for the stability of the ship and the safety of the ship and crew, but also because of the carrier's obligation to care for other cargo.

[25]   For example, during road or rail transfer and handling at the ports and warehouses.

[26]   Williams, "The Implications of the ISM Code for the Transport of Packaged Dangerous Goods by Sea", in *International Symposium on the Transport of Dangerous Goods by Sea and Inland Waterways* (1998), 117, 119.

governing the process.[27] Chapter 7 of the IMDG Code contains detailed provisions regarding stowage and segregation of incompatible dangerous goods. Two substances or articles are considered mutually incompatible when their stowage together may result in undue hazards in the case of leakage or spillage, or any other accident. The extent of the hazard arising from possible reactions between incompatible dangerous goods may vary and so may the segregation. Such segregation is obtained by maintaining certain distances between incompatible goods or by requiring the presence of one or more steel bulkheads or decks between them, or a combination thereof. For the purpose of segregation, dangerous goods having certain similar chemical properties have been grouped together in segregation groups.[28] Whenever dangerous goods are stowed together, the segregation of such dangerous goods from others must be done in accordance with the most stringent provisions for any of the dangerous goods concerned.

When the shipper packs the container, he should do so properly. It is admitted that the carrier's due diligence obligation does not require him to inspect the inside of the container. While the carrier has the authority to open any container once it is on board, the practice of carriers is to leave containers sealed unless a specific reason to open them arises. The carrier has no obligation to open the container, inspect it and inform the shipper how to stow the said cargo.[29] However, some carriers adopt a system of randomly inspecting the contents of containers listed as carrying the IMDG cargo.[30] Generally, containers are sealed and the purpose of sealing them is to ensure the integrity of their contents. It is practice and desirable in shipping that the seal be intact upon arrival at the port of destination. Moreover, it is almost impossible to inspect every sealed container.[31] Conse-

---

[27] IMO/ILO Guidelines for Packing Cargo in Freight Containers or Vehicles.

[28] Segregation groups referred to in the DGL are: acids, ammonium compounds, bromates, cholorates, chlorites, cyanides, heavy metals and their salts, hypochlorites, lead and lead compounds, liquid halogenated hydrocarbons, mercury and mercury compounds, nitrites, perchlorates, permanganates, powdered metals, peroxides and azides. For more information see IMDG Code Chap. 7. 2.

[29] Edgcomb, "The Trojan Horse Sets Sail: Carrier Defences against Hazmat Cargoes" [2000-01] 13 *U.S.F. Mar. L.J.* 31, 39.

[30] Williams, "The Implications of Shipping Dangerous Cargo", in *Pursuit and Defence of Cargo Claims,* 10th & 11th May 1999 1, 13.

[31] Therefore, the carrier does not fail to exercise due diligence in neglecting to have the interior stowage of containers inspected prior to sailing. In *Poliskie Line Oceniczne v. Hooker Chemical Corp.*, 499 F.Supp. 94, 99, a cargo of sulphur dichloride was carried. The shipper stowed drums of sulphur dichloride in the container in violation of the Code of Federal Regulations where no dunnage or bracing was placed in the spaces between the drums or between the drums or door of the container and where the drums were not properly placed in the container to eliminate slack space. The court concluded that the shipper's stowage negligence *per se* was based on numerous violations of the Hazardous Materials Regulations. The shipper asserted that the carrier had a duty to open the container, inspect it and inform the shipper how to stow the drums properly. The court rejected this argument, noting that the shipper had improperly certified that the cargo in the container was packaged in compliance with the HMR. The court found that the carrier was entitled to rely on this certification, relieving him of any duty to

quently, improper stowage of cargo in containers by the shipper, who stuffs and seals the containers, is the responsibility of the shipper.[32]

Bulk cargoes also be loaded and trimmed properly and in accordance with the regulations.

IV. Marking of dangerous goods

In practice, it is the shipper who supplies the details that are written on the bill of lading. Thus, the bill of lading terms usually require the shipper to mark and describe the cargo accurately, to guarantee the accuracy of any such description and to undertake to indemnify the carrier against losses and third-party liability arising from inaccurate descriptions and markings.[33] These obligations are of particular relevance with respect to the receipt function of the bill of lading and the carrier's liability for the goods shipped as described on the bill of lading.

The Hague/Hague-Visby Rules require the shipper to mark the goods and indicate the leading marks necessary for identification of the goods if the shipper wants the carrier to issue a bill of lading in accordance with Art. III.3. The bill will be prima facie or conclusive evidence of the shipment in good order or condition of the indicated quantity of goods, identified by leading marks.[34] According to Art. III.5 "the shipper shall be deemed to have guaranteed to the carrier the accuracy ... of the marks, number, quantity and weight, as furnished by him and... shall indemnify the carrier against all loss, damages and expenses arising or resulting from inaccuracies in such particulars".[35] The obligation of the shipper to mark and describe the goods accurately is of central importance, as the liability of the carrier will relate to goods the particulars of which will have been furnished by the shipper. The shipper is liable for inaccurate particulars furnished by him, irrespective of fault. This rule gives the carrier a right of action for indemnity for any loss he suffers as a result of inaccuracy in the particulars contained in bills of lading derived from the shipment of the goods in respect of marks, quantity and weight, though not apparently as to the *nature* of the goods themselves, nor as to their apparent order and condition.

---

inspect. The court found that, in any case, the carrier's due diligence obligation did not require him to inspect the inside of the container.

[32]   Tetley, *Marine Cargo* Claims (1988), 546.

[33]   See ANL Australia/South East Asia Service Bill Cl. 18; "K" Line Bill of Lading Cl. 19.

[34]   Hague/Hague Visby Rules Art. III.4.

[35]   Under Art. 17.1 of the Hamburg Rules, the shipper "is deemed to have guaranteed to the carrier the accuracy of particulars relating to the general nature of the goods, their marks, number, weight, and quantity as furnished by him for insertion in the bill of lading. This article follows Art. III.5 of the Hague/Hague-Visby Rules, except that the "nature" of the goods is also guaranteed in the Hamburg Rules.

In practice, clauses requiring the shipper to mark the goods in a certain manner are relatively rare.[36] This may be due to the fact that a clause that imposes more extensive obligations on the shipper, thus limiting the carrier's liability beyond the exceptions to the Hague/Hague-Visby Rules, would be void under Art. III.8.[37] However, the shipper is under statutory duty to mark dangerous goods in a certain manner and this is generally contained in the bills of lading.

V. Marking, labelling and placarding in accordance with IMDG Code

The purpose of marking and labelling dangerous goods, in general, is to inform related persons about the contents of packages to allow them to handle the goods appropriately for safety and health purposes and to ensure that the substances can be readily identifiable during the transport. This is particularly important in the case of an accident involving these goods, in order to determine what emergency procedures are necessary to deal properly with the situation and, in the case of marine pollutants, for the master to comply with the reporting requirements of MARPOL 73/78.[38] In that respect marking and labelling serve the same objective as the documentation. However, marking and labelling give more clear and immediate notice of dangerous characteristics of the goods in proximity with the packages who may not be aware of the same by virtue of the dangerous goods documentation.

Chapter 5 of the IMDG Code contains detailed provisions concerning marking and labelling. By means of a symbol, the nature of the risk can be indicated to all concerned; no matter what language they speak. Differently coloured labels make it easier to distinguish the goods, thus providing a very useful guide for handling and storing operations.[39] And by their general appearance they are easily recognizable from a distance as indicating dangerous goods.

For marking in accordance with the IMDG Code, the Proper Shipping Name and the corresponding UN Number, preceded by the letters "UN", shall be displayed on each package.[40] All marking on packages[41] must be:

– readily visible and legible

---

[36] For instance, in order to allow identification of the contract goods by the carrier for the purpose of safe and economic stowage of the cargo and delivery to the contractual consignee. Gaskell/Asariotis/Baatz, *Bills of Lading: Law and Contracts* (2000), 457 f.

[37] *Ibid.*

[38] Protocol I.

[39] Kervella, "IMO Roles on the Transport of Dangerous Goods in Ships and the Work of the International Bodies in the UN system, the Harmonization Issues with Regard to Classification, Criteria, Labelling and Placarding, Data Information, Emergency Response and Training", in *the 11th International Symposium on the Transport of Dangerous Goods by Sea and Inland Waterways* (1992), 74, 81.

[40] In the case of unpackaged articles, the marking shall be displayed on the article, on its cradle or on its handling, storage or storage device.

[41] There are special marking provisions for Class 7 radioactive materials and marine pollutants. See IMDG Code, Chapter 5.2.

- such that this information will still be identifiable on packages surviving at least three month's immersion in the sea.
- displayed on a background of contrasting colour on the external surface of the package and
- must not be located with other package marking that could substantially reduce their effectiveness.

Enlarged labels (placards) will be affixed on cargo transport units to provide a warning that the contents of the unit are dangerous goods and present risks, unless the labels and/or marks affixed to the packages are clearly visible from the exterior of the cargo.[42]

The shipper should correctly and conspicuously mark and label dangerous goods, since this will be part of his duty to give notice of dangerous cargo.[43] If the shipper falsely, incorrectly or recklessly labels or describes goods, his liability is unavoidable.

VI. Notice of dangerous goods

1. In general

When looking at the implications of the carriage of dangerous goods, there are various factors to be considered, the most important of which is information. It is highly important for all involved in a voyage, and especially for the master and crew, to be fully aware of the nature and properties of the cargoes they are to carry. This is not only related to cargo protection but also to the safety of human lives and property. Due to the need for quick on- and off-loading and for security reasons, it is unrealistic to expect shipowners to inspect the contents of all or even many of these containers. A shipowner and crew, as well as P&I and hull insurers, rely on the representations of shippers as to the contents of containers. If a shipper does not tell what has been loaded, the carrier will not be able to take the appropriate precautions. The consequences are obvious: a ship at sea cannot jettison a burning container if it is located deep down in stow. Cargoes in other containers stowed nearby can also catch fire and, if they happen to be undeclared hazardous cargo, the consequences can be very serious.[44] Therefore, it is essential to have accurate and reliable information about the cargo, as decision-making on partial or unreliable information is inherently dangerous.[45]

---

[42] The IMDG Code Chap. 5.3. See *Colormaster Printing Ink. Co. v. S.S. Asiafreighter*, 1991 U.S. Dist. LEXIS 4644 (S.D.N.Y), where the freight consolidator failed to placard the shipping container holding poisonous gas and provide the required written notification to the carrier.

[43] Wong K.K, "Packing Dangerous Goods" [1976-77] 8 *J. Mar L. & Com.* 387, 395 f.

[44] Webster, "Managing Risk" [2004] 18(9) *M.R.I.* 15.

[45] Williams, "The Implications of Shipping Dangerous Cargo", in *Pursuit and Defence of Cargo Claims* (1999), 1.

Common law imposes an obligation on a shipper of cargo not to ship dangerous goods without giving information to the carrier so as to enable him to take precautions to ensure that the goods can be carried without causing damage.[46] In *Brass v. Maitland*,[47] the rights and duties of the shipper concerning dangerous goods were discussed in detail. "Bleaching powder" was carried in casks which mainly consisted of "chloride lime", a highly corrosive substance which damaged other goods for which the shipowner was responsible. The shipowner had no knowledge either of the corrosive nature of the contents or of the defective condition of the casks. It was held that the shipowner's claim for damages was justified, unless the state of the casks and the dangerous and corrosive nature of their contents ought to have been known to the master. The court agreed that it is the duty of a shipper to pack sufficiently and to give warning of any danger.[48]

This is also true in American law. Under general maritime law, a shipper is under a duty to advise the carrier of any dangers in the cargo of which he is or ought to be aware and of which the carrier is not and cannot reasonably be expected to be aware.[49]

It must, however, be noted that it is a fact of life that dangerous goods are not always declared by shippers and it is not unreasonable to assume that at times this is for commercial reasons rather than oversight.[50]

2. Hague/Hague-Visby Rules

Art. IV.6 of the Hague-Hague Visby Rules states that "goods of inflammable, explosive or dangerous nature to the shipment whereof the carrier, master, or agent of the carrier has not consented with knowledge of their nature and character...."[51] It follows from this wording that under the Rules it may not be correct to say that the shipper is under a duty to warn the carrier of the dangerous nature of the cargo of which the carrier was unaware and could not reasonably be expected to become aware.[52] Rather, it can be said that a shipper guarantees that no goods of inflammable, explosive or dangerous nature shall be shipped unless notice is given.[53] However, the provision requires the consent of the carrier, master, or

---

[46] Cooke/Young/Taylor, *Voyage Charters* (2001), 149.

[47] (1856) 6 E&B.470. See also *Acatos v. Burns* (1878) 3 Ex.D.282; *Bamfeld v. Goole, & Sheffield Transport* [1910] 2 K.B. 94; *Great Northern Railway v. L.E.P. Transport* [1922] 2 K.B. 742; *Ministry of Food v. Lamport & Holt* [1952] 2 Lloyd's Rep.371.

[48] However, the court waived this obligation when the shipper himself is unaware of the nature of the goods. See *supra* Part 4.

[49] Wilford/Coghlin/Kimball, *Time Charters* (2003), 187; *International Mercantile Marine Co. v. Fels* 170 F. 275, 277 (2d. Cir. 1909).

[50] Williams, "The Implications of Shipping Dangerous Cargo", in *Pursuit and Defence of Cargo Claims* (1999), 1, 4.

[51] HGB § 564b corresponds to this provision of the Hague Rules accept nature of liability expressly stated. See *infra* Part 4.

[52] In fact, it was intended that shipper should give a declaration to the carrier. Sturley, *Legislative History of the Carriage of Goods by Sea Act* (1990) Vol. 1, 272 f.

[53] Astle, *Shipowner's Cargo Liabilities and Immunities* (1981), 182.

agent of the carrier and this consent could be given upon knowledge communicated by the person having the knowledge. Therefore, the duty of notice of the dangerous nature and character of the goods is implied in the liability.

On the other hand Hamburg Rules expressly states the duty of notice of the shipper. According to Art. 13.2 "Where the shipper hands over dangerous goods to the carrier or an actual carrier, the shipper must inform him of the dangerous character of the goods and, if necessary, of the precautions."

3. Nature of the notice

The necessary notice is the information regarding the nature and character of the goods which is unknown to the carrier. The guiding principle is that the information to be given by the shipper or charterer must be such that an ordinarily experienced and skilful carrier will be able to appreciate the nature of the risks involved in the carriage and to guard against them.[54] All relevant information necessary to warn the carrier of foreseeable hazards inherent in the shipment of the particular cargo should be provided.[55] The shipper should assume that the carrier lacks knowledge of the hazardous characteristics of any dangerous cargo which the shipper offers for transport and that the carrier relies solely on the shipper to disclose any such characteristics to the carrier prior to shipment.[56]

On the other hand, the carrier has no right to expect any communication in respect of the nature of the goods when he himself may easily discover the same.[57] The carrier is not expected to be, or to call in, an expert chemist or to resort to investigations inconsistent with the usual course of commercial business;[58] however, he would be expected to consult IMO publications and cargo-handling manuals, and other sources of information regarding the characteristics of cargoes which are normally consulted by interested parties.[59]

---

[54] Cooke/Young/Taylor, *Voyage Charters* (2001), 151.

[55] In an Australian case *Hoey v. Hardie* (1912) 29 30 N.S.W, only the name of the substance – bichrome of potash – was provided. The chemical was shipped in bags and badly packed. The notice was found insufficient to warn of the actual nature of the goods and so prevented steps being taken to avoid or prevent the damage.

[56] Edgcomb, "The Trojan Horse Sets Sail: Carrier Defences against Hazmat Cargoes" [2000-01] 13 *U.S.F. Mar. L.J.* 31, 51.

[57] Cooke/Young/Taylor, *Voyage Charters* (2001), 151. It was also asserted that a carrier does have some obligation to ascertain the dangerous characteristics of the cargo he carries. *Borgships, Inc.v. Olin Chems. Group*, 1997 U.S. Dist LEXIS 3065 (S.D.N.Y). Furthermore, it was held that a carrier must have knowledge of dangerous characteristics, such as film scrap, when he undertakes to carry such cargo. If he does not have actual knowledge, then the carrier is under an obligation to seek it. *Remington Rand, Inc. v. American Export Lines, Inc.* 132 F.Supp.261 (D.Md. 1951).

[58] Cooke/Young/Taylor, *Voyage Charters* (2001), 151.

[59] In a Canadian case *Elders Co. v. Ralf Misener* [2005] 3 F.C.R. 367, a cargo of alfalfa pellets caught fire when being discharged. The owner of the cargo sued the shipowner, who counterclaimed for the damage caused by the fire. The court considered whether the shipowner ought to have known of that dangerous condition. The IMDG Code current at the time of the shipment categorized alfalfa pellets as dangerous, while

4. "Nature and character" of the goods

While in common law the shipper needs to give to the carrier notice of the fact that such goods are dangerous,[60] under the Hague/Hague-Visby Rules the shipper must give to the carrier notice of the "nature and character" of the dangerous goods. However, the difference between "nature" and "character" of the goods is not explained in the Rules. It is said that "nature" deals with general aspects of the goods carried and "character" specific aspects thereof.[61] For example, concentrates have a propensity to shift, but the risk of shifting greatly increases once the moisture level reaches a particular percentage of the bulk. A carrier may consent to the carriage of concentrate knowing its nature but if he does not know that its moisture content is over the critical level, he does not have the relevant knowledge of the true "character" of the cargo. Therefore, it is often common knowledge that goods have dangerous characteristics, but it is contended that a notice of some special hazard involved in their carriage should be given.[62] It is true that a special notice is necessary if any specific characteristics of goods render the risks in the carriage of the given cargo different in degree from those generally involved in such cargo or if some special precautions are necessary beyond those required in the carriage of such cargo.[63] Accordingly, both elements have to be within the knowledge of the carrier so that he can give or withhold his consent in an informed way.

5. Sufficient notice

The notice should be sufficient to show the dangers of the goods. The question of whether sufficient notice has been given is a question of fact and like all findings

---

Canadian regulations published three years earlier available to the vessel's master did not. The court held that the master could rely on the Canadian Regulations and was under no duty to refer to the IMO Code. So the shipowner did not have knowledge of the dangerous condition.

[60] Abdul Hamid, *Loss or Damage from the Shipment of Goods, Rights and Liabilities of the Parties to the Maritime Adventure* (Diss. Southampton 1996), 223.

[61] Cooke/Young/Taylor, *Voyage Charters* (2001), 1008.

[62] *The Athanasia Comninos and Georges Chr. Lemos* [1990] 1 Lloyd's Rep.277; *The Atlantic Oil Carriers v. British Petroleum (The "Atlantic Duchess")* [1957] 2 Lloyd's Rep. 55.

[63] *The Mediterranean Freight Services v. BP Oil International (The "Fiona")* [1994] 2 Lloyd's Rep. 506. (C.A); *aff'd* [1993] 1 Lloyd's Rep.257, where the cargo was described with complete accuracy as "fuel oil", but this was held to be inadequate notice on the grounds that, at the time of the carriage (1988), the risks attendant on the carriage of certain types of fuel oil were not generally known. Similarly it has been held that although shipowners were generally aware of certain dangerous characteristics of sulphur, the highly corrosive properties of wet sulphur were not generally known to shipowners in the early 1980s and that therefore a special notice of characteristics should have been given. See also *International Mercantile Marine Co. v. Fels* 170 F.275 (2nd Cir. 1909).

of fact is to be decided in the light of the surrounding circumstances of the case.[64] In this respect, the courts seek generally to balance the duty of notice with commercial practicability.[65] The notice need not be a lecture on physics and chemistry. However, as the notice must be such that an ordinarily competent carrier will be able to appreciate the nature of the risks involved in the carriage and to guard against them, it would therefore not be sufficient for the shipper to give notice that the cargo was "oxygen water" when perhydrol, was shipped,[66] or "rock salt" when bisulfate[67] was carried or stating "fireworks" without mentioning that it is listed in Class 1 of the IMDG Code. Therefore, it is submitted that the proper shipping name according to the IMDG Code as well as the chemical name should be submitted to the carrier. Furthermore, in some cases proper labelling would suffice the duty of notice unless particular goods have unknown characteristics.[68] In the

---

[64]   For instance, in *the Athanasia Comminos and Georges Chr. Lemos* [1990] 1 Lloyd's Rep. 277, oral and written warnings were given to the carrier to the effect that "there is gas in our coal, watch your ventilation" and "all coal gassy … treat her like a tanker". They were also warned about the use of naked lights and supplied with an extract from "Thomas on Stowage". However, it was said that these warnings were not intended to constitute warnings as to the specially gassy nature of the coal, since it did not bring to the attention of the carriers that the coal was specially gassy. The warnings were merely intended to remind the master of the correct method of carrying coal. The carriers could not, by virtue of warnings, appreciate the true nature of the risks of the cargo shipped. Hence such warnings would not suffice to discharge the obligation to notify under common law. Similarly in *The Fiona* [1993] 1 Lloyd's Rep. 257, 269. the description on the bill of lading "IMCO Class 3 inflammable liquids" was found to be weak. It was said that the notation endorsed on the bill of lading should have conveyed a warning to the owners that the cargo was flammable and therefore required special precautions to be taken when para. 1.2.4 of the IMDG Code expressly provides that 2 substances which have a flashpoint above 61°C (141°F) are not considered dangerous by virtue of their fire hazard and when the certificates of quality placed on board the ship at the time of shipment showed that the flashpoint was 88°C and 98° respectively. However, the marking "Sodium Chlorite" was found sufficient notice, since everyone concerned knew that it was classified as dangerous goods. *Shaw Savill & Albion Company Ltd. v. Electric Reduction Co. ("The Mahia")* [1995] 1 Lloyd's Rep. 264. In *Ionmar Compania Naviera, S.A.. v. Olin Corp.* 666 F.2d 897, 1982 A.M.C. 19 (5th Cir.1982), steel drums of pool chlorine were carried. All necessary labeling was fulfilled by the shipper. However, warning was found to be insufficient as to the specific propensity of chlorine to ignite when in the presence of sawdust or other fine, organic material.

[65]   Abdul Hamid, *Loss or Damage from the Shipment of Goods, Rights and Liabilities of the Parties to the Maritime Adventure* (Diss. Southampton 1996), 65.

[66]   *Great Northern Railway v. L.E.P Transport and Depository* [1922] 2 K.B. 742. On the other hand, the use of the word "naphtha" in the description of a cargo of soap was notice of its dangerous propensity to give off naphtha vapor. *International Mercantile Marine v. Fels* 164 F.337 (S.D.N.Y 1908), aff'd, 170 F. 275 (2d Cir. 1909).

[67]   RGZ 93 S.163.

[68]   In *Pitria Star Navigation Co. v. Monsanto Co.* 1984 WL 3636, claims were asserted by the shipowner against a voyage charterer and the manufacturer and shipper of the cargo of parathion, a poison, liquid insecticide. This cargo is generally carried in gallon drums and it has been calculated that one drum contains sufficient quantity to kill about

case of bulk cargo, the present state of the cargo, such as humidity or temperature, should be provided.

6. To whom is the notice to be given?

In general, notice to the master or other person in control of the ship of the dangerous characteristics of the cargo will suffice to discharge the shipper's duty to inform.[69] The Hague/Hague-Visby Rules Art. IV.6 provides for the "…dangerous nature to the shipment whereof the carrier, master, or agent of the carrier has not consented…" Naturally, the servant or agent must be in position of authority to receive orders or warnings. A warning to an ordinary seaman does not suffice such a requirement.

---

10 million people. A dose of 25 mg is likely to be fatal on inhalation. Three seamen died during the voyage as a result of an accident which resulted in the parathion contaminating the ship's bilges. Each drum was painted green and lithographed in contrasting white with 14 large skull and crossbones. Each drum also contained warnings and information regarding the insecticide's characteristics, drum handling, spill clean-up, decontamination and the type of medical treatment required in case of accidental exposure. In addition, each drum was labelled with printed four-language warnings and information, unique pictorial warnings against touching, breathing and swallowing and the IMO poison warning. As a result each drum in 14 different places and in various languages, contained labels with short, prominent displayed warnings of "can kill you", "stop", "danger" and "poison". The master testified that he had no means of knowing how dangerous the commodity was and that they had not been given any notice or warning as to its potentially lethal characteristics. However, the court found that there was no negligence with respect to giving required warning of dangerous nature of the cargo. Rather uncontroversial evidence indicated that at no time did any officer of the vessel read these warning labels carefully or thoroughly.

[69] Colinvaux, *Carriage by Sea* (1982) Vol. 2, 844. In *Portsmouth Steamship Co. v. Liverpool & Glasgow Salvage Ass.* (1929) 34 Ll. L. Rep. 459, the plaintiff's vessel was hired for salvage services to carry cargo from another vessel which had run ashore. The cargo salvaged included palm oil in barrels. A large number were broken and their contents leaked into the ship's holds and caused serious damage. The decision was based on the shipowner's indemnity against complying with the charterer's orders. Nevertheless, it was held that as regards the implied common law obligation, any warning with regard to the cargo ought to have been given to the shipowner at home. Having regard to the position of the master during the salvage operations, it was not sufficient to say that the captain knew as much about the palm-oil barrels as the charterer's salvage officer. The owner was the proper person to have the opportunity of deciding what should be done if warning was given. It is said that that the observation should not be extended beyond the particular facts of the case, i.e. that the ship was used for salvage purposes. Abdul Hamid, *Loss or Damage from the Shipment of Goods, Rights and Liabilities of the Parties to the Maritime Adventure* (Diss. Southampton 1996), 67.

7. Effect of carrier's, agents' or master's knowledge

a) Notice not required where carrier, agent or master has actual or constructive knowledge

If the specific nature or character of the goods is within the sphere of a reasonably competent and experienced carrier's knowledge, the shipper has no duty to notify him.[70] The relevant knowledge of the carrier, his agents or master may be either actual or constructive.[71] In other words, duty arises only if the carrier or master did not know or ought not to have known of the danger that the goods posed.[72]

However, the question is complicated in the case of the carrier's constructive knowledge. This raises the question of what sort of knowledge the carrier or actual carrier should posses.[73] Such knowledge is material in considering the rights and obligations of the parties. In *Brass v. Maitland,* the cargo was described as "bleaching powder" containing chloride of lime, which was highly corrosive. It was held a good defence that from such a description, the master had the means of knowing and reasonably might and could and ought to have known that it contained chloride of lime and that he had the means of judging the state and sufficiency of the casks and of the packing of the contents. However, a mere allegation of "means of knowledge" would not have been sufficient, as this might be satisfied by calling in a competent chemist and resorting to investigations inconsistent with the usual course of commercial business.[74] It has been suggested that "a means of knowledge" is not tantamount to constructive knowledge unless the nature of the cargo is universally known.[75]

It is clear that the issue of constructive knowledge is one of facts and its parameters are defined by what the courts opine as reasonable. A delicate balance

---

[70] In the case of companies, one may safely assume that it is only the knowledge of present employees that is relevant and the relevant employees are presumably those whose task is to obtain and have at hand relevant information about cargoes to be carried.

[71] Wilford/Coghlin/Kimball, *Time Charters* (2003), 184; Cooke/Young/Taylor, *Voyage Charters* (2001), 1007; Abdul Hamid, *Loss or Damage from the Shipment of Goods, Rights and Liabilities of the Parties to the Maritime Adventure* (Diss. Southampton 1996), 76. In the *Athanasia Comminos and Georges Chr. Lemos* [1990] 1 Lloyd's Rep.277, it was said obiter that "... the words quoted (with the knowledge of their nature and character) must include knowledge which the carrier and crew ought to have, as well as that which they actually have; otherwise there would be premium ignorance". The Hamburg Rules expressly set out the carrier's constructive knowledge in Art. 13 (2) "... If the shipper fails to do so and such carrier or actual carrier does not have otherwise have knowledge of their dangerous character".

[72] DuClos, "Liability for Losses Caused by Inherently Dangerous Goods Shipped by Sea and the Determinative Competing Degrees of Knowledge" <www.duclosduclos.org/LiabilityforLossesCausedByInherently.pdf> 10 f. (visited 13.7.2007) (to be published in *U.S.F. Mar. L. J* Vol. 20 No. 1).

[73] Astle, *The Hamburg Rules* (1981), 122.

[74] Abdul Hamid, *Loss or Damage from the Shipment of Goods, Rights and Liabilities of the Parties to the Maritime Adventure* (Diss. Southampton 1996), 69.

[75] *Ibid.*

must therefore be struck between science and commercial norms and usages. A carrier is only expected to know what an ordinary, experienced, competent and prudent carrier would know.[76] If the state of knowledge in relation to a particular cargo is undetermined or equivocal, it would not be reasonable to impute anything to the carrier. Moreover, a carrier's knowledge also depends on how often he carries those goods and what opportunities he has of observing their condition.[77] Generally, a carrier is expected to be familiar with the usual and expectable characteristics of a cargo.

Consequently, knowledge is a relative concept and the carrier is expected to keep abreast of recent developments and consult when necessary. What was once something that only a "skilful chemist" could know may now be a matter of common knowledge. Accordingly, the creative ways to impute constructive knowledge on both the shipper and the carrier are asserted to be the existence of published regulations and a history of involvement in losses involving dangerous goods.[78]

b) Actual knowledge required in German law

In German law, if the master or another person whose knowledge is equivalent to the master's knows the dangerous nature of the goods, no notice is necessary.[79]

---

[76]   In *Acatos v. Burns* [1878] 3 Ex. D. 282, a cargo of maize was found to be in a dangerous condition at an intermediate port. In an action against him for conversation, the carrier alleged a breach of common law undertaking by the shipper. In rejecting this allegation, the court held that it was reasonable to impute knowledge of the dangerous characteristics of the goods where the carrier has full opportunities of observing the dangerous character of the good. Maize was the consignment and everybody knows that maize may sprout. The shipowner, seeing the maize coming on board and accepting it, could not claim against the shipper on a guarantee that maize would not sprout during the voyage. Where, however, it is impracticable for the carrier to scrutinize and analyze the nature and characteristics of the cargo, no imputation will be made. In *Health Steel Mines, Ltd. v. The 'Erwin Schroeder'(The Erwin Schroeder)* [1969] 1 Lloyd's Rep. 370, because of the moisture content of the cargo of copper concentrate, shifting boards were fitted. However, the cargo shifted and the vessel took list. It was said that there was much uncertainty regarding concentrates and the question of shifting cargoes at the time when the shipment was made. Therefore, the owners could not be expected to know of the dangers involved in carrying such a shifting cargo, having regard to the state of expert knowledge at the time. Imputation of that knowledge would be unreasonable.

[77]   DuClos, "Liability for Losses Caused by Inherently Dangerous Goods Shipped by Sea and the Determinative Competing Degrees of Knowledge", <www.duclosduclos.org/ LiabilityforLossesCausedByInherently.pdf> 7 f. (visited 13.7.2007) (to be published in *U.S.F. Mar.L.J.* Vol. 20 No. 1); Wong K.K, "Packing Dangerous Goods" [1976-77] 8 *J. Mar. L. & Com.* 387, 390.

[78]   DuClos, "Liability for Losses Caused by Inherently Dangerous Goods Shipped by Sea and the Determinative Competing Degrees of Knowledge", <www.duclosduclos.org/ LiabilityforLossesCausedByInherently.pdf> 12. (visited 13.7.2007) (to be published in *U.S.F. Mar.L.J.* Vol. 20 No. 1).

[79]   Rabe, *Seehandelsrecht* (2000), 447.

Knowledge must be "actual" knowledge; "constructive" knowledge is not sufficient.[80] However, in some cases, for instance when straw is carried – which is
listed in the IMDG Code Class 4.1 – it is not necessary from the shipper's viewpoint to inform the carrier of the easy inflammability of straw. Therefore, the
actual knowledge requirement, in contrast to constructive knowledge, is to be
understood in the way that in obvious cases evidence of knowledge *prima facie* is
to be presumed.[81] However, in the case of doubt, constructive knowledge is not
sufficient.[82]

8. Consent to the shipment of dangerous goods

Consent is a declaration of intent which becomes complete upon receipt by the
other party.[83] It is an assent and approval to agree to do something especially after
thoughtful consideration and implies something different from a contractual
agreement. Therefore, a shipper may ship a dangerous cargo in breach of contract
but with the consent of the master.[84] Moreover, a cargo may comply with its contractual description but nevertheless have particular dangerous characteristics, of
which the carrier, master or agent are ignorant and to the carriage of which no
consent, external to the contract, has been given.

Some cargoes are customarily loaded after designated precautions have been
taken, e.g. the treatment of fishmeal with antioxidant.[85] Consent to the loading of
such cargoes would usually be regarded as subject to the taking of such precautions and the absence of such precautions might nullify any apparent consent.[86]

Art. IV.6 of Hague/Hague-Visby Rules designates persons whose consent is
relevant as being "the carrier, master, or agent of the carrier." Normally the person
consenting to the shipment will also have to possess the relevant knowledge.
However, it may not always be so. For instance, a carrier's agent knows or has the
reasonable means of knowledge of the particular nature and character of the goods
shipped because of specific local information, but the carrier and the master are
both ignorant of them. In such a case, it is asserted that the carrier and master may
still be held to give their (implied) consent by accepting the shipment. Otherwise,
there would be a premium on bad communication.[87]

In terms of formulation of the consent, the Hague/Hague-Visby Rules are silent
on this issue. However, it would be wise to obtain express documented notice and
consent. Implied consent will probably be found only where there is both clear
knowledge of the nature and character of the goods and where the act of accepting

---

[80]   *Ibid.*
[81]   Gündisch, *Die Absenderhaftung im Land- und Seetransportrecht* (1999), 197.
[82]   *Ibid.*
[83]   Schwampe, *Charterers' Liability Insurance* (1988), 57.
[84]   *Chandris v. Isbrandtsen-Moller* [1951] 1 K.B.240.
[85]   *General Feeds Inc. v. Burnham Shipping Corp. (The Amphion)* [1991] 2 Lloyd's
       Rep.101; *Islamic Investment Co. v. Transorient Shipping Ltd. (The Nour)* [1999]
       1 Lloyd's Rep. 1.
[86]   Cooke/Young/Taylor, *Voyage Charters* (2001), 1007.
[87]   *Ibid.* at 1008.

shipment is consistent with consent.[88] To ensure that there is no doubt as to whether consent has been given upon knowledge communicated by the shipper consent should be given in writing.[89] Dangerous goods clauses of bills of lading are worded in such a way that they require written notice of the dangerous nature or consent of the carrier.[90]

9. Master's authority to consent

If the contract excludes certain cargoes the master has no discretion to consent to the shipment of excluded cargoes, because the master has no authority to change the contract's terms. Therefore, the acceptance of excluded cargo by the master does not amount to affirmation of the contract.[91] He is the owner's agent for the voyage defined by the charterparty and has no power to change it for another. Although a time charterparty puts the master under the orders of the charterers as regards employment, it is not be construed as compelling the master to obey orders which the charterers have no power to give.[92]

On the other hand, the position is different under the Hague/Hague-Visby Rules. Art. IV.6 of the Rules provides for "goods of inflammable, explosive, or dangerous nature to the shipment where of the carrier, master or agent of the carrier has not consented with knowledge of their nature and character...." Where the contract does not include cargo exclusion clause knowledge of the master is

---

[88] *Ibid.* at 1007.
[89] It was suggested that the nature and character of the dangerous goods must be declared in writing by the shipper to the carrier. Sturley, *Legislative History of the Carriage of Goods by Sea Act* (1990) Vol. 1, 272 f.
[90] Mitsui OSK Lines Combined Transport Bill 1992 Cl.22(1); Ellerman East Africa/Mauritius Service Bill Cl. 19(1); P&O Nedloyd Bill Cl.19(1).
[91] Colinvaux, *Carver Carriage by Sea* (1982) Vol. 1, 32. In *Chandris v. Isbrandtsen-Moller Co. Inc.* [1951] 1 K.B. 240 a charter excluded shipment of "acids, explosives and ammunition or other dangerous cargo" and contained a clause paramount incorporating the provisions of S.4, sub-s.6, of the United States Carriage of Goods by Sea Act, 1936, which is identical with Art. IV.6 of the Hague-Visy Rules". The vessel was delayed for 22 days beyond the lay-days at the discharging port, and all except 6.5 days of that delay was caused by the shipment of the dangerous cargo. Upon learning of the shipment, the owner did not rescind the contract, but claimed and recovered freight at the charter rate, and demurrage for 6.5 days. The owner than claimed damages in general for the balance of the delay at the discharge port, on the grounds that this delay had resulted from the charterer's breach in shipping dangerous goods. It was held that the master's consent was irrelevant, since he had no authority to vary the charter nor did he purport to do so. The incorporation of the Hague Rules made no difference, since it was impossible to imply in Art. IV.6 a provision that, if the master had consented to the shipment of dangerous cargo with knowledge of its nature and character, the carrier should not be entitled to damages for loss resulting from the shipment.
[92] Wilford/Coghlin/Kimball, *Time Charters* (2003), 177.

knowledge of the carrier and knowingly accepting dangerous goods for carriage overrides any prohibition in the contract of the carriage of dangerous goods.[93]

It is to be noted that HGB, the German Commercial Code, § 564c expressly states in that case "… the knowledge of the captain is deemed equivalent to the knowledge of the carrier or of the ship's agent."

10. Notice by freight forwarder

a) Role of freight forwarder

The freight forwarder is any person who offers himself to the general public as a provider and arranger of the transportation of property, for compensation, and who may assemble and consolidate shipments of such property, and performs or provides for the performance of break bulk and distribution operations with respect to such consolidated shipments and assumes responsibility for the transportation of such property from the point of receipt to the point of destination and utilizes for the whole or any part of the transportation of such shipments the services of a carrier or carriers, by sea, land or air, or any combination thereof.[94] Freight forwarders are prominent links in modern transportation systems in a buyer's market. The forwarder might take different roles at different stages. The legal responsibility of freight forwarders often seems mysterious, because freight forwarders have assumed two different legal roles: agents and principal contractors.[95]

Although the freight forwarder traditionally acts as an agent who arranges for the shipment of goods belonging to his client/the shipper, the freight forwarder sometimes acts as principal contractor arranging the carriage in his own name.[96] Whether the freight forwarder is an agent or a principal contractor will depend upon the facts of each case and upon the law in the particular jurisdiction in question.[97]

A freight forwarder acts as a "travel agent" for the shipper, making the transport, documentation, customs and payment arrangements for the shipper. A freight forwarder, like any professional, must exercise the reasonable care and skill of his profession. The specific duties which the forwarding agent assumes vis-à-vis the customer who retains his services are determined principally by the contract between them.[98] The freight forwarder who acts purely as an agent is not a carrier

---

[93] Tetley, *Marine Cargo Claims* (1988), 466f. *Ente Nazionale v. Baliwag Navigation* 605 F.Supp. 355, 363 f.; *Skibs v. A/S Gylfe v. Hyman-Michaels Co.* 304 F.Supp.1204, 1221.

[94] Hill, *Freight Forwarders* (1972), 16.

[95] *Ibid.* at 133 ff; Glass, *Freight Forwarding and Multimodal Transport Contracts* (2004), 48 ff.

[96] Tetley, *Marine Cargo Claims* (1988), 692.

[97] *Ibid.* at 693 ff.; Tetley, "Responsibility of Freight Forwarders" [1987] 22 *E.T.L* 79 ff.

[98] See standard printed terms and conditions: British International Freight Association (BIFA) Standard Trading Condition; German Freight Forwarders' Standard Terms and Conditions, (ADSp); Standard Trade Conditions of The Canadian International Freight

and is not subject to the Hague/Hague-Visby Rules or national enactments of them, unless he issues a bill of lading with a clause paramount making one of those conventions or laws applicable to the bill or unless he incorporates the ocean carrier's bill of lading by reference into his own.[99]

b) Duties of freight forwarder with regard to dangerous goods

The freight forwarder has responsibilities when it comes to dealing with shipments of dangerous goods. A freight forwarder may be involved in any or all aspects of dangerous goods, such as handling or offering for transport. He is responsible for ensuring that the shipper's dangerous goods documentation and packing is complete, correct and meets national and international regulations. Likewise, when he delivers the dangerous cargo, he should give notice of the dangerous nature of the cargo. It is to be noted that a shipper has the main responsibilities for dangerous goods, i.e. ensuring the shipment is properly identified, classified, packed, marked, labelled, documented, in the proper condition for transport and meets all national and international regulations. Upon the information provided by the shipper to the freight forwarder, the latter communicates all necessary information to the carrier.[100] However, there are differences of opinion on the case when the freight forwarder does not know the dangerous nature of the cargo.

c) Notice of dangerous goods by forwarder

aa) English law

In English law, the forwarder is placed in the position of requiring exact knowledge of the contents of every shipment he handles on behalf of his client and must ensure that the documents accurately describe the contents.[101] Although, in practice, a forwarder will rely on his client, the shipper, to a considerable extent for such information, if the latter either deliberately or inadvertently misinforms him, the forwarder will still be liable for breach of warranty.[102] However, it should be

---

Forwarders Association, Inc. Glass, *Freight Forwarding and Multimodal Contracts* (2004), 74 ff.

[99]   Tetley, *Marine Cargo Claims* (1988), 692 f.

[100]   FIATA Model Rules for Freight Forwarding Art. 16 states that: The customer shall be deemed to have guaranteed to the Freight Forwarder the accuracy, at the time the Goods were taken in charge by the freight forwarder, of all particulars relating to the general nature of the Goods, their marks, number, weight, volume and quantity, and if applicable, to the dangerous character of the Goods, as furnished by him or on his behalf. When the freight forwarder was found liable to the third parties, he can have recourse to the customer (shipper).

[101]   Hill, *Freight Forwarders* (1972), 132.

[102]   *Brass v. Maitland* (1856) 6 E & B 470, was a forwarder case in which it was held that there is implied warranty by the shipper. In *Great Northern Railway Com. V. L.E.P Limited* [1922] 2 K.B 742, it was held that forwarding agents who deliver goods which are in fact dangerous to the carrier without informing him of their danger are liable for

clearly understood that liability will only attach to a forwarder where he actually act as a forwarder.[103]

bb) American law

In American law, the freight forwarder does not have the primary responsibility for the proper labelling of dangerous cargo, nor does he have any obligation to make an independent investigation of the cargo to determine if it is dangerous.[104] The obligation exists only where the forwarder knows or should know of the dangerous nature of the cargo. The primary responsibility for the proper labelling of the cargo rests with the shipper and there is no duty on the part of the freight forwarder to independently investigate each cargo to determine whether and to what extent the cargo may be dangerous.[105] When a freight forwarder knows or should know that a cargo is dangerous, he is obliged to note that fact on the bill of lading. Otherwise, it is the principal's (customer's) duty to inform the freight forwarder regarding the nature of the cargo.

However, if an intermediary is acting as non vessel operating common carrier (NVOCC) he is considered shippers in his relationship to the carrier and is bound to provide information as to dangerous nature of goods.[106]

cc) German law

German law distinguishes *Befrachter* (shipper) and *Ablader* (actual shipper). The later might the person acting in the name of the (shipper) as an independent agent.[107] The internal relationship between *Befrachter* and *Ablader* can be a char-

---

consequent damage sustained through that danger. See also *Bamfield v. Goole Transport Co.* [1910] 2 K.B. 94.

[103] Hill, *Freight Forwarders* (1972), 133.

[104] *Ward v. Baltimore Stevedoring Co.,* 437 F.Supp. 941, 1978 AMC. 965, a longshoreman who was injured while discharging highly toxic cargo, brought a suit against the freight forwarder. It was held that in the absence of any allegation that the freight forwarder was told that the cargo was highly toxic and could cause harm if improperly packaged or handled and where the freight forwarder's only connection with the cargo was that he had prepared the bill of lading for shipment, the freight forwarder was not liable to the longshoreman. Likewise, when the weight of cargo has been supplied to the freight forwarder by the shipper, the freight forwarder is not responsible to the carrier for improperly declared cargo weights, because the forwarder agent is under no duty to the carrier to provide accurate shipping documents. Accordingly, when the carrier was fined $65,520 by customs in the Ivory Coast for improper weights, it was held that the carrier could recover this sum from the shipper under Section 3(5) of COGSA, but not from the freight forwarder. *Atlantic Overseas Corp. v. Feder,* 452 F.Supp.347, 1978 A.M.C. 1203 (S.D.N.Y); see also *Scholastic Inc. v. M/V Kitano*, 362 F.Supp.2d 449, 2005 A.M.C. 1049.

[105] Schoenbaum, *Admiralty and Maritime Law* (2004) Vol. 1, 609.

[106] *Scholastic Inc. v. M/V Kitano* 362 F.Supp.2d 449.

[107] See *infra* p. 180 f.

ter, forwarding or purchase contract. Thus, a freight forwarder who is an *actual* shipper is bound to give notice of dangerous nature under HGB 564b.[108]

VII. Documentation of dangerous goods

The SOLAS 1974 Convention requires a certificate or a declaration, as well as a container/vehicle packing certificate to be provided along with the dangerous goods shipping documents.[109] One of the primary requirements of a transport document for dangerous goods is to convey the fundamental information relating to the hazards of the goods. It is, therefore, necessary to include certain basic information on the document for the consignment of dangerous goods.

1. Dangerous goods transport document and declaration

In accordance with the regulations of the SOLAS Convention, the IMDG Code requires that the consignor who offers dangerous goods for transport must describe the dangerous goods in a transport document and provide additional information and documentation as specified.[110] The dangerous goods transport document may be in any form, provided it contains all the information required by the IMDG Code.[111]

In addition, the dangerous goods transport document shall include a certification or declaration that the consignment is acceptable for transport and that the

---

[108]   Rabe, *Seehandelsrecht* (2000), 303.
[109]   SOLAS Convention Chap.VI.A. Reg.4.
[110]   The IMDG Code Chap.5.4.1.2.
[111]   The transport document shall include:
    1. consignor, consignee and date
    2. dangerous goods description
      a. the UN number preceded by the letters "UN"
      b. the proper shipping name
      c. the class or, when assigned, the division of the goods
      d. where assigned, the packing group for the substance or article which may be preceded by "PG".
In addition to the dangerous goods description:
    1. total quantity of dangerous goods
    2. limited quantities
    3. salvage packing
    4. substances stabilized by temperature control
    5. self-reactive substances and organic peroxides
    6. infectious substances
    7. radioactive material
    8. aerosols
    9. explosives
    10. viscous substances
shall be included.

goods are properly packed, marked and labelled, and in a proper condition for transport in accordance with the applicable regulations.[112]

The "Multi-Modal Dangerous Goods Form" contained in Part 5 of the IMDG Code may be used as a dangerous goods declaration, as it meets the requirements of the above-mentioned regulations. The information on this form, as well as that contained in Chapter 5 of the IMDG Code, is mandatory but the layout of the form is not. The declaration of dangerous goods imposes formidable but necessary responsibilities upon the party making the declaration. If the facts in the declaration are not accurate, that party is liable for the outcome.[113]

Description of dangerous cargoes in bulk must be fulfilled in accordance with the particular code.

2. Container/Vehicle packing certificate

When dangerous goods are packed or loaded into any container, those responsible for packing the container must provide a "container/vehicle packing certificate" specifying the container/vehicle identification numbers and certifying that the operation has been carried out in accordance with the conditions provided in the IMDG Code.[114]

VIII. Is compliance with dangerous goods regulations sufficient?

When a shipper has complied with dangerous goods regulations, will the shipper be deemed to have met all his duties to the carrier as a matter of law? It seems that the answer is no. Giving some warning about the danger and labelling the cargo as indicated cannot be evaluated in a vacuum. It can only be assessed in the light of the knowledge and expertise of the carrier, ship's master and other relevant persons.[115]

---

[112] The text for this certification is "I hereby declare that the contents of this consignment are fully and accurately described above by the Proper Shipping Name, and are classified, packed, marked and labeled/placarded, and are in all respects in proper condition for transport according to applicable international and national government regulations."

[113] In Germany, GGBefG § 10 sets out cases which constitute violation and imposes fines of up to 50,000 euros. The U.S. Hazardous Materials Regulations impose both civil and criminal penalties in §§ 5123 f. Merchant Shipping (Dangerous Goods and Marine Pollution) Act Art. 24 sets out fine for infringement.

[114] The information required in the dangerous goods transport document and the container/vehicle packing certificate may be incorporated into a single document. If not, these documents will be attached to each other. If the information is incorporated into a single document, the document shall include a signed declaration such as "It is declared that the packing of the goods into the container/vehicle has been carried out in accordance with the applicable provisions."

[115] In *Ionmar Compania Naviera S.A. v. Olin Corp.* 666 F.2d 897, 1982 A.M.C. 1489, the defendant Olin shipped steel drums of pool chlorine by vessel. Chlorine, a highly

The IMDG Code and other codes provide lists of dangerous goods; however, these lists are not exhaustive. Particular cargo may not be listed by name but by its properties.[116] Therefore, the shipper is not discharged his duty to give notice by

---

flammable substance, is subject to extensive regulation by the HMR. Olin met those requirements, including properly marking and labeling the chlorine drums, packing chlorine in approved steel drums and providing a written description of the chlorine's characteristics to the shipowner. The shipowner duly noted the presence of the chlorine on his required hazardous cargo manifest. Nonetheless, when a fire erupted in the area where the chlorine was stored, causing substantial damage to the vessel and other cargo, the shipowner sued Olin for failure to properly warn him of all dangers of chlorine. An investigation determined that a reaction between the chlorine and sawdust left in the area by longshoremen caused the fire. The District Court found that, despite its compliance with the HMR, Olin was 85% responsible for the damage due to its negligence in failing to give the stevedore adequate warning about the specific propensity of chlorine to ignite when in the presence of sawdust or other fine, organic material. The stevedore was assigned the other 15% of liability. Olin appealed. The Fifth Circuit reversed and remanded the case, concluding that the District Court failed to make the necessary findings with respect to critical issues, including whether Olin properly warned the carrier. The Fifth Circuit provided the District Court with the following guidance on the "sufficiency of warning" issue: "Olin as the manufacturer of the chlorine had a duty to warn the stevedore and the shipowner of the foreseeable hazards inherent in the HTH shipment of which the stevedore and the ship's master could not reasonably have been expected to be aware. The Appellate court agreed that Olin had complied with the HMR by placing cautionary yellow labels on each drum and making the required statements on the bills of lading describing the chlorine as an oxidizing agent (highly flammable). While these steps gave the stevedore and shipowner some warning about the dangers associated with HTH, the Court concluded that this was not necessarily enough. Based on its conclusion that the findings do not indicate what the stevedore knew or should have known about stowing this particular cargo, it directed the District Court to inquire as to "what knowledge, aside from that disclosed by the labels and the bill of lading, the vessel actually had about the cargo prior to its stowage. In respect to Ionmar, compliance with the HMR regulations does not mean as a matter of law that shippers have met their entire pre-shipment duties to carriers, especially if their cargo has dangerous characteristics with potentially serious consequences that are not likely foreseeable by the carriers. See also *Borgships. Inc. v. Olin Chemicals Group* 1997 U.S. Dist. LEXIS 3065 (S.D.N.Y 1997), where, although the cargo of SDIC was exclusively excluded from the list of HMR, the court, following *Ionmar*, concluded that compliance with Department of Transport (DOT) regulations does not satisfy as a matter of law the shipper's duty to warn.

[116] Some cases illustrate very well how improper consultation of the IMDG Code may cause a disaster. In the *Asian Gem* the vessel contracted to carry some low-grade powdered zinc dross from Long Beach to Japan, where at the time they were less careful about pollution when smelting low-grade zinc ores. The ship asked whether the material was dangerous and was told: "It is not in the IMDG Code". This was a bulk cargo and was transported to the ship in trucks and loaded. The ship was told to keep the material dry, despite it having been stored outside for up to two years and despite the trucks having been sprayed with water to reduce dusting problems. When the ship sailed, the diligent crew began to apply Ram-neck bitumen tape to the hatch covers to keep water out, but it was cold, so the seaman doing the work used a paraffin blow

simply stating that the cargo is not listed in the IMDG Code.[117] It is unfortunate that a new substance is only given a UN number and an entry in the Orange Book and the IMDG Code if there is accident experience, new information or a new requirement to ship.[118] Moreover, there may be cases where instructions provided in the IMDG Code are insufficient.[119]

IX. The 24-Hour Advance Manifest Rule or the 24-Hour Rule and its implication for the shipper's duties

In the past forty years, containerization has helped lower the cost of moving goods and created a relatively transparent and sophisticated logistics network for the more efficient delivery of goods. Unfortunately, these gains have a downside because of their very transparency and ease of access. As a result, over the past twenty years, government authorities have had to impose an ever-widening array of restrictions and demands upon anyone using this elaborate system. Drug-smuggling, stowaways and cargo theft were early problems for which special measures had to be introduced, and those have become more elaborate. Now concerns about the deployment of weapons of mass destruction by terrorists, as well as marine pollution, have prompted stricter measures tailored to these "extra-commercial" activities.[120]

Following the events of 11 September 2001, safety and security considerations have been at the forefront of international concerns. The need to enhance security

---

lamp to heat the tape and metal to get good adhesion. The inevitable happened. There was an explosion, the hatch covers were blown up and one removed the head of the unfortunate seamen. Investigations revealed that it was not a self-heating problem like direct reduced iron or iron scrap but was simply a low-temperature reaction between zinc, dust and water producing hydrogen gas which continued to burn on the surface of the stowage. Despite the knowledge of the shipper that it produced hydrogen, he decided to say "It is not listed in the IMDG Code". While at the time zinc dross and zinc ashes were not specifically listed in the IMDG Code, it was effectively present as a Class 4.3 "water reactive substance N.O.S" or "not otherwise specified" material. Therefore, the shipper should know that there are often NOS catch-all categories in the IMDG Code. Following the incident, zinc ash was specifically incorporated into the codes as a hazardous material generating flammable gas when wet. Watt/Burgoyne, "Know Your Cargo" [1999] 13(5) *P&I Int'l* 102.

[117] *Senator Linie GmbH v. Sunway Line* 291 F.3d. 145.

[118] Compton, "Dangerous Goods" 2004 (January) *Cargo Systems*, 34, 35.

[119] "New IMDG Code 'dangerous' says club", 2000 (14 December) *Fairplay* 7; *In re M/V Harmony and Consolidated Cases*, 393 F.Supp.2d 649. In that case the M/V Harmony stowed the containers in accordance with the IMDG Code. Neither the manufacturer/ shipper nor the carrier knew the true risks and dangers of storing and shipping this chemical in the manner utilized by the shipper. Testing subsequent to the accident revealed that the chemicals should have been stored at a lower temperature than provided for by the IMDG Code.

[120] Bulow, "Charter Party Consequences of Maritime Security Initiatives: Potential Disputes and Responsive Clauses" [2006] 37 *J. Mar. L & Com.* 79.

worldwide is recognized by all governments and industry. As world trade is largely dependent on maritime transport, the security of the maritime transport system has received particularly significant attention. The U.S. government, in response to its own analysis of the vulnerability of the maritime transport system, has taken the lead and initiated a considerable number of measures aimed at enhancing the security of maritime traffic, including port, vessel and cargo security. Given that a reported 50% of the value of all U.S. imports arrives in sea containers, much of the focus has been directed at the particular security challenge posed by maritime container shipments of dangerous weapons and dangerous cargoes and a number of specific measures relevant to container security have been implemented in the forms of laws, regulations and voluntary partnership programs. Several international organizations, including the World Customs Organization (WCO) and the International Labour Organization (ILO) have also reacted swiftly to the need for strengthened security measures to enhance maritime transport security.

The worry that a vessel or her cargo could be used as a weapon of terror led to the passage in the U.S. Congress of the Maritime Transportation Security Act (MTSA) of 2002. This law, along with a concomitant international effort to address the same concern, led to the adoption in the same year of the International Ship and Port Facility Security (ISPS) Code by the IMO.[121]

New security measures obviously cost time and money. Delays caused by security have started fairly new areas of dispute.

1. U.S. cargo security initiatives

The main U.S. initiatives relevant to maritime container security are the Customs Trade Partnership Against Terrorism (C-TPAT) and the Container Security Initiative (CSI), which focus on establishing partnership relations with industrial actors and ports, as well as the so-called "24-Hour Rule" and recent regulation under the U.S. Trade Act of 2002, which amend U.S. customs regulations (19 CFR) and are aimed more specifically at obtaining and monitoring information on cargo. The U.S. Customs and Border Protection Service (CBC) is the relevant government agency in charge of the administration and enforcement of these programs and regulations.

a) Customs Trade Partnership against Terrorism

The Customs Trade Partnership Against Terrorism (C-TPAT) is a joint government/business initiative aimed at building a co-operative relationship that strengthens overall supply-chain and border security. It is intended to enhance the joint efforts of both entities in developing a more secure border environment, by improving and expanding existing security practices. C-TPAT is a non-contractual voluntary agreement, terminable at any time by written notice by either party. Initially, importers, carriers, as well as U.S. port authorities/terminal operators and

---

[121]   The ISPS Code is incorporated by reference in the MTSA.

certain foreign manufacturers are eligible to participate in the program. However, it is envisaged to broaden participation to include actors in all international supply chain categories.[122]

U.S. Customs, on their part, mainly undertake to assist the carrier in his efforts to enhance security and to expedite clearance of cargo at the U.S. border. Once a company becomes a C-TPAT member, its risk score in the Automated Targeting System (ATS) is partially reduced.[123] U.S. Customs also undertake to conduct initial and periodic surveys to assess the security in place and suggest improvements.

C-TPAT operates on the basis of individual "non-contractual voluntary agreement" to implement certain recommendations.[124] The parties are thus expected to use their best endeavour to comply with the C-TPAT recommendations and to enhance security throughout their supply chain, without, however, incurring liability in the case of errors or non-compliance.[125]

b) Container Security Initiative

The Container Security Initiative (CSI) is another program concerning ocean-going sea containers which was developed shortly after 11 September 2001. CSI is based on the premise that the security of the world's maritime trading system needs to be enhanced and that it will be more secure if high-risk cargo containers are targeted and screened before they are loaded. The initiative aims at facilitating detection of potential problems at the earliest possible opportunity and is designed to prevent the smuggling of terrorist or terrorist weapons in ocean-going cargo containers. CSI consists of four core elements:[126]

1. Establish security criteria for identifying high-risk containers based on advance information.
2. Pre-screen containers at the earliest possible point.
3. Use technology to quickly pre-screen high-risk containers.
4. Develop secure and smart containers.

---

[122] Applicants wishing to participate need to fill in a C-TPAT Supply Chain Security Profile Questionnaire and to sign a C-TPAT Agreement to Voluntary Participation.

[123] As a result, the likelihood of inspections for Weapons of Mass Destruction is decreased.

[124] C-TPAT partners receive these benefits: a reduced number of inspections (reduced border times); an assigned account manager (if one is not already assigned); access to the C-TPAT membership list; eligibility for account-based processes (bimonthly/ monthly payments, e.g.); an emphasis on self-policing, not Customs verification. Bishop, "A 'Secure' Package? Maritime Cargo Container Security After 9/11" [2002] 29 *Transp. L.J.* 313, 321.

[125] U.S. Customs may remove a company from C-TPAT membership if they determine that its commitment is not serious or that it has intentionally misled Customs.

[126] Bishop, "A 'Secure' Package? Maritime Cargo Container Security After 9/11" [2002] 29 *Transp. L.J.* 313, 319.

Increased security is critical in the area of containerized shipping, since about ninety percent of the world's trade is transported in cargo containers.[127] The goal of CSI is to improve security without slowing down the movement of legitimate trade. Thus, wherever possible container screenings are to be carried out during periods of down time, when containers sit on the docks waiting to be loaded on a vessel and screenings should not have to be carried out again in the U.S.

c) The 24-Hour Rule

Whereas the C-TPAT and CSI are partnership-oriented programs, other security initiatives focus on the collection of information, in particular cargo-related information. The main such initiative or relevance to maritime container transport is the so-called 24-Hour Rule, which is closely connected to CSI. U.S. customs regulations now require detailed manifest information on U.S.-bound cargo to be provided 24 hours before loading at the foreign port.

U.S. law and customs regulations impose certain documentary requirements upon vessels bound for the United States. Inter alia, vessels destined for the U.S. and required to make entry must have a manifest meeting certain requirements.[128] U.S. Customs are the competent authority to specify the form and data content of the vessel manifest, as well as the manner of production and delivery or electronic transmittal of the vessel manifest. Prior to 2 December 2002, the relevant customs regulations[129] simply required the master of every vessel arriving in the U.S. to have the manifest on board the vessel. The vessel manifest had to include a cargo declaration listing all the inward foreign cargo on board the vessel, regardless of the intended U.S. port of discharge of the cargo.[130] No merchandise would be unloaded until U.S. Customs had issued a permit for its discharge.[131] In cases where the master of a vessel had committed any violation of customs laws, for example by presenting or transmitting a forged, altered or false manifest, he was liable to pay a civil penalty.[132] Following the events of September 2001, new regulations have been adopted with the aim of enabling U.S. Customs to evaluate the terrorist risk of cargo containers.[133]

d) Shipments affected by the 24-Hour Rule

The new regulations require ocean carriers to transmit the cargo manifest for cargo being shipped on a container vessel to the U.S. 24 hours in advance of loading at

---

[127] *Ibid.* at 320.
[128] 19 U.S.C. 1431, 1434.
[129] 19 CFR 4.
[130] 19 CFR 4.7a (c)(1)
[131] 19 U.S.C. 1448.
[132] 19 U.S.C. 1436(b).
[133] The final rule, leading to changes to 19 CFR 4, has been published in 67 FR 66318. Note that some provisions have been revised and others added as a result of new regulations under the Trade Act of 2002, which entered into force on 5 December 2003.

foreign ports. Transit containers, so-called FROB,[134] bound for destinations outside the U.S. are equally affected by the Rule. Bulk shipments are exempted from the requirements. As for break-bulk cargo, exceptions may be made on a case-by-case basis.[135] It should be emphasized that any container which is transhipped before reaching its final U.S. destination will have to fulfil the 24-hour requirement at the last transhipment port. Thus, if a consignment is cleared under the Rule and loaded onboard a vessel bound for a specific destination but the vessel later diverts to an intermediate port for shipment, the carrier will have to comply once again with the Rule. As far as empty containers are concerned, it appears that notification of relevant information needs to be provided to U.S. Customs 24 hours before arrival of the vessel.

e) Required information

For each container, the manifest must provide a large number of data elements including, *inter alia*:

- detailed and precise description of the cargo or the 6 digit HTSUS (Harmonized Tariff Schedule of the United States)
- numbers and quantities of the lowest external packing units as per bill of lading
- container number and, if applicable, seal number,[136]
- accurate weight of the cargo[137]
- the foreign port where the cargo is loaded, the last foreign port before the vessel departs for the U.S. and the first foreign port where the carrier takes possession of the cargo
- the full names and complete, accurate and valid addresses of the consignee and the shipper of the cargo; alternatively, unique "identification number" for shipper and consignee to be "assigned by Custom Border Protection (CBP) upon completion of the Automated Commercial Environment."

As the Rule seeks to establish precisely what is carried in every container, the description of the cargo must be precise enough to enable identification of the shapes, physical characteristics and likely packing of the manifest cargo so U.S. Customs can identify any anomalies in the cargo when a container is run through imaging equipment. Generic descriptions, such as "FAK" (freight of all kinds), "general cargo" and "STC" (said to contain) are not acceptable, as they do not provide adequate information regarding the merchandise. Descriptive clauses which were commonly used and accepted until recently are no longer acceptable

---

[134] The so-called "Foreign Cargo Remaining on Board", ("FROB"), refers to the cargo loaded in a foreign port and to be unloaded at another foreign port which remains on board during one or several intervening stops in U.S. ports.

[135] Customers wishing to obtain an exemption must send an exemption request in writing to U.S.Customs. 67 FR 66318, 66321.

[136] In the case of containers, the smallest packing unit inside a container is relevant, 19 CFR 4.7a (c) (4) (v).

[137] For sealed containers, the weight as declared by the shipper may be provided, 19 CFR 4.7a (c) (4) (vii).

and have to be replaced by more specific clauses.[138] Indirectly, the new requirements affect also bills of lading and other transport documents used in trade, as carriers need to relate a number of data elements from the relevant shipping documents, including the identity of the shipper and consignee.

f) The 24-Hour Rule in carriage contracts

The information needs to be provided to U.S. Customs by the carrier, not the shipper of the cargo. In practice, however, this means that shippers must provide the necessary information several days ahead of sailing, whereas in the past manifests were invariably submitted long after the vessel had departed.

Failure to provide the required information within 24 hours prior to loading may result in the delay of a permit being issued to discharge the cargo in the U.S. and/or the assessment of penalties or the level of claims for liquidated damages against the carrier by the U.S. Customs. Since bill of lading terms are heavily influenced by charterparty terms, more thought has been put into the charterparty rather than the bill of lading.[139]

For instance, BIMCO has produced two standard clauses for incorporation into voyage and charterparty respectively.[140] Acknowledging the fact that the charterers are usually in a better position than the owners to obtain and assess the correctness of the information provided for the cargo, the sub-clause (a) applies the principle that the charterers shall provide all necessary cargo information to the owners to enable them to submit a timely and accurate cargo declaration. Ideally it would be more expedient if the charterers were themselves to submit the cargo declaration directly to U.S. Customs. However, given the wording of the regulations, the charterers are only allowed to do so in a very limited number of circumstances.[141] To cater for these situations, sub-clause (a) (ii) stipulates that if permitted by the regulations the charterers must submit the cargo declarations. Sub-clause (b) sets out the legal consequences for the charterers if they do not comply with the provisions of sub-clause (a).

2. European Union developments and the 24-Hour Rule

Customs work in the EU has typically evolved in the context of financial and commercial controls. Broadly speaking, the fight against fraud has traditionally been seen as part of the task of supervising the flow of goods. Work has been organized on this basis and Community instruments such as customs modernization programs have been used to identify best practice for financial and commercial controls. In the light of the growing threats from dangerous goods, organized crime or terrorist organizations, this approach has been found to be inadequate protection for the Community. Current controls ensure the financial interests of

---

[138]   67 FR 66318, 66324.
[139]   Todd, "ISPS Clauses in Charterparties" [2005] *J.B.L.* 2005 372, 374.
[140]   See Appendix.
[141]   19 CFR 4.7.

the Community and its member states are protected, particularly as any problems arising once goods have entered the Community can be rectified through post-clearance audits. But they do not provide adequate countermeasures or prevent terrorist action and they are no longer able to guarantee a high level of protection for citizens against the risks from dangerous or defective goods. More recently, the focus has been on the threat to public security from dangerous goods.

In 2003, the European Union made several proposals to amend the Community Customs Code in order to simplify administration and strengthen security at its external borders to meet the need for safety and security in relation to goods crossing international borders, including requirements linked to the U.S. Container Security Initiative. The proposals were adopted by the European Parliament and Council.[142] Among other measures, the new regulation requires traders to supply customs authorities with advance information on goods brought into, or out of, the customs territory of the European Community. This will provide for better risk analysis, but, at the same time, for quicker process and release upon arrival, resulting in a benefit for traders that should be equal to, if not exceed, any cost or disadvantage of providing information earlier than at present.

The measures will not apply, however, until the necessary implementing provisions have also come into force. These provisions will define

- the data elements to be included in the pre-arrival and pre-departure declarations;
- the time limit for the provision of this information;
- the rules for variations and exceptions to this limit; and
- the framework for the exchange of risk information between member states;

Discussions with member states on the draft implementing provisions commenced in the Custom Code Committee in July 2005. It is anticipated that a 24-hour deadline for prior declaration will apply to goods brought into the customs territory of the EU by sea[143] and, in contrast to the U.S. Regulations, will be applied both bulk and containerized cargo. When the Code is promulgated and comes into force, probably a clause requiring compliance with the EU 24-hour rule will be produced for carriage contracts.

C. Duties of the carrier

The safe carriage of dangerous goods cannot be attained if duties were imposed only on the cargo interests. Some of the requirements of the IMDG Code and other codes must be observed by the carrier and his employees. In fact, the duties of shipper and carrier are interrelated. There are some duties which are exclusively imposed on the carrier as well as duties which serve as a back-up mechanism.

---

[142] Regulation 648/2005, OJ L 117, 4.5.2005, 13.
[143] But in most other cases, prior notification will probably need to be given just 2 hours, if electronically, or 4 hours, if on paper, before the goods are brought into or out of the customs territory of the EU/1250/2005-Rev.6, 13 July 2006.

I. Seaworthiness of the ship

In common law, the provision of a seaworthy vessel by the carrier at the beginning of the voyage was implied in every contract to ship goods by water. This warranty was absolute and was not dependent on a finding of knowledge or negligent conduct.[144] That means the shipowner is liable in common law for failure to make the ship seaworthy in fact, although he may have taken all reasonable pains and precautions to do so.

However, the Hague-Visby Rules modified common law by reducing the obligation to use due diligence. Accordingly, the Hague/Hague-Visby Rules Art. III.1 set out:[145]

> The carrier shall be bound before and at the beginning of the voyage to exercise due diligence to
> (a) make the ship seaworthy
> (b) properly man, equip, and supply the ship
> (c) make the holds, refrigerating and cooling chambers and all other parts of the ship in which goods are carried fit and safe for their reception, carriage and preservation.

This article is paired with Art. IV.1, which provides that

> neither the carrier nor the ship shall be liable for loss or damage arising or resulting from unseaworthiness unless caused by want of due diligence on the part of carrier to make the ship seaworthy, and to secure that the ship is properly manned, equipped and supplied, and to make the holds refrigerating and cool chambers, and all other parts of the ship in which goods are carried fit and safe for their reception, carriage and preservation in accordance with the provisions of 1 of Art. III. ...

Seaworthiness means the ship is fit in design, structure, condition, manning and equipment to encounter the ordinary perils of the voyage.[146] She must also have a competent master, and a competent and sufficient crew.[147] The carrier must per-

---

[144] Colinvaux, *Carver Carriage by Sea* (1982) Vol. 1, 109.

[145] U.S. COGSA sec. 3(1); British COGSA Art. III; HGB § 559.

[146] Colinvaux, *Carver Carriage by Sea* (1982) Vol. 1, 114.

[147] *Ibid.* Chap.1.3 of the IMDG Code comprises provisions in relation to training which is recommendable. A state may decide its mandatory application. When a carrier is contracted to carry dangerous cargo, the crew must be qualified to handle dangerous cargoes. It may be that an incompetent crew that does not know how to handle dangerous goods may result in the vessel being found unseaworthy. For instance, U.S. Hazmat Regulations require training of all employees who perform work functions covered by the Hazardous Material Regulations (49 CFR Parts 171-180). Any employee who works in shipping, receiving or material-handling areas or who may be involved in preparing or transporting hazardous materials is required to have training. A carrier may not transport a hazardous material by vessel unless each of its hazmat employees involved in that transportation is trained as required. The record of training for a crew member who is a hazmat employee subject to training requirements must be kept on board the vessel while the crew member is in service on board the vessel (49 CFR 176). Tugman, "US and International Hazardous Material Regulations" [1995] (8) *P&I Int'l* 154.

form *due diligence* to make the ship seaworthy, i.e. a genuine, competent and reasonable effort to fulfil his obligations. Due diligence is equivalent to reasonable diligence, having regard to the circumstances known, or fairly to be expected, and to the nature of voyage and the cargo to be carried.[148] It will suffice to satisfy the condition if such diligence has been exercised at the loading port. The fitness of the ship at that time must be considered with reference to the cargo and to the intended course of the voyage. Obviously a test of seaworthiness can be applied only on a case-by-case basis, considering the particular and relevant facts of the situation at hand.

The duty to provide a seaworthy ship at the beginning of the voyage is often delegated to the master, the crew or some other agent. However, this duty is non-delegable and the carrier is accordingly responsible for the acts of any agent or shipyards he uses to fulfil this duty.[149]

When the carrier agrees to carry dangerous cargo, depending on the type of the dangerous cargo, the vessel must be constructed and equipped in accordance with all applicable regulations for the particular cargo. The vessel must be cleaned out after the former voyage and be ready for the said cargo.[150] Dangerous cargo must

---

[148] Colinvaux, *Carver Carriage by Sea* (1982) Vol. 1, 351.

[149] Schoenbaum, *Admiralty and Maritime Law* (2004) Vol. 1, 684.

[150] In the *Northern Shipping Co. v. Deutsche Seerederei (The "Kapitan Sakharov")* [2000] 2 Lloyd's Rep. 255, a container on deck containing dangerous cargo exploded, causing fire on deck which spread below resulting in the sinking of the vessel and the death of two seamen. It was found that the cause of loss of the vessel and its cargo was the isopentane in eight tank containers stowed in the aft part of the hatch catching fire as a result of the explosion and fire on deck. The international standards as to seaworthiness were those embodied in the SOLAS, and, in this case, the U.S.S.R's (now the Russian Federation's) version of the IMDG Code, given the initials "MOPOG" At the time of the accident, SOLAS Chap.VII, Reg.6.3 provided that dangerous goods in packaged form which give off dangerous vapors shall be stowed in a mechanically ventilated space or on deck, and in the IMDG Code, Reg. 5. that Class 3.1 liquids should be carried in a well ventilated space. MOPOG made similar provisions. It was found that the poor natural ventilation in the vessel's hold would not have effectively removed flammable vapour from the cargo spaces below deck and that the availability of fire-fighting equipment in the holds was no adequate substitute. The presence of isopentane below deck, and its combustion by the initial fire and exacerbation of it, was responsible for the heating and explosion of one or both of the tanks, causing, in turn, the rupture of the bulkheads between holds 2 and 3, so allowing both holds to flood with fire-fighting water and the ship to sink. If the isopentane had been stowed on deck, the vessel would not have been lost, and neither SOLAS nor the IMDG Code nor MOPOG nor the vessel's technical certificate permitted stowage of isopentane under deck in the inadequately ventilated hold. It was concluded that the stowage of the isopentane under deck made the vessel unseaworthy and in that respect the carrier failed to exercise due diligence. See also *the Fiona* [1994] 2 Lloyd's Rep. 507, where it was held that fuel oil was an inflammable, explosive and dangerous cargo to the shipment of which the master and owners had not consented, as they knew of its nature and character; however, although the shipment of the fuel oil was a cause of the explosion, the dominant cause was the failure to clean the tanks before loading.

be segregated and stowed in accordance with the regulations. If necessary, crew must be trained for dangerous cargoes.

II. Care of goods

Art. III (2) of the Hague/Hague-Visby Rules provides that "... the carrier shall properly and carefully load, handle stow, carry, keep, care for, and discharge the goods carried."

The carrier is to see that the cargo is loaded safely and loaded without delay. The carrier; through his master; is obliged to be careful and skilful in stowing and protecting each part of the cargo.[151] All reasonable precautions ought to be taken, by pumping, ventilation and so forth, to prevent the goods being damaged either through the action of causes external to them or by their own infirmities.[152] Where damage has happened, all measures which are reasonably available, under the circumstances, ought to be taken to arrest or check the further destruction or deterioration of the goods.

Like the duty of seaworthiness, the duty of caring for the cargo is non-delegable and the carrier is accordingly responsible for the acts of the master, the crew, the stevedore and his other agents. However, unlike the duty of seaworthiness, which applies only before and at the commencement of the voyage, the duty of caring for the cargo applies during the voyage as well. Therefore, the carrier is responsible for the acts of the master and crew while the vessel is underway.[153]

Again, if the carrier agrees to carry dangerous cargo, he has a duty to exercise a much higher degree of care in loading, stowing and caring for those materials.[154]

III. Stowage and segregation by the carrier

The carrier's duty to properly and carefully load, handle, stow, carry and keep the cargo is a stringent obligation though not an absolute one. Particularly, the word "properly" enunciates a very high standard of care.[155] The final responsibility for proper stowage remains in all circumstances with the carrier, even if there is a Free In/Free out (F.I.O) expression in the contract.[156] Questions of stowage are under the absolute control of the master of the vessel and as such he has the final say as to how stowage is to be effected.[157] This is not only because of the carrier's

---

[151] Colinvaux, *Carver Carriage by Sea* (1982) Vol. 1, 132.

[152] *Ibid.*

[153] Schoenbaum, *Admiralty and Maritime Law* (2004) Vol. 1, 687 f.

[154] Edgcomb, "The Trojan Horse Sets Sail: Carrier Defences against Hazmat Cargoes" [2000-01] 13 *U.S.F.Mar.L.J* 31, 48; *Contship Containerlines, Ltd. v. PPG Industries, Inc.* 442 F.3d 74, 77. 7

[155] Tetley, *Marine Cargo Claims* (1988), 527 ff.

[156] This expression refers to the cost of loading and discharging, a cost which is borne by the shipper.

[157] Tetley, *Marine Cargo Claims* (1988), 545.

responsibility for the stability of the ship and the safety of the ship and crew, but also because of the carrier's obligation to care for other cargo.

The carrier is expected to be an expert with respect to ordinary cargo and should stow it properly without having to receive special instructions. However, he cannot be expected to be an expert with regard to cargo which requires special care. The shipper should give the carrier instructions for cargo requiring special care. With the information provided by the shipper, the carrier must stow the cargo in accordance with all applicable regulations. When performed by the carrier, stowage into containers must be done properly and carefully, with proper bracing, blocking and dunnage inside. The stowage of cargo inside containers in such circumstances is part of the loading of the ship.[158]

Generally, improper stowage is an act or omission by the carrier, his agents or servants which implies negligence or want of care and skill in stowing cargo. For instance, stowage of cargo subject to spontaneous combustion at high temperatures without providing adequate ventilation[159] is improper, as is stowage of grain without shifting boards.[160]

The IMDG Code's detailed provisions with regard to stowage and segregation must be strictly followed.[161] The dangerous goods plan should ensure that only valid positions are utilized for dangerous goods when planning and loading.[162] In order to protect dangerous goods from adverse weather and also from the impact of a collision, they must not be stowed in outboard stacks.[163] Likewise, the dangerous goods plan must be duly filled in with information pertaining to the dangerous goods on board. This allows a quick and easy location of dangerous goods.

IV. Obtaining information on the goods

It is unrealistic to expect operators of container vessels to know or to acquire knowledge of the characteristics of the hazardous material cargo they are asked to carry. It is the shipper that knows the most about the cargo being shipped and, therefore, he should have the initial and higher obligation to provide the information relevant for any hazardous materials.[164]

However, the carrier's right to receive and obtain information about the cargo is implicit in Art. IV.6 of the Hague Rules.[165] The carrier has the duty to use all

---

[158] Tetley, *Marine Cargo Claims* (1988), 541, 546.
[159] *Old Colony Ins. V. S.S. Southern Star,* 280 F.Supp.189, 1967 A.M.C. 1641.
[160] *The Standale* (1938) 61 Ll. L.Rep. 223.
[161] IMDG Code Chap.7.
[162] Peermohamed, "Dangerous Cargo" [2002] *P&I Int'l* 17, 19.
[163] *Ibid.* Furthermore, during stowage and segregation procedures it may be necessary to keep dangerous goods under constant surveillance in order to detect any leaks at an early stage and to cool the goods with water spray in case a fire breaks out on board.
[164] Edgcomb, "The Trojan Horse Sets Sail: Carrier Defences against Hazmat Cargoes" [2000-01] 13 *U.S.F. Mar. L.J.* 31, 46.
[165] Tetley, *Marine Cargo Claims* (1988), 554; Bulow, "'Dangerous' Cargoes: the Responsibilities and Liabilities of the Various Parties" [1989] *LMCLQ* 342, 353.

reasonable means to ascertain the nature and characteristics of the goods tendered for shipment and to exercise due care in their handling.[166] That is to say if he is informed of the particular cargo, then he is expected to ascertain the characteristics of it for him to properly handle it. For instance if the cargo is ore he should ascertain moisture content or if the cargo is coal specific type or the cargo or temperature if he is not provided any information on these issues by the shipper. The safe carriage of goods depends on the information available regarding the nature of the cargo and handling conditions.

V. Documentation, marking and labelling packaging

If there is due cause to suspect that dangerous goods are not packed, marked, labelled and documented in accordance with the regulations, the carrier should not take the goods on board.[167] This duty of carrier is a back-up mechanism to ensure that the cargo interest has fulfilled his duties.

VI. Dangerous goods manifest

The carrier's initial responsibility is to make sure the shipper has provided the correct paperwork and to create his own records upon receival of certification from the shipper concerning dangerous goods. The carrier, through the master, must prepare a dangerous goods list or manifest containing, among other information, the name, packaging and stowage location of each hazardous material carried aboard the vessel. A detailed stowage plan, which identifies by class and sets out the location of all dangerous goods and marine pollutants, may be used in place of such a special list or manifest. This dangerous goods or marine pollutants list or manifest shall be based on the documentation and certification required in the IMDG Code.[168] When the manifests have been received on board, they must be carefully checked by comparing the information submitted with the IMDG Code and the vessel's dangerous goods cargo check-list.[169]

---

[166] *Ensley City* 71 F.Supp. 44; *Remington Rand, Inc. v. American Export Lines, Inc.* 132 F.Supp. 129, 136.

[167] SOLAS Reg. 4.

[168] The master must ensure that only such dangerous goods are loaded which are included in the dangerous goods manifest. A copy of this document must be available before departure to the person or organization designated by the port state authority

[169] Peermohamed, "Dangerous Cargo" [2002] (7) *P&I Int'l* 17, 19.

VII. The International Safety Management (ISM) Code and dangerous goods

1. The ISM Code

The ISM Code is a major development in safety in the marine context. It heralds a period of significant change in the marine industry and its repercussions are likely to be felt right up to the very top of the marine food chain. The ISM Code sets international standards for the safe management operation of ships and particularly has a significant impact on standards of seaworthiness.[170]

It is the first mandatory international instrument to tackle operational management all the way from the ship to the boardroom. If properly implemented, there is no doubt that it should have a significant impact on the general quality of shipping operations, making it much harder for sub-standard shipping to prosper. The safe transport of dangerous goods by sea is governed by comprehensive and relatively complex regulations which call for a high degree of mutual reliance on the key players in the transport chain. They all have important responsibilities. The ISM Code serves to place further emphasis on these responsibilities and to increase transparency within this transport chain whilst promoting, within shipping, the ideal of self- regulation.[171]

It would not be an overstatement to claim that what the ISM Code is all about is the development and implementation of a safety management system (SMS).[172] The Code has a significant impact on the concept of carrier-liability assessment.[173] What in the past may have been categorized as crew negligence may now be the owner's fault.

2. Implications of the ISM Code for the carriage of dangerous goods

The Safety Management System (SMS) provides the framework for compliance with the Code. It is a written system of safety and environmental protection policies and procedures to be followed by vessel- and shore-based personnel, with

---

[170] For more information on the ISM Code, see Philip Anderson, *ISM Code, A practical guide to the legal and insurance implications* (1998); Terry Ogg, "IMO's International Safety Management Code (The ISM Code)" [1996] *IJOSL* 143, 145 f.; Edelman, "The Maritime Industry and the ISM Code" [1999] *8-WTR Currents:Int'l Trade L.J* 43, 44; Rodriguez/Hubbard, "The International Safety Management (ISM) Code: A New Level of Uniformity" [1999] *73 Tul.L.Rew.* 1585, 1595 ff.

[171] Williams, "The Implications of the ISM Code for the Transport of Packaged Dangerous Goods by Sea", in *International Symposium on the Transport of Dangerous Goods by Sea and Inland Waterways* (1998), 117.

[172] Anderson, *ISM Code, A Practical Guide to the Legal and Insurance Implications* (1998) 21; Rodriguez/Hubbard, "The International Safety Management (ISM) Code: A New Level of Uniformity" [1999] *73 Tul.L.Rew.* 1585, 1595 ff.; Edelman, "The Maritime Industry and the ISM Code" [1999] *8-WTR Currents: Int'l Trade L.J.* 43.

[173] Rodriguez/Hubbard, "The International Safety Management (ISM) Code: A New Level of Uniformity" [1999] *73 Tul.L.Rew.* 1585, 1599 ff.

specific record keeping, reporting and internal audit requirements which are meant to enable the company to uncover and correct safety deficiencies before they result in an accident. It must ensure compliance with applicable mandatory rules and regulations and must take into account applicable guidelines and recommended standards.[174] This clearly includes the IMDG Code and other codes.

Therefore, under the ISM Code, if the cargo is to be carried on the specific type of vessel, such as gas or chemical cargo, the company has the duty to ensure that the ship is properly constructed and equipped to carry dangerous goods safely. Employees involved with handling stowage and carriage of dangerous goods in the ship must be properly informed, instructed, trained and supervised.[175] Surveys point particularly to improper stowage and securing of packages within containers, inappropriate packaging, incorrect marking and labeling of goods, stowage of incompatible goods together and non-declaration of dangerous goods. However, it is unclear what the shipping company can do to independently assess whether potential hazards have been identified, such as correct classification of the goods with the IMDG Code, and if adequate precautions have been taken. It may well be that ship operators should exercise some form of vetting process, for instance on packing companies, in order to safeguard their duties and responsibilities.[176] In any case, master and operator are obliged not to accept dangerous goods which do not comply with the IMDG Code.[177]

In order to manage the risks inherent in carrying dangerous cargoes, the carrier needs to ensure that his team, including ship officers, are advised of the risks involved in advance and the carrier's team should follow the necessary procedure.

D. Is the carrier obliged to carry dangerous goods?

In general, a carrier is not obliged to accept the cargo if he cannot give it proper care during the voyage. Instead, he should refuse the cargo or advise the shipper

---

[174] ISM Code Art. 1.4 provides that "Every company should develop, implement and maintain a safety management system (SMS) which includes the following functional requirements:
1. a safety and environmental-protection policy,
2. instructions and procedures to ensure the safe operation of ships and protection of the environment in compliance with relevant international and flag-state legislation
3. defined levels of authority and lines of communication between, and amongst, shore and shipboard personnel
4. procedures for reporting accidents and non-conformities with the provisions of this Code
5. procedures to prepare for and respond to emergency situations and
6. procedures for international audits and management reviews.
[175] Williams, "The Implications of the ISM Code for the Transport of Packaged Dangerous Goods by Sea", in *International Symposium on the Transport of Dangerous Goods by Sea and Inland Waterways* (1998) 117, 118.
[176] *Ibid.*, at 121.
[177] *Ibid.*

that he cannot provide proper care.[178] However, the issue may be complicated under certain circumstances.

With regard to dangerous goods, the purpose of the notification of the dangerous characteristics of the cargo is to enable the carrier to take the necessary precautions to ensure safe carriage of the cargo or to reject it, if he is not contractually obliged to carry it. The question then arises as to the circumstances in which the carrier is entitled to refuse the cargo.

I. Common law

In common law, the common carrier is the insurer of the goods and is bound to carry goods he publicly professes to carry unless there is reasonable excuse.[179] He may, however, refuse the goods where there is no possibility to carry the goods in safety or where the goods are insufficiently protected.[180] Based on these notions, it has been said that there are no common carriers of dangerous goods.[181]

1. Goods expressly prohibited

Under a charter contract, the right granted to charterers, particularly by most standard form time charters, is to give orders regarding the "employment" of the vessel. Employment means employment of the ship to carry out the purposes for which the charterers wish to use her. The master remains responsible for the safety of the vessel, her crew and cargo. If an order is issued which exposes the vessel to a risk which the owners have not agreed to bear, the master is entitled to refuse to obey it.[182] Accordingly, in the absence of a particular undertaking, a private carrier is generally not compelled to receive goods for shipment, dangerous or not. Therefore, a carrier may refuse to accept goods which fall within the scope of an express prohibition of dangerous goods.[183]

---

[178] Tetley, *Marine Cargo Claims* (1988), 555 f; *The Ensley City* 71 F.Supp.444.

[179] Abdul Hamid, *Loss or Damage from the Shipment of Goods, Rights and Liabilities of the Parties to the Maritime Adventure* (Diss. Southampton 1996), 117.

[180] *Jackson v. Rodgers* (1683) 2 Show. 327; *Batson v. Donovan* (1820) 4 B.&Ald.21, 33; *Edwards v. Sherrat* (1801) 1 East 604.

[181] Abdul Hamid, *Loss or Damage from the Shipment of Goods, Rights and Liabilities of the Parties to the Maritime Adventure* (Diss. Southampton 1996), 117; Kahn-Freund, *The Law of Inland Transport* (1956), 314.

[182] Wilford/Coghlin/Kimball, *Time Charters* (2003), 315 ff.; Hamblen & Jones, "Charterers Orders –"to obey or not to obey" [2001] 26 *Tul. Mar. L. J.*, 105, 113-114.

[183] Cooke/Young/Taylor, *Voyage Charters* (2001), 155. The loading of excluded cargo is a breach of the contract. If the charterer or shipper does so, the shipowner or carrier may elect to rescind the contract, or he may affirm the contract, reserving his right to damages. If he affirms it, the old contract is not displaced, but its terms will become applicable to the cargo in fact loaded. Alternatively, the carrier may treat the shipment as an offer to ship goods for carriage at the current rate of freight for those goods, but

2. Goods not conforming to contractual description

Even where the cargo is not prohibited by an express prohibition, the shipowner may correctly refuse its shipment if the cargo actually provided by the charterer does not correspond with its contractual description, as it is not the cargo which the shipowner contracted to carry.[184] A carrier has a duty to transport cargo which he has contracted to carry pursuant to the terms and conditions of the contract of carriage. However, the duty to transport a cargo in accordance with the contract of carriage is not an unqualified one. A carrier is not obliged to load a cargo which, through no fault of the carrier, cannot be carried without danger to the vessel, her crew and her cargo. The determination of whether a cargo can be carried safely is committed to the sound discretion of the master, upon whom the responsibility for safe carriage rests.[185]

It has also been said that the carrier could also refuse to accept goods if their shipment is prohibited by law or if any statutory requirements as to packing or otherwise have not been complied with.[186]

The purpose of the notification of the dangerous characteristics of the cargo is to enable the carrier to take the necessary precautions to ensure safe carriage of the cargo or to reject it, if he is not contractually obliged to carry it. The issue is a bit more complicated where the cargo has been described specifically in the contract but presents unusual risks which are different in kind or different in degree, approximating to a difference in kind, from those usually associated with the cargo specified. The question involves two distinct inquiries: whether the cargo conforms to its description and, even if it does conform, whether the cargo presents dangers different from those which the carrier contracted to bear.[187]

Then the question arises of whether the carrier is entitled to refuse the goods on the grounds that they fall outside the contract description or, appropriate notice having been received, he is obliged to carry them. The former view was taken in *The Amphion.*[188] Here the view indicates that the shipowner is entitled to refuse a cargo which has unusual risks as described above. Particularly, the issue was considered as one of conformity, i.e. whether the particular cargo conformed to the contractual description. If it does not, then there is a breach of contract on the part of the charterer. In this view, this arises in the case where there are unusual risks which are different in kind or degree proximate to that in kind. Thus, the shipowner will be entitled to rescind the contract and discharge himself of further performance. According to another view, if the danger can be avoided by expense

---

otherwise on the terms of the charterparty, and he may accept that offer. In this case, a new contract will be implied. Colinvaux, *Carver Carriage by Sea* (1982) Vol. 2, 821.

[184] Wilford/Coghlin/Kimball, *Time Charters* (2003), 177.

[185] *A&D. Properties v. M.V. Volta River* 1983 WL 637, 1984 A.M.C. 464; *Boyd v. Moses* 74 U.S. 316 (1869); *The Ensley City* 71 F.Supp. 444; *Birth v. Hardie*, 132 F.61 (S.D.N.Y 194).

[186] Abbot, *Treatise of the Law Relative to Ships and Seamen* (1901), 643.

[187] Abdul Hamid, *Loss or Damage from the Shipment of Goods, Rights and Liabilities of the Parties to the Maritime Adventure* (Diss. Southampton 1996), 119.

[188] *General Feeds v. Burnham Shipping*, [1991] 2 Lloyd's Rep. 101.

and care on the carrier's part, the carrier must take them, but if it cannot be so avoided, he is not bound to carry them.[189]

It is obvious that no simple answer can be given. The issue would depend on the manner in which the cargo is described in the contract.[190] Therefore, the proper approach should be to consider whether the cargo tendered for shipment is a reasonable cargo having regard to the terms of the carriage contract and all the circumstances.[191]

II. German law

If the cargo is not specifically but only generally identified in the contract, the carrier is, in principle, obliged to carry all type of cargo.[192] However, the carrier can refuse loading of cargo which endangers the ship or cargo according to the principle set out in HGB § 564.V.[193]

If the cargo is specifically identified in the contract, the shipper should deliver the agreed cargo. In this case, in principle, he has no right to substitute, i.e. to load

---

[189] Boyd/Burrows/Foxton, *Scrutton on Charterparties* (1996), 106. This view is also in accord with *The Atlantic Duchess* [1957] 2 Lloyd's Rep. 55, 95 and *The Fiona* [1993] 1 Lloyd's Rep. 257.

[190] If the cargo is described only in general terms, it has been suggested that the carrier is entitled to refuse it if the extra precautions required to ensure safe carriage will cause unreasonable delay or expense. Cooke/Young/Taylor, *Voyage Charters* (2001), 155.

[191] In *Internav Ltd. v. Scanbulk Ltd. (Wismar)* 1980 S.M.A. No:1454, under a time charterparty, the actual owner of the vessel refused to load a cargo of Direct Reduced Iron (D.R.I.) because it could be subject to the emission of highly flammable gas if it had catalyzed with water. D.R.I. are iron-ore pellets which are produced in a process which removes 25%of the oxygen naturally linked to raw iron ore. The pellets thus produced have a purity content of 92%-93% iron, making them far more economically efficient to transport. The cargo-exclusion clause in the NYPE did not specifically exclude D.R.I., but stated that "no injuries, dangerous or inflammable cargo" should be loaded "unless packed, labeled, loaded, stowed and discharged according to IMCO Regulations/U.S. Coast Guard Regulations". It was ruled that the cargo was not inherently dangerous as it had not been listed as "dangerous" in the IMDG Code, although it was classified as "Materials Hazardous in bulk by the IMO". Furthermore, the exclusion clause specifically allowed the loading of "dangerous cargo" provided IMCO and USCG regulations were followed. The arbitrators concluded that the owner was wrong in refusing to load the cargo and that the cargo could have been safely transported had all the instructions prepared by the charterer's experts been followed. On the other hand, in *Tramp Shipping Co. Inc. v. Steamship Co. Ltd. (The Agia Erini II)* (1974) S.M.A No. 875, the owner refused to load a cargo of sulphur under an exclusion clause in the NYPE charterparty relating to "injuries, dangerous or inflammable cargoes". The panel ruled that the owner was justified in refusing to allow the vessel to load this type of cargo, which is included in Title 46 of the Federal Regulations.

[192] HGB § 562(1).

[193] § 564.V provides that if the goods endanger the ship or the rest of the cargo, the master is authorized to land the goods or, in urgent cases, jettison them.

any other cargo.[194] However, on the other hand, refusal of loading of equal or similar cargo by the carrier for the same destination is not justified based on good faith, unless such cargo aggravates his situation and endangers the ship or its cargo.

Likewise, the carrier is justified to refuse cargo which does not conform to laws and regulations, for instance if it is insufficiently packed or not certified as required. This is not only the right of the carrier but also his legal duty.[195]

III. Refusal of dangerous goods not accompanied by the declaration

The SOLAS Convention imposes a duty not to accept the cargo transport unit where there is due causes to suspect that a cargo transport unit in which dangerous goods are packed is not accompanied by either a certificate or declaration or container/vehicle packing certificate.[196] Accepting cargo not complying with the requirements would constitute violation of dangerous goods regulations and results in the carrier's liability. Therefore, the carrier has not only the right but also the duty to refuse loading of such cargo.[197]

IV. Dangerous goods declaration including indemnity clause

Some carriers attempt to turn the declaration into a contractual commitment by the person completing the declaration. The form of declaration that these carriers demand includes an indemnity in favour of the carrier against damages resulting from a failure to comply with the IMDG Code requirements for proper packing, marking etc.[198]

Although the Hague/Hague-Visby Rules give carriers protection against the risks attached to the transport of dangerous goods, ocean carriers do not regard the protection under international law as adequate, as they often include the following term in their bills of lading.

> Whether or not the Merchant was aware of the nature of the Goods, the Merchant shall indemnify the Carrier against all claims, losses, damages, liabilities or expenses arising in consequence of the Carriage of such Goods.
> Nothing contained in this Clause shall deprive the carrier of any of his rights provided for elsewhere.[199]

However, such a clause makes the obligation to indemnify the carrier absolute and independent of the consent or knowledge of the carrier. It is doubtful whether the

---

[194] HGB § 562 (2).
[195] GGVSee § 9(5).
[196] SOLAS Chap. VII. Reg.4.4.
[197] GGVSee, § 9(5). U.S HMR §176(24). MSA (Dangerous Goods and Marine Pollutants) Regulations 1997 Art. 7.
[198] Ellerman East Africa Service Bill cl. 19; P&O Nedloyd Bill cl. 19.
[199] *Ibid.* See also Mitsui OSK Lines Combined Transport Bill 1993 cl. 19.3.

entire provision will be valid where the Hague/Hague-Visby or Hamburg Rules apply, as the indemnity provided for in these Rules does not cover the situation where the carrier was properly advised and has knowledge of the dangerous nature of the cargo.[200] According to Art. III.8 of the Hague/Hague-Visby Rules and Art. 23.1 of the Hamburg Rules such a contractual clause would be null and void.[201] So carriers incorporate a short form of this indemnity into the declaration, where the need for the signature imposes the obligation inescapably on the shipper.[202]

E. Duties under the CMI/UNCITRAL draft instrument

I. Duties of the shipper

Comparing the Hague-Visby Rules and Hamburg Rules, the draft instrument contains a more detailed, complete and perhaps precise regulation on the shipper's obligations.[203]

1. Delivery for carriage

In the draft instrument, a number of positive obligations affecting the shipper are regulated. Chapter 7 of the draft instrument is entitled *Obligations of Shipper* and sets out in considerable detail both the shipper's obligation and liability for breach of these obligations. In particular, the obligations of the shipper arising from the contract of carriage towards the carrier are described in a detailed manner:[204]

> Subject to the provisions of the contract of carriage the shipper shall deliver the goods ready for carriage in such condition that they will withstand the intended carriage, including their loading, handling, stowage, lashing and securing, and discharge, and that they will not cause injury or damage. In the event the goods are delivered in or on a container or trailer packed by the shipper, the shipper must stow, lash and secure the

---

[200]   Hague/Hague-Visby Rules Art. IV.6. HGB § 564b, COGSA § 1304(6). In *Apl Co. Pte. Ltd. v. UK Aerosols Ltd.* 2006 WL 2792875 (N.D. Cal.), 2006 A.M.C. 2418 a clause in the bill of lading providing that "regardless of the Merchant's knowledge about the 'nature of the Goods, the Merchant shall indemnify the Carrier against all claims, losses, damages, liabilities or expenses arising in consequence of the Carriage of such Goods'" said to violating COGSA. Where a carrier knows that cargo "poses a danger and requires gingerly handling or stowage" it cannot invoke strict liability under sec, 1304(6). *Ibid.* at 9. See *infra* Part 4.

[201]   HGB § 662 corresponds to this article. It is because the carrier may himself be at fault and such a clause may lower his liability.

[202]   Jones, "Carriers Include Indemnity Undertaking in Dangerous Goods Declaration that They Require Shippers Execute" <www.forwarderlaw.com/library/view.php?article_id =225> (visited 10.11.2006).

[203]   Asariotis, "Main Obligations and Liabilities of the Shipper", 2004 *TranspR* 284, 285; Zunarelli, "The Liability of the Shipper" [2002] *LMCLQ* 350.

[204]   Art. 25. It ssubsequently became Art. 28, A/CN.9/WG.III/WP.56.

goods in or on the container or trailer packed by the shipper, the shipper must stow, lash and secure the goods in or on the container or trailer in such a way that the goods will withstand the intended carriage, including loading, handling and discharge of the container or trailer, and that they will not cause injury or damage.

Generally, obligations of the shipper represent a break from previous practice in the field of maritime transport, since neither the Hague/Hague-Visby Rules nor the Hamburg Rules have such extensive provisions relating to the shipper's obligations. Some doubts were expressed as to whether the chapter was in fact needed at all, because it was viewed that the chapter placed a heavy responsibility on shippers and it was suggested that small shippers, particularly those from developing countries, could find it difficult to meet the requirements of the draft instrument.[205] However, there was in general support for the context in which the contract of carriage required the shipper and carrier to cooperate to prevent loss of or damage to the goods or to the vessel. Obligations in the contract of carriage had evolved over the years beyond mere acceptance to carry goods and make payment for such carriage. It is obvious that that there was a need to strike an overall balance between the obligations of the shipper and the carrier.[206]

2. Obligation to provide information, instructions and documents

The draft instrument provides that the shipper is obliged to provide the carrier with information, instructions and documents that are reasonably necessary for:[207]

(a) the handling and carriage of the goods, including precautions to be taken by the carrier or a performing party;

(b) compliance with rules, regulations, and other requirements of authorities in connection with the intended carriage, including filings, application, and licenses related to the goods;

(c) The compilation of the contract particulars and the issuance of the transport documents or electronic records, including the particulars referred in art. 34(1) (b) and (c), the name of the party to be identified as the shipper in the contract particulars, and the name of the consignee or order, unless the shipper may reasonably assume that such information is already known to the carrier.

These obligations of the shipper were thought to be especially important in the light of contemporary transport practice, in which the carrier seldom sees the goods he is transporting, even when they are non-containerized goods. In this context, the flow of reliable information between the shipper and the carrier was said to be of utmost importance for the successful completion of a contract of carriage, particularly with respect to dangerous goods.[208]

---

[205] A/CN.9/591 para.106.

[206] The caution expressed was that unnecessarily detailed shipper's obligations could result in creating hurdles for the ratification of the draft instrument.

[207] Art. 27 subsequently became Art. 30. A/CN.9/WG.III/WP.56.

[208] A/CN.9/591 para.129.

3. Special rules on dangerous goods

As noted in Part 2, the specific provision concerning the shipment of dangerous goods disappeared and the general regime of shipper's liability provided for all kinds of cargo; however, subsequent discussion revealed the necessity of maintaining special rules on dangerous goods. Consequently, a special provision on dangerous goods was drafted providing that:[209]

> 2. The shipper must mark or label dangerous goods in accordance with any rules, regulations or other requirements of authorities that apply during any stage of the intended carriage of the goods. If the shipper fails to do so, it is liable to the carrier and any performing party for all loss, damages, delay and expenses directly or indirectly arising out of or resulting from such failure.
> 3. The shipper must inform the carrier of the dangerous nature or character of the goods in timely manner before the consignor delivers them to the carrier or performing party. If the shipper fails to do so and the carrier or performing party does not otherwise have knowledge of their dangerous nature of character, the shipper is liable to the carrier and any performing party for all loss, damages, delay and expenses directly or indirectly arising out of or resulting from such shipment.

Obviously this provision is clearer and more precise than Art. IV.6 of the Hague/Hague-Visby Rules. The shipper's duty to mark is expressly stated, as well as his duty to inform of the dangerous nature of the cargo. However, it was said to be that the obligation of the shipper to mark or label dangerous goods in accordance with the applicable local rules depending on the stage of the carriage could place too heavy a burden on a shipper if he was not aware of the intended route of the voyage.[210]

II. Duties of the carrier

The draft instrument sets out the two fundamental obligations of the carrier, as contained in the Hague-Visby Rules arts. III.1 and III.2, and adds some supplementary provisions. Firstly, the aforementioned provisions are preceded by a general statement of the duty of the carrier to carry the goods to the destination and deliver them to the consignee. The addition of this provision, of which there is no equivalent in the either the Hague/Hague-Visby Rules or the Hamburg Rules, constitutes the general frame within which the rules on the duties and responsibilities of the carrier operate.

1. Seaworthiness of the ship

According to draft instrument:[211] "The carrier shall be bound before, at the beginning of and during the voyage by sea to exercise due diligence to:

---

[209] Art. 33. A/CN.9/WG.III/WP.56.

[210] Berlingieri, "Basis of liability and exclusions of liability" [2002] *LMCLQ* 336, 337 ff.

[211] Art. 13 subsequently became 16. A/CN.9/WG.III/WP.56.

(a) make [and keep] the ship seaworthy
(b) properly man, equip and supply the ship
(c) make [and keep] the holds and all other parts of the ship in which the goods are carried, including containers where supplied by the carrier, in or upon which goods are carried fit and safe for their reception, carriage and preservation.

It was agreed that the carrier's obligation of due diligence in respect of seaworthiness should be continued. Making this obligation a continuing one affects the balance of risk between the carrier and cargo interests in the Draft.

2. Care of goods

The duties of the carrier set out in respect of the care of the cargo set out in the draft instrument are same as and in the Hague-Visby Rules Art. III.2.[212] However, whereas the Hague-Visby Rules Art. III.2 provides that the duty of care of the goods is subject to the provisions of Art. IV, where the excepted perils are set out, a similar link did not exist in Art. 11 of the draft instrument with Art. 14.[213]

3. Obligation to provide information and instructions

In the Draft the carrier is obliged to provide the shipper, on his request, with such information as is within the carrier's knowledge and instructions that are reasonably necessary or of importance to the shipper in order to comply with his obligations.[214] This obligation is placed in the chapter on shipper's obligations.[215] This provision was thought to be necessary to balance the obligations of the shipper in Art. 25 (28).

The provision reflects the duty of cooperation between the parties with respect to the exchange of information necessary for the performance of the contract of carriage. The purpose of the article is not so much to establish the independent liability of the carrier for his failure to provide the shipper with necessary information as to deny the carrier the ability to rely on his failure in defending a cargo claim.[216]

---

[212] Art. 11 subsequently became 14. A/CN.9/WG.III/WP.56.
[213] Berlingieri, "Basis of Liability and Exclusions of Liability" [2002] *LMCLQ* 336, 339.
[214] Art. 26 subsequently became Art. 29 A/CN.9.WG.III/WP.56.
[215] Doubts were expressed regarding the placement of the draft article in a chapter dealing with the obligations of shipper. However, it was viewed that the said article was appropriately located as a logical complement to the shipper's duties.
[216] A/CN.9/591 para.122.

Part 4:   Rights and liabilities of the parties

A. In general

Generally speaking, in the field of carriage of goods, goods are the object of loss or damage. The system is mainly based on the concept of the loss of or damages or delay to goods. However, it is not always so. Goods can cause various damages in different situations while in transit. In particular, the carriage of dangerous goods by sea carries with it substantial risks of damage both to the vessel and to other cargoes. In this case, goods become the subject rather than the object. Dangerous goods can, for example, cause an explosion or fire, corrode the cargo holds or the hull, or liquids in the cargo holds can solidify. Furthermore, the carrier may incur extra expenses such as unloading and reloading of cargoes, extra bunker expenses due to the damage to the ship since he may have to deviate. The vessel may be detained or quarantined, so the carrier may suffer economic loss or the vessel may need to be cleaned. Moreover, cargo may damage other cargo onboard or may injure the crew. In accordance with this fact, Art. IV.6 of the Hague/ Hague-Visby Rules provides that "Goods of inflammable, explosive, or dangerous nature to the shipment whereof the carrier, master or agent of the carrier has not consented with knowledge of their nature and character ... the shipper of such goods shall be liable for all damage and expenses directly or indirectly arising from such shipment."

The carriage of dangerous goods raises first of all the question of the shipper's liability to the ship and the carrier. Moreover, it gives rise to the question of liability of the shipper to third parties, e.g. seamen, stevedores and owners of the other cargoes. This liability is, however, in tort; therefore, it does not fall principally under the carriage contract. However, if the carrier incurs a liability to third parties due to the shipment of dangerous goods, he can recover the amount of liability in a recourse action against the shipper. Therefore, claims for third parties might fall indirectly under the contract of carriage and/or the Hague-Hague/Visby Rules.

If the carrier is made aware of the dangerous characteristics of the cargo prior to loading, as mentioned in the previous part, precautions can be taken to reduce those risks. However, if the shipper fails to notify the carrier of the dangerous characteristics of the goods, the carrier will be denied the opportunity to take the precautions necessary for the safe carriage of the cargo. For this reason, the shipper will be held responsible for the consequences of his failure to give the necessary notification. However, this basic formulation provides only a partial delineation of the respective obligations of the shipper and the carrier in respect of the carriage of dangerous cargo. The formulation will need modification if the carrier has voluntarily accepted those risks that have resulted in damage or if his own breach of contract has contributed to the damage.[1] Thus, the rights and responsibilities of the shipper and carrier vary and greatly depend on whether the carrier knew or should have known of the dangerous nature of the cargo.

B. Shipper's liability for dangerous goods

The dispute between shippers and carriers on the point of liability for the shipment of dangerous goods has been the subject-matter of much controversy in common law. Particularly, where cargo has been shipped in circumstances where neither the shipper nor the carrier has the means of knowledge of the dangerous nature of the cargo, the question arises as to who should bear the risk.

I. English law

1. Early principles

In the first edition of his pioneering *Treatise of the Law Relative to Ships and Seamen,* Abbot speaks of the general duties of the merchant, stating the general principles: (i) that the hirer of anything is required to use it in a lawful manner and according to the purpose for which it was let, and (ii) that the merchant must load no prohibited or unaccustomed goods, by which the ship may be subjected to detention or forfeiture.[2] He cited no direct English authorities for either of these propositions, as there were clearly none, and referred to Justinian's Digest 19.2.61.1.[3] and the French Ordinance, liv.3, tit.3, Fret. Art. 9.[4]

In the same year as the publication of Abbot's Treatise, the case of *Williams v. East India Co.*[5] was argued before the King's Bench. In this case, the plaintiffs had chartered their vessel, the Princess Amelia, to the East India Company and, on one of the voyages, a package of flammable oil/varnish was received on board the vessel by the chief mate, since deceased, allegedly with no declaration of its dangerous qualities. Had notice of the dangerous character of the goods been given to the owners? It was held that in order to make the putting on board wrongful, the defendant shippers had to be "conusant of the dangerous quality of the article put onboard; if being so, they yet gave no notice, considering the probable danger thereby occasioned to the lives of those on board,... for which

---

[1]   Baughen/Campbell, "Apportionment of Risk and the Carriage of Dangerous Cargo" [2001] 1 *IntML* 3.
[2]   Abbot, *Treatise of the Law Relative to Ships and Seamen* (1901), 270.
[3]   *Ibid.*
[4]   *Ibid.*
[5]   (1802) 3 East 192.

[defendants] were criminally liable, and punishable as for misdemeanors at least."[6] Thus this case clearly limited the shipper's liability to those instances where the shipper had knowledge of the dangerous nature of the goods.[7]

2. Brass v. Maitland

The leading case of *Brass v. Maitland & Ewing,* which was argued before the Queen's Bench, cited both Abbot and Williams with approval but also went further on the standard of the shipper's liability.[8] In this case, a consignment of bleaching powder, which consisted largely of chloride of lime, was shipped on board the plaintiff's "general ship", the *Regina,* in casks. During the voyage, the chloride of lime corroded, bursting the casks and mixing with and damaging other cargo. The majority of the court held that *ex contractu* the shipper should be *strictly* liable to the carrier. They had an obligation not to ship goods which were so dangerously packed that: (1) the carrier could not by reasonable knowledge and diligence be aware of their dangerous character, and (2) the carrier had not been so notified by the shipper or charterer. The court agreed that it is the duty of a shipper to pack sufficiently for safety and to give warning of any danger, when he is himself aware of it; but they differed as to his obligations when he is himself ignorant. However, ignorance was not accepted as an excuse for putting on board without notice the dangerous goods which were insufficiently packed. The underlying basis for this reasoning was expressed as follows:[9]

> If the plaintiffs and those employed by them did not know and had no means of knowing the dangerous quality of the goods which caused the calamity, it seems most unjust and inexpedient to say that they have no remedy against those who might easily have prevented it. ... It seems much more just and expedient that, although they were ignorant[10] of the dangerous qualities of the goods, of or the insufficiency of the packing, the loss occasioned by the dangerous quality of goods and the insufficient packing should be cast upon the shippers than upon the shipowners.

---

[6] *Ibid.* at 200.

[7] Girvin, "Shipper's Liability for the Carriage of Dangerous Goods by Sea" [1996] *LMCLQ* 487, 492.

[8] (1856) 6 E. & B.470.

[9] *Ibid.* at 483. It is argued that in *Brass v. Maitland,* the facts of the case had nothing to do with the situation of neither party having no knowledge of the dangerous nature of the goods. The shipowner knew that the casks contained bleaching powder and had the means of knowing that it contained chloride of lime, which could thereby be dangerous to other cargo. Panesar, *"The Shipment of Dangerous Goods and Strict Liability"* [1998] *I.C.C.L.R* 136, 139.

[10] It is asserted that the word "ignorance" indicates a deliberate course of action, where the shipper simply fails to pay any attention to the dangerous nature of the cargo. Thus it is doubtful whether Lord Campbell was considering those cases where the shipper simply has no means of knowing that they are dangerous. *Ibid.*

Dissenting opinion[11] strongly disagreed with this approach, principally because it reasoned that there was no decisive authority to support an absolute obligation on the part of the shipper:

What then is the nature and extent of this duty or engagement on the part of shippers of goods? On the one hand it is clear a tortuous act for the consequences of which the shipper are responsible, to ships goods apparently safe and fit to be carried, and from which the shipowner is ignorant that any danger is likely to arise, without notice of such goods being dangerous. If the shipper is aware of such danger, such shipment, when the scienter is made out, is clearly wrongful and tortuous, and perhaps an action on a contract to give notice in such a case might be supported, though it would seem rather to be the subject of an action of tort. On the other hand, I cannot agree with the doctrine contended for on the part of the plaintiffs, that there is an absolute engagement on the part of the shipper that the goods are safe and fit to be carried on the voyage. Such a warranty would include the cases where the goods may be openly seen, and are known by the shipowner to be dangerous. It does not seem that there is any authority decisive on the point as to whether the shipper is liable for shipping dangerous goods without communication of their nature, when neither he nor the shipowners are aware of the danger. It seems very difficult that the shipowner [? shipper] can be liable for not communicating what he does not know. Supposing that hay or cotton should be shipped, apparently in a fit state, and not dangerous to the knowledge of the shippers or shipowners, but really being then in a dangerous state, from a tendency to heat, are the shippers to be liable for the consequences of fire from heating of such goods? ... I entertain great doubt whether either the duty or the warranty extends beyond the cases where the shipper has knowledge, or means of knowledge of the dangerous nature of the goods when shipped or where he has been guilty of some negligence as shipper, as by shipping without communicating danger which he had the means of knowing, and ought to have communicated. Probably an engagement or duty may be implied that the shipper will use and take due and proper care and diligence not to deliver goods apparently safe, but really dangerous, without giving notice thereof, and any want of care in the course of the shipment in not communicating what he ought to communicate might be negligence for which he would be liable ; but where no negligence is alleged, or where plea negatives any alleged negligence, I doubt extremely, whether any right of action can exist.

3. Reaction to Brass v. Maitland

The dissenting opinion in *Brass v. Maitland* found support in certain first-instance decisions. In *Hutchison v. Guion*,[12] where salt-cake caused damage to the vessel, the court, although finding for shipowners, who did not know that salt-cake was capable of corrosion and destruction, applied the reasoning of the dissenting opinion in *Brass v. Maitland*. The same principle was taken "in the narrowest and most limited way" in *Farrant v. Barnes*.[13] The same reasoning was also applied in

---

[11]   (1856) 6 E. &B. 470, 490 ff.
[12]   (1858) 5 C.B. (N.S) 149; 141 E.R. 912, 916.
[13]   (1862) 11 C.B. (N.S) 553, 563, 142 E.R. 912, 916.

Mitchell Cotts & Co. v. Steel Bros & Co. Ltd.,[14] although this was not case of goods which were physically dangerous.[15]

In cases decided this century, the judgment of *Brass v. Maitland* has been strongly endorsed both in the first instance and in the Court of Appeal. It received guarded support in *Acatos v. Burns*,[16] although it was not directly relevant because that case concerned the carriage of a cargo of maize. The majority judges in *Bamfield & Goole v. Sheffield Transport Co. Ltd*[17] both accepted the existence of an absolute obligation in a case in which ferro-silicon, carried in casks, had given off poisonous gases which killed the plaintiff's husband. This majority view was also endorsed in *Great Northern Railway Co. v. L.E.P. Transport & Depository.*[18]

4. Implied absolute warranty not to ship dangerous goods without notice

Two recent decisions have also endorsed the existence of an absolute obligation by the shipper. In *Athanasia Comninos & Georges Chr. Lemos*,[19] it was felt that the weight of authority supported the absolute obligation. Finally, this view was affirmed by the House of Lords in *Effort Shipping Co. Ltd. v. Linden Management S.A. (The Giannis NK)*.[20] Although the claim in *Giannis NK* was governed by the Hague Rules, the end of the controversy was determined by stating that "the liability of a shipper for shipping dangerous goods at common law, when it arises, does not depend on his knowledge or means of knowledge that the goods are dangerous."[21]

Consequently, in common law there is an implied term[22] in a contract of affreightment that the shipper of goods will not ship goods of such a dangerous character that the carrier or his agent could not by reasonable knowledge and diligence be aware of their dangerous character; and the shipper is therefore strictly liable for damage resulting from the shipment of such dangerous goods.[23]

---

[14] 1916 2 K.B.610,
[15] See *supra* Part 2.
[16] (1878) 3 Ex.D.282, 288.
[17] [1910] 2 K.B. 94, 105 f.
[18] [1922] 2 K.B. 742, 764.
[19] [1990] 1 Lloyd's Rep.277.
[20] [1998] 1 Lloyd's Rep. 337 (H.L.).
[21] *Ibid.* at 345.
[22] A term will be implied in the contract if it is necessary to do so in order to give business efficacy to the transaction, but not otherwise. *The Moorcock* (1889) 14 P.D. 64, 68. The terms of a contract in English law fall under three classes: (1) conditions (2) warranties (3) intermediate terms. A warranty is a term of the contract of minor importance such that no breach of it will give rise to any right to the innocent party to terminate the contract. Cooke/Young/Taylor, *Voyage Charters* (2001), 37 ff.
[23] See also *Williams v. East India Company* (1802) 3 East 192. *Mediterranean Freight Services Ltd. v. Bp Oil International Ltd. (The Fiona)* [1993] 1 Lloyd's Rep. 257; Colinvaux, *Carver Carriage by Sea* (1982) Vol. 2. 841; Cooke/Young/Taylor, *Voyage Charters* (2001) 149; Wilford/Coghlin/Kimball, *Time Charters* (2003) 180 f.; Boyd/Burrows/Foxton, *Scrutton on Charterparties,* (1996) 105; Gaskell/Asariotis/Baatz, *Bills*

In other words, unless the carrier knows or ought to know the dangerous character of the goods, there will be an implied warranty by the shipper that the goods are fit for carriage in the ordinary way and are not dangerous.

5. Liability under the Hague/Hague-Visby Rules

a) General rule: negligence

As noted previously, the Hague/Hague-Visby Rules contain very few rules on the shipper's obligations and liabilities. Art. IV.3 of the Hague/Hague-Visby Rules sets out a general rule to the effect that the shipper shall only be liable to the carrier in cases of fault on the part of the shipper or his servants or agents. This general principle is supplemented by two more specific rules contained in Art. III.5 and IV.6.

b) Strict liability of shipper for carrying dangerous goods without knowledge of carrier

According to Art. IV.6 the shipper of dangerous goods which were shipped without the carrier's knowing consent shall be liable for all damage or expenses directly or indirectly arising out of or resulting from such shipment. On the other hand, Art. IV.3 provides that the shipper shall not be responsible for loss or damage sustained by the carrier or the ship arising or resulting from any cause without the act, fault or neglect of the shipper, his agent, or his servants.

The principal issue the House of Lords faced in *Effort Shipping Co. Ltd. v. Linden Management S.A.* (The "*Giannis NK*")[24] was how this was to be reconciled with the specific provision in Art. IV.6, in particular whether the liability of a shipper of goods who does not know they are dangerous is fault-based or strict. The House of Lords favoured strict liability for a number of reasons.

In the absence of clear wording indicating whether or not Art. IV.6 was subject to Art. IV.3, the House of Lords made its decision by taking into account a number of factors. First, Art. IV.6 was read as a whole. Cumulatively, its first and third sentences provide that, whether or not goods are shipped with the carrier's knowledge of and consent to their dangerous character, the carrier has a right to land or destroy them or render them innocuous; and this right could not depend on the shipper's knowledge of their character.[25] It was therefore concluded that the rider in the first sentence that "the shipper of such goods shall be liable" was not confined to cases where he is at fault.[26]

Secondly, the breadth of Art. IV.3 was such as to make it unlikely that it was intended to qualify the specific provisions of Art. IV.6: *generalia specialibus non*

---

of Lading: Law and Contracts, (2000) 470 ff.; Treitel/Reynolds, *Carver on Bills of Lading* (2005), 512.

[24] [1998] 1 Lloyd's Rep. 337.
[25] *Ibid.* at 342, 346.
[26] *Ibid.* at 342.

derogant.[27] The opinion was that it was a common-sense consideration, equally applicable to international conventions, that "in the construction of documents it may be proceeded on an initial premise that a general provision does not necessarily qualify a specific provision in the same document."[28]

It was pointed out that the method used elsewhere in the Hague Rules to make one rule subject to another[29] was not employed to subject Art. IV.6 to Art. IV.3.[30] It was viewed that the absence of a qualifying word in either rule had a neutral effect.[31] Attention was drawn to Art. III.5, which provides that the shipper shall be deemed to have guaranteed to the carrier the accuracy at the time of shipment of the marks, numbers, quantity and weight, as furnished by him. It was pointed out that Art. IV.3 is not expressly made subject to Art. III.5 and it was held that this factor supports the argument that Art. IV.3 does not qualify Art. IV.6[32] either.

The House spent little time on the assistance to be derived from the now readily available *travaux préparatoires* on the ground that they provided no assistance for the question at issue.[33] Accordingly, *travaux préparatoires* could only assist where truly feasible alternative interpretations existed and the court was satisfied that the *travaux préparatoires* "clearly and indisputably point to a definite legal intention."[34]

Of obviously greater importance was the effect of authority, particularly given the accepted desirability of an internationally uniform approach to the interpretation of international conventions. In this respect, common law was considered to be important, on the ground that it must have been known when the Hague Rules were drafted that the dominant theory in a very large part of the world was that shippers were under an absolute liability not to ship dangerous goods and that there was an alternative theory that shippers ought to be liable only for want of due diligence, yet neither the express words of the Rules nor the *travaux préparatoires* revealed an intention to change the position.[35]

The majority of the House therefore concluded that Art. IV.6 is a free-standing provision dealing with a specific subject-matter; it was neither expressly nor by implication subject to Art. IV.3 and it imposed strict liability on shippers in relation to the shipment of dangerous goods irrespective of fault or neglect on their part.[36] However, the adjective "free-standing" was condemned on the basis that the Hague Rules had to be read as a whole, and that, Art. IV.3 and Art. IV.6 were

---

[27] *Ibid.*

[28] *Ibid.* at 347.

[29] Art. II is expressly made subject to Art. VI and Art. VI applies "notwithstanding the provisions of the preceding articles".

[30] *Giannis NK* [1998] 1 Lloyd's Rep. 337, 342.

[31] *Ibid.* at 346.

[32] *Ibid.* This argument, however, was rejected as it was rather an insubstantial point in the interpretation of a multilateral trade convention.

[33] *Ibid.* at 347 f.

[34] *Ibid.*

[35] *Ibid.* at 348 f.

[36] *Ibid.* at 342 f, 349.

reconcilable, Art. IV.6 not being free-standing but *taking priority over Art. IV.3.*[37] Whichever description appears preferable, however, it seems to have no practical significance.[38]

Consequently, there is no requirement that the shipper or his agent should have the relevant knowledge, either actual or constructive, of the nature and character of the goods if the provisions of the rule are to take effect. In other words, the liability of the shipper does not depend on whether he himself knew or should have known of the dangerous nature of the relevant cargo; even if he did not know and had no means of knowing the dangerous nature of the goods, he is liable for the consequences of shipping those goods.

II. American law

1. Early principles: strict liability

The first American case to address this issue of the shipper's liability to the carrier for offending cargo was *Pierce v. Winsor*.[39] In this case, the shipowner sued the charter/shipper for a cargo of mastic, an article new to commerce. Neither party knew that on a long voyage mastic had a tendency to melt, set and then bind everything near to it, thereby becoming extremely difficult to discharge. During the voyage, the cakes melted together and spilled onto other goods on the ship, hardened, and caused loss to the cargo as well as damage, including the cost of cleaning the ship.[40] The vessel owner had been held liable under bills of lading for other cargo damaged by mastic and had himself suffered damage in breaking up the mastic to discharge it.[41]

The court looked at the *Brass* majority decision for the general rule that a shipper need not have scienter to be held strictly liable.[42] The court voiced several policy reasons for deciding in favour of the carrier.[43] Counsel for the carrier compared a shipper's duty of absolute warranty of goods with the carrier's duty of absolute warranty that his ship is seaworthy.[44] The court agreed with the carrier and proposed that strict liability for the shipper is the only just outcome.[45] The court reasoned that if the shipper is not held strictly liable, fraud may be encouraged or shippers may try to experiment with shipping profitable, yet volatile goods at the total expense of the carrier.[46] Citing *Brass v. Maitland,* particularly com-

---

[37]   *Ibid.* at 349 f.
[38]   Rose, "Liability for Dangerous Goods" [1998] *LMCLQ* 480, 484.
[39]   19 F. Cas.646 (C.C.D Mass) (No:11, 150), aff'd, 19 F.Cas. 652 (C.C.D Mass, 1861) (No:11, 151).
[40]   *Ibid.*
[41]   *Ibid.* at 650.
[42]   *Ibid.* at 651.
[43]   *Ibid.*
[44]   *Ibid.* at 648.
[45]   *Ibid.* at 650 f.
[46]   *Ibid.* at 651.

ments on equally ignorant shippers and carriers,[47] it was held that the shippers, not the shipowners, must suffer if, owing to the ignorance of the shipper, notice of the defect was not given to the carrier.[48] The reasoning was that the loss should fall upon the shipper, who generally has the better means of determining the condition of the article to be shipped.[49]

2. Reactions: unclarity, no cargo warranty?

In 1910, a Second Circuit decision assailed *Pierce* as an overly expansive reading of *Brass*. In the *William J. Quillan*[50] case, the schooner *Wm. J. Quillan* was chartered to transport "tankage" from New York to Georgia. Tankage is a dry powder, the result of the boiling, drying and pressing of street garbage, and is packed in bags. It was assumed that everyone knew the dangerous nature of the tankage, as it had been regularly transported for over twenty-five years. A fire broke out during transport, causing damage to the ship, loss of cargo and loss of profit as a result of the inability to sell the destroyed or damaged bags.[51] The shipper claimed general average benefits, notwithstanding that his cargo caused the peril. The court used this case as an opportunity to criticize *Brass* and *Pierce* for their policy of strict liability for shippers and made its decision in favour of the shipper. Citing another British decision in *Greenshields, Cowie & Co. v. Stephens & Sons*,[52] where the court held that the general rights of a cargo owner, in the absence of any negligence or wrongdoing, are not affected because of inherent vice or defect in the cargo, it was held that a shipper ignorant of the cargo's dangerous nature does not owe a duty to the carrier pursuant to an implied absolute warranty of the fitness of the cargo and, therefore, is not liable for ensuing damage.[53]

It is said that the *William J. Quillan* opinion gives the erroneous impression that the court in *Greenshields* not only failed to read *Brass* as expansively as it was in *Pierce* but also failed even to mention the *Brass* decision.[54] In fact, the court in *Greenshields* did discuss *Brass* and, particularly, the issue of the shipper's absolute warranty where the shipper and carrier were equally ignorant of the hazardous nature of the cargo as shipped.[55] Reaching a conclusion similar to that of *Brass*, the *Greenshields* court decided that the absolute warranty would not arise in situations where the carrier took on goods openly carried and known by the carrier to be by nature possibly productive of danger.[56] Therefore, since the cargo was trans-

---

[47] *Ibid.*
[48] *Ibid.*
[49] *Ibid.*
[50] 180 F.681 (2d Cir.), cert. denied, 218 U.S.682, rev'g 175 F.207(S.D.N.Y 1910)
[51] 175 F. at 210.
[52] [1908] 1 K.B. 51 (1907), where coal heated and ignited on the voyage, but the carrier made no attempt to prove that it had been in a dangerous condition on shipment.
[53] 180 F. at 210.
[54] Maloof/Krauzlis, "Shipper's Potential Liabilities in Transit" [1980] 5 *Mar. Law* 175, 178 f.
[55] *Ibid.* at 179.
[56] *Ibid.*

ported openly and both carrier and shipper were aware of its potential hazards, the *Greenshields* court concluded in part that the shipper should still be entitled to his general average benefits, notwithstanding that his cargo created the peril.[57]

Accordingly, it has been contended that the *Greenshields* decision should not be interpreted as concluding that no absolute warranty exists where, as in *William J. Quillan,* both the shipper and carrier are unaware of the dangers or defects of the cargo.[58] Taken together, these cases supported the existence of an absolute warranty on the part of the shipper of defective goods to give the carrier notice of the dangers of his goods.[59] Such absolute warranty existed even if the shipper was ignorant of any defects; the shipper would be liable for damage caused by a breach of the warranty. On the other hand, if the carrier was aware of the defects or reasonably should have been aware, then the shipper would not owe such a warranty, there being no need to inform a knowledgeable carrier.[60]

A number of cases that followed these decisions dealt with the question of whether the shipper had given proper notice to the carrier. The cases basically held that a carrier could not recover damages from a shipper who failed to give notice of the cargo's characteristics if the nature of the cargo was readily ascertainable from the inscriptions on the packing.[61]

3. Enactment of COGSA: no strict liability

In 1936, the Carriage of Goods by Sea Act (COGSA) was enacted to re-define the rights and liabilities between carrier and shipper. Although even a strict reading of Section 1304(3)[62] compels the conclusion that a shipper cannot be held liable to the carrier or vessel unless there is an incident of fault or negligence, the idea that the shipper owed an implied absolute warranty to the carrier was slow in dying.

In *Sucrest v. Jennifer,*[63] the vessel developed a serious list while carrying a bulk cargo of raw sugar which had been damaged by salt water because of a transverse

---

[57]   *Ibid.*

[58]   *Ibid.* at 179; Roark, "Explosion on the High Seas! The Second Circuit Promotes International Uniformity with Strict Liability for the Shipment of Dangerous Goods: Senator v. Sunway" [2003] 33 *Sw. U. L. Rev.* 139, 177.

[59]   Maloof/Krauzlis, "Shipper's Potential Liabilities in Transit" [1980] 5 *Mar. Law* 175, 179.

[60]   *Ibid.*

[61]   In *The Ragoon Maru,* 27 F.2d 722 (2d Cir. 1928), the court found the carrier had full notice of the characteristics of the cargo of bleaching powder from the bill of lading and held that the shipper would not be liable for the costs of transshipment when the cargo had to be discharged because of the chlorine fumes emanating from the bleach. Similarly in *Hansen v. E.I. Du Pont de Nemours & Co.* 8 F.2d 552 (W.D.N.Y. 1925), a carrier of cordite, a slow burning powder, was denied the right to recover against the cargo owner for the loss of vessel on a theory of misrepresentation of the cargo's characteristics because the nature of the cargo was found to be readily ascertainable from inscriptions on the boxes.

[62]   Corresponds to Art. IV.3 of the Hague/Hague-Visby Rules.

[63]   455 F.Supp. 371.

shift of cargo. The vessel was purposely stranded for the safety of the vessel, its crew and the cargo. The cargo-owner sued the vessel and its owner for damage to the cargo, for salvage expenses and for cargo contribution to general average. The ship and the owner denied liability and claimed for hull damage, salvage expenses, damages for lost use of the vessel and the vessel's contribution to general average. Normally, raw bulk sugar is a passive, stable cargo, not a dangerous one. Because it had previously been damaged by salt water, however, it had become semi-lique-fied, or "thixotropic", from the vibration of the vessel during the voyage. The cargo therefore shifted, causing the vessel to list. The court found that the ship-per/charterer was not at fault, because he could not have been expected to know of the unique hazards of this cargo. This case was the first instance in which the maritime or scientific community learned of the thixotropic properties of raw sugar.[64] The court dismissed both the suit and the counterclaim and held that the cargo owner had neither actual nor constructive knowledge of the inherent dangers of the cargo and was not liable for the accident.[65]

It is very clear that in a series of decisions, courts, including the Second Circuit, have held that a shipper of goods does not give an absolute warranty that the cargo is safe and instead can be liable for loss or damage caused by dangerous goods only if he had actual or constructive knowledge that the cargo was dangerous.[66] Consequently, it was considered that, as a result of the COGSA, implied warran-ties of transportability do not exist; negligence is the only basis of shipper's liabil-ity.[67]

4. Senator v. Sunway: strict liability of shipper for the shipment of dangerous goods

a) Facts of the case

Since the adoption of COGSA, the shipper was not considered to warrant abso-lutely that the cargo contains no inherent dangers and that the cargo owner's duty is based on either its actual or constructive knowledge danger, although the deci-

---

[64] Bulow, "''Dangerous' Cargoes: the Responsibilities and Liabilities of the Various Parties" [1989] *LMCLQ* 342, 350.

[65] *Sucrest v. Jennifer* 455 F. Supp. 371, 386.

[66] Wilford/Coghlin/Kimball, *Time Charters* (2003), 187. In *Akt. Fido v. Lloyd Brazileiro*, 267 F.733, 738, the court ruled that charterers under several voyage charters were not liable to owners for expenses and delay incurred in discharging cargoes of coal which had overheated. The charters expressly provided for the loading of the coal. The court found that neither the owners nor the charterers were at fault and declined to impose any liability on the charterers. As stated by the court: "In our opinion the heating of the coal was such an extraordinary and unusual occurrence that it could not have been anticipated when the charters were made and should not be regarded as a "default" within the terms of the charters." However, the opinion contradicts preceding explana-tions of the courts, where it is said that "… the tendency of coal to heat was as well known to the masters as it was to charterers".

[67] Maloof/Krauzlis, "Shipper's Potential Liabilities in Transit" [1980] 5 *Mar. Law* 175, 181.

sion of the United States Court of Appeals for the Second Circuit in *Senator Linie G.m.b.H & Co. K.G v. Sunway Line, Inc.*[68] raised the question of whether that statement was correct and takes the position in dictum that there is a clear division of opinion on the point.[69]

In 1994, the shipper Easter Sunway Line hired the carrier Tokyo Senator to carry a cargo of 300 drums of thiourea dioxide (TDO), a white, odorless powder used as a reducing agent and in the bleaching of protein fibers. At the time of the shipment, TDO was not named as a hazardous or dangerous cargo either in the IMDG Code or in the Department of Transportation of Hazardous Materials Table.[70] It was not until 1998 that TDO was named as a hazardous or dangerous material in the IMDG Code and in 1999 in the Code of Federal Regulations. A fire broke out aboard the Senator and investigators later discovered that one of the drums of TDO had spontaneously ignited. After the incident, there was extensive litigation between various parties, but the important issue was the one between the carrier, Senator, and the shipper. The trial court held in favour of the shipper Sunway and interpreted Section 1304(6) of COGSA as imposing a scienter requirement upon shippers in order to hold them liable for shipping dangerous goods.[71] The Second Circuit reversed that decision and held that under Section 6, the shipper has *strict liability* for damage and expenses arising out of the shipment of dangerous cargo even if he had no actual or constructive knowledge of the dangerous nature of the cargo.

---

[68]   291 F.3d 145.

[69]   Wohlfeld, "The Senator Linie: Shipper's Strict Liability for Inherently Dangerous Goods" [2003] 27 *Tul. Mar. L.J.* 663; ;Roark, "Explosion on the High Seas! The Second Circuit Promotes International Uniformity with Strict Liability for the Shipment of Dangerous Goods: Senator v. Sunway" [2003] 33 *Sw. U. L. Rev.* 139; Cho/Huynh/McKee/Tanner/Walsh, "2002-2003 Survey of International Law in the Second Circuit" [2003] 30 *Syracuse J. Int'l L. & Com.* 75.

[70]   It was listed in the IMDG Code in 1998 and in the Code of Federal Regulations in 1999.

[71]   The District Court determined that the fire was caused by an exothermic (heat-releasing) reaction within the container holding the TDO drums. The District Court also concluded that, although an exothermic reaction may result from a number of factors such as excess heat or moisture, the plaintiffs had failed to establish the actual case of the exothermic reaction, or that any particular party was responsible. The District Court granted the shipper's motion for summary judgment with respect to the carrier's claims, holding that the COGSA "does not impose liability on a shipper of inherently dangerous goods unless it can be shown that the shipper actually or constructively knew of the dangerous nature of the cargo prior to shipment and failed to disclose that the nature to the carrier." Additionally, the District Court held that the general maritime law in the United States Court of Appeals for the Second Circuit and the United Kingdom did not impose an absolute warranty on the part of the shipper that its cargo was not hazardous. 2001 WL 238293 at 3.

b) Plaintiff's argument

On appeal, the carrier Senator argued that past case decisions requiring the shipper's knowledge for the shipper to incur liability were either incorrectly decided or clearly distinguishable from the facts in this case.[72] The carrier also argued that the COGSA statute had been misinterpreted.[73] Senator claimed that although COGSA seemed to be a negligence-based statute, specific sections such as 1304(6) were exceptions that pointed to strict liability.[74] Senator further argued that even if general maritime law applied, it did not prohibit the absolute warranty of the shipper, and in fact, if the goods were dangerous, then general maritime law stated that the shipper would be strictly liable.[75]

c) The defendant's argument

The shipper Sunway claimed that because it did not know, and could not have known, that the TDO was dangerous and therefore could not have warned Senator, it should not be held liable under COGSA.[76] Sunway then pointed out to various dangerous goods cases, all of which held that shippers must have scienter in order to be held liable.[77] Sunway further argued that COGSA is not only a codification of general maritime law, but that general maritime law does not impose strict liability upon a shipper and does not hold a shipper to an absolute warranty of his goods.[78]

d) Standard of review

To determine whether shipper liability under § 1304(6) is strict or knowledge-based, the second Circuit began by examining: 1) the plain meaning of the statutory provision together with the cases interpreting it; 2) the legislative purpose and history of COGSA; and 3) the general maritime law of the shipper's liability existing in pre-COGSA American and British case law: It then inquired whether in 1936 federal maritime common law had firmly settled the question of whether § 1340(6) codified general maritime law concerning shipper liability for dangerous goods.[79]

e) Plain meaning and the COGSA case law

It was found that the only reference in 1304(6) implicates the carrier.[80] A plain meaning approach would suggest that it is the carrier's knowledge of the goods'

---

[72] *Senator Linie v. Sunway Line* 291 F.3d 145, 148.
[73] *Ibid.* at 158.
[74] *Ibid.* at 158 f.
[75] *Ibid.*
[76] *Ibid.* at 148.
[77] *Ibid.* at 151, 154 f.
[78] *Ibid.*
[79] *Ibid.* at 153.
[80] *Ibid.* at 154.

dangerous nature, not the shipper's, that conditions shipper liability.[81] The Second Circuit observed that the District Court did not apply the plain meaning approach in reading § 1304(6).[82] Instead, the District Court concluded that a shipper would be liable for damage under § 1304(6) if the shipper *knowingly* shipped hazardous cargo without informing the carrier of the hazardous nature of the cargo.[83] To support its conclusion, the District Court relied on the only other case in the Second Circuit, addressing § 1304(6).[84] In *Borgships Inc. v. Olin Chemical Group*,[85] the court ruled that the shipper's compliance with certain Department of Transportation regulations did not, as a matter of law, satisfy the shipper's duty to warn and therefore the carrier had stated a cause of action under COGSA. The Second Circuit explained that the Borgships court did not employ COGSA in interpreting the scope of a shipper's duty under the bills of lading.[86] Instead, the Borgships court relied on general maritime law to support the position that a shipper was not held to guarantee absolutely the safe nature of its cargo and was liable only for what was actually or constructively known.[87] Hence, although the Borgships decision implied that COGSA § 1304(6) incorporated general maritime law, it did not directly answer the question of whether § 1304(6) codified pre-existing maritime common law on a strict liability standard for a shipper with respect to the dangerous nature of his cargo.[88] In essence, although a few courts have addressed § 1304(6), none has dealt with the specific question of strict liability.

In this case, the District Court further concluded that under the general rule of COGSA, a shipper was not liable for loss or damage sustained by the carrier or to the ship without the shipper's act, fault or neglect.[89] This general rule was based on § 1304(3) of COGSA, which sets forth a fault- or negligence-based theory of shipper liability. Surprisingly, U.S. courts have not specifically addressed the relationship between § 1304(3) and § 1304(6) when considering cases involving dangerous cargo.

Conversely, the Second Circuit discovered that in *Effort Shipping Co. v. Linden Management SA ("The Giannnis NK")*, the British House of Lords had examined the relationship between Articles 4(3) and 4(6) of the Hague Rules, the identical counterparts to §§ 1304(3) and 1304(6) of COGSA.[90] The Second Circuit found that the House of Lords' analysis in *Giannis NK* to be highly persuasive and applicable to the COGSA counterpart provision.[91] The Second Circuit concluded that the specific language and subject matter of § 1304(6) expressed an exception to,

---

[81]   *Ibid.*
[82]   *Ibid.*
[83]   *Ibid.*
[84]   *Ibid.*
[85]   1997 WL 124127 (S.D.N.Y)
[86]   *Senator Linie v. Sunway Line* 291 F.3d 145, 154.
[87]   *Ibid.* at 154 f.
[88]   *Ibid.*
[89]   2001 WL 238293 at 3.
[90]   *Supra* p. 142 f.
[91]   *Senator Linie v. Sunway Line* 291 F, 3d. 145, 157.

not a special application of, the fault-based standard of § 1304(6).[92] The Second Circuit favoured the above interpretation because by recognizing that §§ 1304(3) and 1304(6) apply different types of liability, the meaning of § 1304(6) was not made superfluous within the design of COGSA.[93] That is, since § 1304(3) already imposed liability on a shipper for his act, fault or neglect, § 1304(6) would be redundant if the purpose of the provision was merely to denote that the shipper required knowledge of the dangerous nature of the cargo.[94] Thus the Second Circuit found that the plain-meaning analysis strongly suggested that § 1304(3) and § 1304(6) play distinct roles in COGSA's allocation of risk between shippers and carriers.[95] The Second Circuit further found that the above analysis supported the proposition that § 1304(6) set forth a rule of strict liability for a shipper of inherently dangerous goods when neither the shipper nor the carrier had actual or constructive pre-shipment knowledge of the dangerous nature of the cargo.[96]

f) Legislative purpose and history of COGSA and the Hague Rules

The legislative history of COGSA shows that the Act is the identical twin to the Hague Rules of 1921.[97] The purpose of the Hague Rules was "to establish uniform ocean bills of lading to govern the rights and liabilities of carriers and shippers *inter se* in international trade."[98] COGSA on the other hand, was designed to provide a uniform law in governing the transport of goods by sea. In its objective to conform to international consensus, COGSA could not diverge significantly from the Hague Rules. The Second Circuit noted that the COGSA legislators had debated little on the terms and language of COGSA and that the legislators had focused mainly on preserving the international consensus contained in the language of the Hague Rules rather than on codifying particular rules of general maritime law expressed in U.S. case law.[99]

The Second Circuit traced the evolution of COGSA from the nineteenth century to its enactment in 1936. In doing so, the Second Circuit discovered that "while the language concerning carrier's right to land or destroy dangerous goods underwent various alterations in form and emphasis, the language describing shipper's liability remained virtually unchanged."[100] However, the Second Circuit noted that certain pre-Hague versions of language similar to that in § 1304(6) appeared in the Liverpool Conference Form ("Liverpool Form") and the Hamburg Rules of Affreightment ("Hamburg Rules"), promulgated by the International Law Association in 1882 and 1885 respectively. The Liverpool Form was a model bill of lading offered for voluntary adoption by agreement of shippers and carriers; the

---

[92] *Ibid.*
[93] *Ibid.*
[94] *Ibid.*
[95] *Ibid.*
[96] *Ibid.*
[97] *Ibid.* at 158.
[98] *Ibid.*
[99] *Ibid.* at 158 f.
[100] *Ibid.* at 159.

Hamburg Rules were model provisions that parties could voluntarily incorporate by reference into their bills of lading. These model documents made it expressly clear that shippers who agreed to adopt them as or within bills of lading would be subject to strict liability for damage resulting from inherently dangerous goods.[101]

Reviewing the record of the legislators' debate, the Second Circuit found nearly no comment directed at the language of shipper liability, but did find debates concentrating on the carrier's knowledge of dangerous goods and the problem of adequate notice to the carrier of a cargo's dangerous nature. The Second Circuit emphasized that in the ten years of Congressional deliberations over COGSA, Congress repeatedly stressed the fact that COGSA "is an important step in a world-wide movement for uniformity in ocean bills of lading".[102]

Although the Second Circuit found the legislative history of COGSA and the Hague Rules to be largely silent on the type of liability intended by the legislators to be imposed by § 1304(6), the Second Circuit found that Congress was mainly concerned with preserving international consensus in the context of the Hague Rules and securing American shipper and carrier interests under COGSA.[103] Alternatively, the shippers suggested that the lack of Congressional debate in enacting COGSA showed that COGSA legislators did not consider strict liability for shippers of dangerous goods.[104] The shippers advanced this suggestion by contending that § 1304(6) codified the well-established American maritime common law rule that shippers were not strictly liable.[105] For the shippers to make such a contention, the well-established American maritime common law rule had to exist; however, the Second Circuit found no such rule.[106]

In reaching its conclusion about American maritime common law, the District Court followed the Second Circuit holding in the *William J. Quillan* case, which rejected the absolute guarantee theory for shippers.[107] The shippers asserted that *Quillan*'s holding constituted, and still represented, American maritime common law.[108] However, the Second Circuit in the present case noted that it was doubtful that the *Quillan* decision was representative of general maritime law in 1936. The Second Circuit observed that the *Pierce* court adopted in 1861 a rule of strict liability for shippers of inherently dangerous goods and that *Pierce* was still persua-

---

[101] Hamburg Rules of Affreightment Art. 4 provides "Shipper accountable for any loss or damage to ship or cargo caused by inflammable, explosive, or dangerous goods, shipped without full disclosure of their nature, whether such shipper shall have been aware of it or not and, and whether such shipper be principal or agent: such goods may be thrown overboard or destroyed by the master or owner of the ship at any time without compensation." reprinted in Sturley, *The Legislative History of the Carriage of Goods by Sea Act and the Travaux Préparatoires of the Hague Rules* (1990) Vol. 2, 123. Liverpool Form 1882 is the same. *Ibid.* at 62.

[102] *Senator Linie v. Sunway Line* 291 F.3d, 145, 161.

[103] *Ibid.*

[104] *Ibid.*

[105] *Ibid.*

[106] *Ibid.*

[107] *Supra* p. 145.

[108] *Senator Linie v. Sunway Line* 291 F.3d 145, 161.

sive authority when *Quillan* was decided. The *Pierce* court relied heavily on the English case of *Brass v. Maitland*, which was regarded as the leading authority on maritime common law for imposing strict liability on shippers.[109] The *Quillan* court concluded that *Brass v. Maitland* did not set forth a general rule of strict liability for shippers of inherently dangerous goods.

In *The Giannis NK*, the House of Lords reaffirmed *Brass* when they unanimously held that both *Brass* and the British counterpart to § 1304(6) imposed strict liability on a shipper of dangerous goods where the carrier did not have knowledge of the dangerous nature, whether or not the shipper had knowledge of the danger. The Second Circuit in the present case found the decision in *The Giannis NK* to suggest that the decision in *Pierce* properly applied the strict liability rule for shippers of inherently dangerous goods.[110] Furthermore, the Second Circuit differentiated between the courts of *Pierce* and *Quillan* as possibly constituting a circuit split over the liability of shippers of inherently dangerous goods. Hence, the Second Circuit was unconvinced that the COGSA legislators ignored the conflicting rules between *Quillan* and *Pierce*. Moreover, the Second Circuit discovered that twelve years after the *Quillan* decision, another pre-1936 Second Circuit decision, in *the Santa Clara*[111] case, approvingly cited the strict liability holding in *Pierce* and *Brass*, although *Santa Clara* did not involve strict liability for shippers of inherently dangerous goods.[112]

g) Whether COGSA § 1304(6) codified general maritime law concerning shipper liability in the dangerous goods context

The Second Circuit concluded that the nature of a shipper's dangerous goods liability under general maritime law in the United States was not firmly settled in 1936.[113] The Second Circuit's conclusion rested on the fact that, although decisions for shipper's liability for inherently dangerous goods conflicted in *Quillan* and *Pierce*, the *Santa Clara* court approvingly cited *Pierce* and *Brass* for the policy behind their strict liability decisions twelve years after *Quillan*.[114] Likewise, the Second Circuit held that, by enacting § 1304(6), Congress had not codified a general maritime rule that required proof of a shipper's knowledge of the inherently dangerous nature of goods to impose liability on the shipper.[115]

Because the Second Circuit concluded that § 1304(6) did not codify general maritime law regarding liability for shippers of inherently dangerous goods, the Second Circuit held that § 1304(6) displaced inconsistent maritime common law.[116] As the Second Circuit stated there was a question as to whether federal

---

[109] *Supra* p. 139 ff.
[110] *Senator Linie v. Sunway Line* 281 F.3d 145,163.
[111] 281 F.725, 736, where the charterer loaded copper concentrates instead of copper ore and the court ruled that the charterer was in breach of contract.
[112] *Senator Linie v. Sunway Line* 281 F.3d 145, 165.
[113] *Ibid.* at 166.
[114] *Ibid.*
[115] *Ibid.*
[116] *Ibid.* at 166.

statutory law or federal common law governs, the courts must assume that Congress had spoken on the applicable standards as a matter of law. The Second Circuit considered several factors in determining whether the presumption of federal statutory law displaced general maritime law: 1. whether statutory language explicitly preserves or preempts judge-made law, or whether evidence of Congressional intent to achieve such results may be found in legislative history; 2. the scope of the legislation; 3. whether applying judge-made law would entail filling a gap left by Congress in a law it has affirmatively and specifically enacted; and 4. whether judge-made law at issue represents a long-established and familiar principle of common law or general maritime law.[117]

With respect to the first factor, COGSA and its legislative history did not discuss the relationship between § 1304(6) and pre-existing maritime common law.[118] As to the second factor, by setting forth in detail the rights, duties, liabilities, and immunities of carriers, COGSA extensively governs the relations between carriers and shippers with respect to all contracts of carriage of goods by sea to or from ports of the United States.[119] Regarding the third factor, the Second Circuit had to determine whether COGSA explicitly addressed the question of strict liability for shippers of inherently dangerous goods.[120] Having determined that § 1304(6) does not require a shipper scienter requirement, the Second Circuit concluded that § 1304(6) directly and affirmatively answered the question of whether a shipper of inherently dangerous goods may be liable when neither the shipper nor the carrier had actual or constructive pre-shipment knowledge of the danger.[121] Hence, § 1304(6) superseded *Quillan* and other pre-1936 decisions that set forth a rule of general maritime law inconsistent with § 1304(6).[122]

h) Conclusion

Consequently, the Second Circuit concluded that § 1304(6) imposed strict liability and displaced inconsistent general maritime law. In reaching this conclusion, the Second Circuit was mindful of the main purpose of COGSA, which was to allocate the risk of loss and create predictable liability. The Second Circuit expressed the view that by displacing uncertain maritime common law, the strict liability rule of § 1304(6) will help to ensure predictability in the allocation of risk between carrier and shipper. Moreover, the Second Circuit noted that a strict liability rule of § 1304(6) would foster fairness and commercial efficiency. The Second Circuit reasoned that since a shipper can be expected to have greater access to and familiarity with shipped goods, the shipper must be liable for dangerous goods, since he should be in a better position to assess the dangerous nature of the goods. Furthermore, by conforming the construction of § 1304(6) to its British counterpart as expounded in *the Giannis NK* case, the Second Circuit noted that the decision in

---

[117] *Ibid.* at 167.
[118] *Ibid.* at 167.
[119] *Ibid.*
[120] *Ibid.*
[121] *Ibid.* at 168.
[122] *Ibid.*

this case advanced the main purpose of COGSA and the Hague Rules: "international uniformity in the law of carriage by sea".[123]

III. Express language of HGB § 564b in German law: strict liability of shipper/actual shipper if dangerous nature of goods unknown to the carrier

Unlike English and American law, HGB § 564b explicitly states that if dangerous goods are brought on board without the master's knowledge of their dangerous character and properties, the shipper of such goods is liable without fault on his part.[124] Therefore, the problem of determining the nature of liability has not arisen in German law.

C. Is strict liability proper?

I. Strict liability in general

In European civil law systems, tort liability is based on a more or less broadly defined general clause of fault liability. Despite the survival of some specific torts, this is also true for English law, where the principle of negligence has been gradually established. However, faced with an increase in technical and industrial risks, many legal systems have, since the end of the 19th century, introduced liability rules that provide for some form of strict liability.[125] In any discussion on the foundations of tort liability, "strict liability" is generally opposed to "liability for fault". Fault liability is liability for unacceptably unreasonable conduct.[126] To judge that the defendant acted with fault, one first has to determine the relevant standard of conduct, and secondly, to establish that the defendant failed to meet this standard. By contrast, strict liability is concerned with the pecuniary consequences of harm not attributable to a lack of precaution on the part of the defendant, or what may be referred to as "accidental harm". Beyond liability for foreseeable and avoidable harm, it may also be extended to harm that is unforeseeable and/or inevitable. Many strict liability rules are explained on the basis that the defendant is in the best position to monitor an activity under his control and to prevent the occurrence of harm.[127]

---

[123] *Ibid.* at. 170.

[124] Rabe, *Seehandelsrecht* (2000), 447; Abraham, *Das Seerecht in der Bundesrepublik Deutschland* (1978) 1. Teil, 527.

[125] Palmer, "A General Theory of the Inner Structure of Strict Liability: Common Law, Civil Law, and Comparative Law [1988] 62 *Tul .L. Rev.* 1303, 1312; Vandall, *Strict Liability, Legal and Economic Analysis* (1989), 1 ff.

[126] Werro/Palmer (ed), *The Boundaries of Strict Liability in European Tort Law* (2004), 7.

[127] *Ibid.* at 6.

Moreover, economic analysis of law has made tort scholars aware that the choice between different liability rules was predominantly of social efficiency.[128] Once social efficiency was accepted as a policy goal in tort law, the debate over whether losses should lie with the victims or be shifted to the defendant became essentially a matter of overall social welfare rather than of justice between the two parties involved. Consequently, discussions concentrated on the question as to which of the parties could bear the loss most cost-efficiently. The availability of insurance as a mechanism for spreading losses within a wider community received much greater attention.

To explain why strict liability may not only be socially desirable but why it may also constitute an appropriate response, academic writers and legislators have traditionally based themselves on the concept of ultra-hazardous risks. This concept underlies many of the liability rules which exist in civil law systems and which impose strict liability for narrowly defined risks, such as those triggered by cars, railways, pipelines etc. While being firmly well-established in the civil law tradition, the concept of ultra-hazardous risk is also considered to underlie the reasoning in *Rylands v. Fletcher.*[129] The defendant in *Rylands* decided to dig a reservoir on land located on the northwest side of a hill. The plaintiff maintained property on the southeast side of the hill beneath the defendant's land, where he worked coal mines. The coal-mine passages were horizontal, though there were many vertical shafts located intermittently throughout the passage- ways. Unknown to the defendant, five of these vertical shafts reached the portion of land that he was excavating for the reservoir. One morning, water from the newly-dug reservoir burst through one of the vertical shafts, travelling downward into the horizontal mines belonging to the plaintiff. The water caused severe damage to the mines and the plaintiff had to cease work. The defendant did not know that coal mines were beneath the ground or that the plaintiff was working in the mines and thus could not have known that the benign water he brought onto his land would become a hazard should it escape. The question before the court was whether a defendant could be liable when he had no knowledge that the substance he rightfully kept on his own land may cause danger to happen to another or to another's land if it were to escape. *The Rylands* court answered this question by likening the water to a wild animal kept on one's property:

> We think that the true rule of law today is, that the person who for his own purposes brings on his land and collects and keeps there anything likely to do mischief if it escapes, must keep it in at his peril, and if he does not to do so, is prima facie answerable for all the damage which is the natural consequences of its escape. He can excuse himself by showing that the escape was owing to the plaintiff's default; or perhaps that the escape was the consequences of vice-major or the act of God....[130]

In other words, regardless of fault, knowledge or a high degree of care, one who brings on his land something wild or something that can escape and do harm to the

---

[128] *Ibid.* at 13.
[129] (1866) LR 1 Exch 265; aff'd (1868) LR 3 HL 330.
[130] *Ibid.* at 279.

others would be strictly liable for the consequences.[131] The *Rylands v. Fletcher* rule became one of the few examples of strict liability at common law. In the United States, the authors of the Restatement of Torts (Second) used it as the starting point for formulating a general rule of strict liability for ultra-hazardous activities.[132]

To view strict liability as a response to the creation of particular risks, whether they are enumerated in a statute or defined by judges, is an appealing concept. Instead of stretching fault liability beyond the limits of reasonableness, it proposes an alternative justification for shifting losses from the plaintiff to the defendant.

II. Ultra-hazardous activity: dangers of cargo unknown to the carrier

Comparatively speaking, dangerous goods are like dangerous activities in terms of magnitude of harm they can cause and therefore should be similarly analyzed.[133] In a case of carriage of dangerous goods by sea, it may or may not be likely that the cargo would cause damage, but it is clear that if it causes damage, such as an explosion or fire on board a ship, it may be disastrous. Hence, if a shipper ships dangerous cargoes without the knowing consent of the carrier, he should be liable for the consequences, even though he did not know that the cargo was dangerous, dangerously packed or stowed or he had no means of knowing about its dangers.

Other policies behind strict liability include loss-shifting, safety, superior knowledge and the availability of insurance. In the context of shipper and carrier, loss-shifting and insurance are not the best arguments in favour of strict liability, because both are professional merchants and both could equally bear the economic loss. However, the idea of safety and the rationale that the shipper is in the best position to discover the nature of his goods are persuasive arguments favouring strict liability. It is clear that shipping is a dangerous business. Even in this modern age, seafarers die on the job at a rate of three times that of coal-miners, five and a half times that of construction workers, and twenty-five times that of manu-

---

[131] Roark, "Explosion on the High Seas! The Second Circuit Promotes International Uniformity with Strict Liability for the Shipment of Dangerous Goods: Senator v. Sunway" [2003] 33 *Sw. U. L. Rev.* 139, 158.

[132] § 519: "One who carries on an abnormally dangerous activity is subject to liability for harm to the other persons, land or chattels of another resulting from the activity, although he has exercised the utmost care to prevent the harm."

[133] Roark, "Explosion on the High Seas! The Second Circuit Promotes International Uniformity with Strict Liability for the Shipment of Dangerous Goods: Senator v. Sunway" [2003] 33 *Sw. U. L. Rev.* 139, 160. For instance, § 520 of Restatements of Torts (2nd) lists six factors for determining whether an activity is abnormally dangerous. These are: a) existence of a high degree of risk of some harm to the person, land or chattels of others b) likelihood that the harm that results from it will be great c) inability to eliminate the risk by the exercise of reasonable care d) extent to which the activity is not a matter of common usage e) extent to which the activity relates to the place where it is carried on, and f) extent to which its value to the community is outweighed by its dangerous attributes.

facturing employees, totalling over 6,500 lives lost per year.[134] Thus, safety is of great concern for workers aboard carriers, as well as safety of the ship itself and its cargo onboard.

III. How does strict liability address safety?

If a shipper knows he will be held strictly liable for his carrying of dangerous cargoes, he will exercise a high degree of care.[135] The shipper is in the best position to protect people from dangerous goods, because he can conduct tests and ascertain the true character of the shipped goods before despatch, whereas the carrier could not reasonably be expected to do so for every type of cargo he carries.[136] Particularly in terms of containerized goods, should a carrier have to open every stuffed container to see what is in them before he departs or test every cargo to find out their properties? Of course he should not, as that would be inefficient. Furthermore, even if the carrier did open every container, how would he be expected to know whether a substance is dangerous? A carrier should be able to trust that the shipper is not going to load dangerous cargoes onto the ship without informing the carrier. At least when he has a warning, the carrier can make the decision of whether or not to take the risk.

In terms of bulk cargo, the issue rarely arises because bulk cargo properties are generally well known in the trade. The carrier, master or agent can see and is supposed to know what is known about particular cargo. However, what the carrier or master should be informed about is the particular state of the bulk cargo, e.g. moisture content, temperature etc. In that case, the strict liability of the shipper for an unknown danger of a bulk cargo would arise if such a danger were unknown by the scientific world.[137]

As the cases deciding for strict liability expressly mention, the strict liability of the shipper arises when the shipper has not informed the carrier, master or agent of the dangerous nature and properties of the goods. In other words, if the shipper has given the necessary notice to make the carrier or master aware of the goods' particulars or if the carrier, master or agent knew the dangerous nature of the goods, strict liability does not become an issue. However, that is not to say that the shipper is not liable once he has given notice of danger.

---

[134]   Li/Mi Ng, "International Maritime Conventions: Seafarers' Safety and Human Rights" [2002] 33 *J. Mar. L. & Com.* 381.

[135]   Vandall, *Strict Liability* (1989), 21; Verro/Vernon, *The Boundaries of Strict Liability in European Tort Law* (2004), 8.

[136]   *In re M/V "DG Harmony"*, No: 98 Civ.8394 (DC), 2005 U.S. Dist. Lexis 23874: 18 October 2005.

[137]   *Sucrest Corp. v. M.V. Jennifer* 455 F.Supp. 371 (Maine N.D).

D. No strict liability where the carrier has knowledge

If the dangerous nature of the goods and the necessary measures to be taken are unknown by the carrier or master, the shipper of such goods is strictly liable. That means that even if he did not know the dangerous nature of the cargo and the necessary measures[138] to be taken to handle such cargo or he had no means of knowing the dangerous nature of the cargo and measures, he will be liable for damages resulting from such shipment. The issue is understood limited to situations in which neither the shipper nor the carrier knew of the dangerous nature of the goods.[139] However, such an issue arises only on the rare occasion when neither carrier nor shipper knows or ought to be reasonably aware of the dangerous nature of the goods.[140]

Art. IV.6 of the Hague/Hague-Visby Rules in its first sentence makes the shipper of dangerous goods liable "…whereof…. the carrier has not consented with knowledge of their nature and character…", while in the second sentence it grants the carrier rights "…such goods shipped with such knowledge and consent shall become a danger to the ship or cargo, … be landed at any place, or destroyed or rendered innocuous …". That is to say that where the carrier has been given of notice of the dangerous nature of the cargo and handling conditions or the carrier knew or ought to have known about the dangerous nature of the cargo and the measures to be taken for the handling of such cargo, the shipper's obligation has been discharged.[141]

In such a case, it is accepted that when the carrier consents to carry such cargo he agrees to carry such risk and charges extra, i.e. a surcharge, for the carriage of dangerous goods.[142] Acceptance with the knowledge of the dangerous nature and character and measures to be taken means accepting on the basis of available scientific information. If the available information is insufficient or found to be improper, the carrier cannot be said that he accepted the cargo with that knowledge.[143]

The Hamburg Rules, on the other hand, expressly provide in Art. 13 that "…the shipper must inform him of dangerous nature of the goods and, if necessary, of the precautions to be taken. If the shipper fails to do so and such carrier or actual carrier does not otherwise have knowledge of their dangerous character the shipper is liable…"

---

[138] *In Re M/V Harmony*, NO:98 Civ.83954 (DC), 2005 U.S.Dist. Lexis 23874, where cargo was carried in accordance with the dangerous goods regulations but regulations found to be insufficient for that cargo. *Supra* p. 114 fn. 119.

[139] *Contship Containerlines, Ltd. v. PPG Indus* 442 F.3d 74, 77.

[140] Wilson, *Carriage of Goods by Sea* (2001), 38.

[141] Boyd/Burrows/Foxton, *Scrutton on Charterparties* (1996) 105, Tetley, *Marine Cargo Claims* (1988), 465; Wilson, *Carriage of Goods by Sea* (2001), 36.

[142] Gaskell/Asariotis/Baatz, *Bills of Lading: Law and Contracts* (2000), 470.

[143] *In re M/V DG Harmony*, No. 98 Civ.8394 (DC), 2005 U.S.Dist. Lexis 23874, *supra* p. 114 fn. 119.

It is, however, to be noted that acceptance of dangerous cargo with the full knowledge of its nature, character and handling measures for transport does not fully exonerate the shipper from liability. The shipper will be still liable to the extent that his negligence.

I. English law

If the master or person in control of the ship chooses to receive the goods on board, knowing their nature and the manner in which they are packed, the shipper will not be liable[144] In other words, a carrier who consents to carry goods of a particular description contracts to perform the carriage in a manner appropriate to goods of that description and thereby assumes all risks of accidents attributable to a failure to carry in that manner.[145] In *The Atlantic Duchess*,[146] the vessel was chartered for the carriage of a cargo of "crude oil and/or diesel oil and/or gas oil and/or distillate of petroleum." A cargo of butanized crude oil was loaded. During ballasting operations fire and exclusion occurred. The shipowner alleged that the charterers were in breach of contract in shipping an unauthorized cargo or that the additional butane injected into the crude oil involved special hazards of which no warning was given by the charterers, thus the charterers were in breach of the implied term at common law not to load a dangerous cargo without notice. However, it was held that the shipowner had failed to prove the butanized crude oil cargo shipped by the charterers was outside the contractual description or that butanized crude oil involved any special hazards. The shipowner's claim failed as the butanized crude oil was not outside the contractual description, but it was reasonable to infer that there was some failure of safety precautions on their part, and so his claim failed.[147]

---

[144] Colinvaux, *Carver Carriage by Sea* (1982) Vol. 2, 844. In *Brass v. Maitland*, 6 E. & B. 470, 486, it was held to be good defence that the master knew that the casks contained bleaching powder and had the means of knowing and reasonably might and could and ought to have known that it contained chloride of lime, and that he had the means of judging the state and sufficiency of the casks and the packing of the contents. "On this supposition, the loss which has happened is to be imputed to the carelessness and misconduct of the master.... in stowing the casks where they were likely to injure other goods. A mere allegation of "means of knowledge" I think would not have been sufficient, as this might be satisfied by calling in an expert chemist and resorting to investigations inconsistent with the usual course of commercial business. But the shippers were justified in acting upon the supposition that the master to whom the goods are alleged to have been delivered did know what "he reasonably might and could and ought to have known."

[145] *The Athanasia Comninos and Georges Chr. Lemos* [1990] 1 Lloyd's Rep. 277, 283.

[146] [1957] 2 Lloyd's Rep. 55.

[147] See also *Greenshields, Cowie & Co. v. Stephens & Sons* [1998] A.C. 431, *Deutsche Ost-Afrika v. Legent* [1998] 2 Lloyd's Rep. 71; *Shaw Savill & Albion Company Ltd. v. Electric Reduction Sales Company Ltd. ("The Mahia")* [1955] 1 Lloyd's Rep. 264; *Macieo Shipping Ltd. v. Clipper Shipping ("The Clipper Sao Luis")* [2000] 1 Lloyd's

II. American law

In American law too, if the shipper informs the carrier of the dangerous properties of the cargo or the carrier knows or ought to know the dangerous nature of the goods, the shipper can be excused.[148] Under this type of reasoning, a carrier is deemed to have accepted the carrying risk and is responsible for inherent vice.[149]

In *Westchester Fire v. Buffola Housewrecking*,[150] the court held that the implied warranty by the shipper of a cargo of metal turnings and borings was not applicable where the shipowner had been given full opportunity to observe the cargo before it was loaded. It also stated that the rule that there is an implied warranty by the shipper that the goods are fit for carriage in the ordinary way and are not dangerous does not apply where the shipowner knows or ought to know the dangerous character of the goods.[151] Likewise, in *The Stylianos Restis*,[152] the carrier sued for fire damage to his vessel caused by a cargo of fishmeal. The cargo was proven to be in good order. It had been inspected at the loading port and conformed to the quality as described, i.e. with no excess oil or fat content which could cause spontaneous heating. The owner ventilated the cargo as much as possible. He had obviously taken proper care of the cargo. During the voyage, smoke was observed coming from the ventilator and the vessel was forced to seek refuge. The court ruled that the susceptibility of fishmeal to spontaneous combustion is an inherent vice of the commodity and that compliance with any known regulation or requirement does not preclude the possibility of fire emanating due to spontaneous combustion. Under the GENCON charterparty, the court found that the owner had failed to sustain his burden of proving that there was any negligence or breach of warranty on the part of the supplier or its agent. The owner's claim against the shipper of the cargo of fishmeal for fire damage was denied under S.1304(3) of COGSA, as the vessel owner had failed to prove actual negligence on the part of the shipper and as the master was clearly aware of the inherent property of fishmeal to heat.[153]

Rep. 645; *Ministery of Food v. Lamport & Holt Line Ltd.* [1952] 2 Lloyd's Rep. 371, 382.

[148] Bulow, "Dangerous' Cargoes: the Responsibilities and Liabilities of the Various Parties" [1989] *LMCLQ* 342, 350. Maloof/Krauzlis, "Shipper's Potential Liabilities in Transit" [1980] 5 *Mar. Law* 175, 179 f.

[149] *Ibid.*

[150] 40 F.Supp. 378 (W.D.N.Y.).

[151] *Ibid.* at 382.

[152] 1974 A.M.C 2343 (S.D.N.Y. 1972).

[153] In *The Rangoon Maru* 27 F.2d 722, 726 (2d Cir. 1928), the court found that the carrier had full notice of the characteristics of the cargo of bleaching powder from the bill of lading and held that the shipper would not be liable for the costs of transshipment when the cargo had to be discharged because of the chlorine fumes emanating from the bleaching powder. Furthermore, a shipper was found free of liability for an explosion caused by naphtha fumes in *International Mercantile Marine Co. v. Fels*, where the shipper had warned the carrier of the need for proper ventilation and the cargo had markings indicating the presence of naphtha. 164 F.337, 345 ff. (S.D.N.Y 1908). See

Furthermore, in *Pitria Star Navigation Co. v. Monsanto Co.*[154] claims were asserted by the shipowner against a voyage charterer and the manufacturer and shipper of a cargo of parathion, a poison, liquid insecticide. The court found that the death of the seamen did not give rise to a claim against the manufacturer of the cargo because there was no showing of any negligence in the manufacture of packaging of the insecticide. Because the shipowner knew that the parathion was a poison, the court found that the defendants had no reason to believe that the shipowner and the crew would not realize the dangerous condition of the cargo. Recently this position was confirmed. In *Contship Containers, Ltd. v. PPG Industries, Inc.*[155] cargo of Cal Hypo which was listed in the IMDG Code was carried. The carrier sued shipper seeking damages after cargo overheated and caused fire. It was held that a carrier cannot invoke strict liability if it knows that a cargo poses danger and requires gingerly handling or stowage.[156]

III. German law

In German law, as in American and English law, if the carrier, his agent or master knew the dangerous nature of the cargo, the shipper is, principally not liable.[157] HGB § 564b is equivalent to Art. IV.6 and likewise in its second paragraph it provides the carrier with the right to land cargo or render it innocuous at any time but it does not mention the liability of the shipper in such cases. For instance, a ship was loaded with sulfur chips, the cargo corroded its equipment and repair became necessary during the voyage. The ship stayed in a dockyard for a long time with its cargo. Liability of the shipper was refused on the ground that the

---

also *International Marine Development Corp. v. Lakes Shipping and Trading Corp. (The M/V Glia)* (1975) S.M.A. No:931; *Skibs A/S Gylfe v. Hyman-Michaels Co. (The Gyda)* 1970 AMC 84, 304 F.Supp. 1204; *Pitria Star Navigation Co. v. Monsanto Co.* 1986 AMC 2966 (E.D.La.1984).

[154]   1984 WL 3636 (E.D.La).

[155]   442 F.3d 74.

[156]   *Ibid.* at 78; DuClos, "Liability for Losses Caused by Inherently Dangerous Goods Shipped by Sea and the Determinative Competing Degrees of Knowledge", <www. duclosduclos.org/LiabilityforLossesCausedByInherently.pdf> 6 ff. (visited 13.7.2007) (to be published in *U.S.F. Mar.L.J.* Vol. 20 No. 1); Homer, "Second Circuit Limits Cogsa Strict Liability for Shippers of Dangerous Goods in Contship Containerlines, Ltd. V. PPG Industries, Inc., 31 *Tul. Mar. L.J.* 199, 203 ff. It is asserted that the holding in *Contship* arguably punishes a carrier for complying, to the best of its ability, with a regulated temperature maximum for stowage of Cal Hypo, while rewarding the shipper-manufacturer for failing to disclose particular dangers of which he was aware. *Ibid.* at 206.

[157]   Rabe, *Seehandelsrecht* (2000), 447; Abraham, *Das Seerecht in der Bundesrepublik Deutschland* (1978) 1. Teil, 527; Gündisch, *Die Absenderhaftung im Land- und Seetransportrecht* (1999), 197, 199.

cargo was properly declared. The opinion was that the carrier agreed to carry the risk himself.[158]

E. Liability where both parties have knowledge

Having said that when the carrier accepts the dangerous cargo knowing its dangerous nature, he accepts to carry the risk, it has to be pointed out that this does not mean the shipper will not be liable at all. When both the carrier and the shipper know the dangerous nature of the cargo and something goes wrong, the case is like any other cargo case. Giving notice of the dangerous nature does not mean that all necessary precautions for the transport of the dangerous cargo have been taken. Even if the shipper has notified the carrier of the dangerous nature of the cargo, the shipper cannot be exonerated from liability if he is negligent in fulfilling duties on his part and failing to take the necessary precautions.[159]

Under Art. IV.3 of the Hague/Hague-Visby Rules,[160] the shipper is liable for loss or damage arising from fault or neglect of himself, his agents or his servants. Hence, if the damage to the vessel or shipment arises from the shipper's failure to due precautions the shipper is liable for its consequences. The shipper's negligence may lie in improper stowage, lack of securing, improper treatment of cargo,[161] lack of proper shifting boards etc.[162]

---

[158] RG, HansGZ 1916, Nr. 9. Similarly where damaged barley was agreed to be carried, the shipper was not liable for the costs of unloading which became necessary during the voyage. HansGZ 1911 Nr.138.

[159] DuClos, "Liability for Losses Caused by Inherently Dangerous Goods Shipped by Sea and the Determinative Competing Degrees of Knowledge", <www.duclosduclos.org/LiabilityforLossesCausedByInherently.pdf> 8 ff. (visited 13.7.2007) (to be published in *U.S.F. Mar.L.J.* Vol. 20 No. 1).

[160] Corresponding to US COGSA § 1304(6); British COGSA IV.6.

[161] *The Amphion* [1991] 2 Lloyd's Rep. 101; *The Nour* [1999] 1 Lloyd's Rep.1.

[162] In *Poliskie Line Oceanicze v. Hooker Chemical Corp.* 499 F.Supp. 94 (S.D.N.Y), the vessel owner sought damages against the shipper of a hazardous cargo, sulphur dichloride. Although the vessel owner knew the hazards of the cargo, he was entitled to rely on the shipper's certification that the stowage was in accord with the Code of Federal Regulations. The court found that the shipper was negligent in the stowage of the drums in a container and therefore liable for damage. Similarly, in *Hellenic Shipping &Industries Co. Ltd. v. American Marine Navigation Co. (The Mesologi)* (1971) S.M.A No. 642, the vessel loaded cotton bales. Although a normally safe cargo, cotton is listed in Class 4 when it is wet and contaminated. It is then described as being liable to ignite spontaneously unless securely bound in bales. The vessel also loaded steel plates and, later, grain. The vessel experienced heavy weather while sailing. Fire was observed in the cotton cargo and the vessel was forced to seek refuge. The panel found that the fire occurred due to bad stowage and a lack of proper securing of the steel plates. During heavy weather, the plates shifted and struck the wire strapping of the bales. The stowage was the charterer's duty under Clause 8 of the NYPE. The owner's claim was granted. In *MS Neuwarder Sand* BGH 25.3.1974, the charterer found liable under HGB § 564 for not informing carrier of high humidity of zinc

Fishmeal is a good example of this situation. It is listed in the IMDG Code in Class 9. It is customarily loaded after taking designated precautions, e.g. having been treated with antioxidant; consent to the loading of such cargoes would usually be regarded as subject to the taking of such precautions and the absence of such precautions might nullify any apparent consent of the carrier or master It is well known in practice that if fishmeal is improperly treated, it may ignite through spontaneous combustion. While the shipper is strictly liable for not informing of the dangerous nature of the good's character, it is also clear that it is negligent not to treat the fishmeal properly.

German Commercial Code, HGB, does not contain the general provision Hague/Hague-Visby Rules Art. IV.3 on shipper's liability. If a dangerous cargo was carried with the knowing consent of the carrier but caused damage due to negligent stowage, packing, or information by the shipper, the issue would be determined either § 564 or under the general provisions of obligation law.

F. The necessity to distinguish between dangerous goods and non-dangerous goods which cause damage

Strict liability makes a person responsible for the damage and loss caused by their acts and omissions regardless of fault. Applying this principle to dangerous cargo, the fact that damage has been caused by dangerous cargo on board the ship is sufficient to establish the shipper's liability. This harsh nature of strict liability raises the question of the necessity to distinguish between dangerous and non-dangerous cargo which may cause damage.

As Part 2 clearly shows, any cargo might cause damage. Moreover, in English Law Art. IV.6 of the Hague/Hague-Visby Rules is given a broad meaning and not restricted to "inflammable, explosive and other dangerous". Consequently, the shipper might be strictly liable for any cargo, whether actually dangerous or not.

The basis of liability in the Hague/Hague Visby Rules is fault, with the exception of Art. III.5 and IV.6. This is also the basis for the carrier's liability. The Hague Rules were intended to rein in the unbridled freedom of contract of owners to impose terms which were "so unreasonable and unjust in their terms as to exempt from almost every conceivable risk and responsibility."[163] At common law the common carrier was virtually an insurer for the safe delivery of the goods. In

---

residue which became colloidal. In *In re M/V DG Harmony* No. 98 Civ.8394 (DC), 2005 U.S.Dist.Lexis 23874: 18 October 2005, in addition to strict liability, the shipper was found negligent in failing to determine the true nature of the risk in handling the cargo and for failing to warn the vessel about the danger, even though the cargo was carried in accordance with the IMDG Code, because the precautions provided for the given cargo in the IMDG Code were found to be incorrect for the cargo.

[163] Roskill, "Book Review, The Legislative History of the Carriage of Goods by Sea Act and the Travaux Préparatoires of the Hague Rule" [1992] 108 *L.Q.R.* 501, 502.

the 19th century it became the practice of shipowners to enter into contracts of affreightment with shippers that exempted them from responsibility for certain events, even including the carrier's own negligence. The use of these clauses led to legislation in the United States and many countries dealing with the validity of such clauses.[164] Courts were also asked to interpret them and this led to varying interpretations and a distressing lack of uniformity in shipping practice and law.[165] This situation hastened the need for uniform international regulation of the rights and duties of the carriers of ocean cargo. The International Convention for the Unification of Certain Rules of Law relating to Bills of Lading was signed in 1924.[166] The Hague Rules aimed to achieve this by a pragmatic compromise between the interests of owners and shippers and were designed to achieve a partial harmonization of the diverse laws of trading nations, at least in the areas which the convention covered.

Therefore, it is against the objective of the Rules to interpret Art. IV.6 broadly to include *any* cargo causing damage, such as infested cargo or contraband cargo of rice, and to hold the shipper *strictly* liable for damage arising from any shipment. As this outcome would be unreasonable and unjust where the carrier is, by contrast, liable for only his negligence; two situations should be distinguished as a matter of law. It should be understood that the liability of the shipper is, principally, based on fault, mirroring the liability regime in respect of the carrier; and strict liability is an exception. In the leading case *Brass v. Maitland,* it was held that the obligation on the carrier to furnish a seaworthy ship and on the merchant to supply the goods packed in a fit condition for the journey are considered as coextensive.[167] This could be justified in common law where seaworthiness is an absolute warranty. However, under Hague/Hague-Visby Rules seaworthiness is an obligation to use due diligence. Thus strict liability of the shipper for dangerous goods can only be justified if the concept is restricted to its objective, i.e. if "dangerous goods" are be understood as exceptionally dangerous goods that pose significant risk and significant potential damage.

In terms of "legally dangerous goods", e.g. goods causing delay or detention of the cargo, it was said that "it would be most unsatisfactory if different principles governed each type of case... moreover there is no logic in either law or common sense for requiring a shipowner to prove negligence on the part of a shipper when he has suffered delay and expense from the shipment of non-physically dangerous goods but to absolve him from that requirement when delay or expenses has arisen from the shipment of a physically dangerous goods."[168] This is no more unsatisfactory than where one party's liability is greater than other's under an

---

[164] Clarke, *Aspects of the Hague Rules, A Comparative Study in English and French Law* (1976), 3 f.
[165] Schoenbaum, *Admiralty and Maritime Law* (2004) Vol. 1, 636.
[166] The Brussels Protocol of Amendments to the Hague Rules, the so-called Visby Rules, were signed in 1968.
[167] *Brass v. Maitland* 6 E. & B 474, 475.
[168] *The Giannis NK* [1994] 2 Lloyd's Rep.171.

instrument which was said to be aimed at balancing the parties' interests. Therefore, it is doubtful if this approach is viable under the principal aim of the Rules.

Consequently, the extension of the meaning of 'dangerous" to goods which may cause delay is not justified under the Hague/Hague-Visby Rules, as the Rules are supposed to balance the interests of carrier and shipper. Since the liability of the carrier for delay under the Rules is fault-based, there is no good reason to destroy the said balance in favour of the carrier.

I. Separation in German law: fault liability for incorrect information on properties and contraband goods

In addition to the specific provision on dangerous cargo, the shipper's liability for incorrect information and contraband cargo is regulated separately in German law. Accordingly, HGB § 564 provides that:

> Giving the incorrect information as to type and properties of the goods renders shipper or actual shipper liable, if he is negligent, to the carrier and to others specified in § 512(1), for damages arising out of incorrect information as to properties of the goods.
>
> The same applies, if the shipper loads contraband or culpable cargo whose export, import, and transit passage forbidden, or by loading such cargo if he violates statutory provisions, particularly police, tax, and custom regulations. ...

The information provided to the carrier under § 564 is different from that provided under § 563, which corresponds to Art. III.5. of the Hague/Hague-Visby Rules. Information as to the nature and property of the cargo is necessary for the proper, safe and secure carriage of all kind of cargoes. The shipper or actual shipper should provide all necessary information regarding any type of cargo. If he fails to do so, he will be liable for his fault. A fault would also be attributable to the shipper or actual shipper if he ought to have known the correct information as to the properties of the cargo.[169]

For instance, the declaration of bisulphate as rock salt, canons as merchandize, weapons as hardware, or the loading of war contraband or cargo whose export or import is forbidden fall under § 564 rather than the specific provision § 564(b).[170] In the *MS Neuwardersand*,[171] the charterer was found negligent under § 564 for failing inform carrier of high humid state of zinc residue.

Consequently, German law expressly distinguishes between cargoes which are dangerous cargoes and cargoes which, although not dangerous, may cause damage or may be contraband. While the shipper is strictly liable for shipping dangerous goods without the consent of the carrier, in the case of the latter he is liable only for negligence.

---

[169] Rabe, *Seehandelsrecht* (2000), 440.
[170] Abraham, *Das Seerecht in der Bundesrepublik Deutschland* (1978) 1. Teil, 520.
[171] BGH, *VersR* 74, 771.

II. The Hague/Hague-Visby Rules Art. IV.3

Applying the above mentioned principle, under the Hague/Hague Visby Rules Art. IV.3 failing to inform about the shifting properties of an innocuous cargo, such as orange juice[172], known properties or actual state of bulk cargoes, such as humidity or heat, is a fault of the shipper. The shifting problem of a cargo of orange-juice packages does not make a cargo dangerous; the danger is rather in the failure to take necessary precautions to prevent shifting. Similarly, fault liability in a case of illegal, contraband cargo, for the reason explained above, is more just under Hague/ Hague-Visby Rules.[173]

It is clear that even in cases of dangerous cargoes, it would be possible to hold the shipper liable for his fault. Once again, as a policy matter, the division between liability arising from dangerous cargo and rather innocuous cargo but damaging cargo should be drawn. It would not be just to place the burden on the shipper for a innocuous but damaging cargo where the damage well may be created by the carrier's negligent handling. The carrier can still recover his damages but the only difference is that he must prove the shipper's negligence.

G. Liability of transferees of bills of lading

I. In general

The carriage of goods by sea is performed on the basis of a contract of carriage between the carrier and the consignor. In practice, a contract of carriage other than charter contracts is usually evidenced by a transport document issued after the goods have been delivered by the consignor to the carrier. Transport documents contain particulars about the parties to the contract of carriage, the goods and the terms and conditions of the carriage. They may serve as evidence not only of the contract of the carriage, but also of the receipt of goods. However, the role of transport documents can be more complex than simply acting as evidence.

---

[172] *Narcissus Shipping Corp. v. Armada Reefers, Ltd.* 950 F.Supp.1129, 1997 A.M.C 2499 (M.D.Fl 1997).

[173] In *Mitchell Cotts v. Steel,* [1916] 2 K.B. 610, where a cargo of rice could not be discharged without the permission of the government, it was held that this was analogous to one of shipping dangerous cargo. Even in this case there is place for the charterer's liability for negligence. Because at the time of shipment, the goods were fit for carriage without problems to their original destination, to Alexandria; but the destination was changed later to Piraeus. The charterers, but not the owners, were aware that rice could not be discharged at Piraeus without permission. It is obvious that the charterer is at fault for knowingly changing the destination. As mentioned above in such a case, under German law the charterer would be liable for his fault under § 564 for the incorrect particulars of cargo rather than § 564(b), that provides strict liability for dangerous cargo. By contrast, this case was considered to be analogous to a dangerous cargo case in English Law. See *supra* Part 2.

Some transport documents have the function of documents of title, which means that they are able to represent the goods and entitle their holders to demand delivery from the carrier. The phrase "document of title" is a common term used to denote documents issued by a carrier or by a warehouseman acting as bailee. The main purpose of documents of title is to facilitate the transfer of rights to goods while they are in the custody of a carrier or warehouseman.[174] The bill of lading is a typical document of title and is invested with particular attributes of great practical importance commercially. The bill of lading as a document of title can have several functions:

(a) the bill of lading represents the goods, so that possession of a bill of lading is equivalent to possession of goods
(b) under certain conditions, the transfer of the bill of lading may have the effect of transferring the property of the goods
(c) the lawful holder of a bill of lading is entitled to sue the carrier

The effect of the transfer of a bill of lading depends on the intention of the parties and on the applicable law.[175] The only right that is indisputably transfer by the endorsement of a bill of lading is the right to demand and have possession of the goods described in it. This right is guaranteed to a legal holder of the bill of lading in all jurisdictions. However, this is not the case in respect of the property and the transfer of contractual rights.

II. Transfer of contractual rights

1. Common law

Under common law in the past, contracts were not assignable, so that although the transfer of a bill of lading could affect a transfer of property in the goods, it did not transfer the rights and liabilities under the contract of carriage.[176] Consequently, in accordance with the doctrine of privity of contract,[177] the transferee of a bill of lading was not able to sue the carrier on the contract of carriage as he was not a party to it. This problem was partly solved by the Bills of Lading Act 1855 in English law. Section 1 sought to deal with this problem by providing that the transfer of a bill of lading has as a consequence the transfer of the contract of carriage. However, a transferee had a right of action under the bill of lading only if the property in the goods passed to him "upon or by reason of such consignment or endorsement".

This linking of contractual rights with the property has given rise to inconvenient and unfair consequences in cases when the property did not pass by consign-

---

[174] Treitel/Reynolds, *Carver on Bills of Lading* (2005), 267.
[175] Pejovic, "Documents of Title in Carriage of Goods by Sea: Present Status and Possible Future Directions" [2001] *J.B.L.* 461, 467.
[176] Treitel/Reynolds, *Carver on Bills of Lading* (2005), 167 ff.
[177] The doctrine of privity in contract provides that a contract cannot confer rights or impose obligations arising under it on any person or agent except the parties to it.

ment or by endorsement of the bill of lading.[178] In March 1991, the Law Commission reported to Parliament with proposals for the reform of the law relating to rights of suit in respect of the carriage of goods by sea. The report was the culmination of detailed research over the preceding five years, prompted by concern within the international commercial community that English Law was seriously defective: it failed to make provision for the passing of title in an undivided bulk and Sec. 1 of, Bills of Lading Act 1855 only enabled a consignee to assert rights under a bill of lading in circumstances where property in the goods had passed to him by reason of consignment or endorsement of the bill. These deficiencies were addressed in two pieces of legislation: first, the Carriage of Goods by Sea Act 1992 and secondly, the Sale of Goods (Amendment) Act 1995. The carriage of Goods by Sea Act 1992, by which the Bills of Lading Act 1855 was repealed, resolved the above-mentioned problems by abolishing the link between the passing of property and the passing of rights and liabilities.[179]

This problem does not seem to exist in the United States, where the Federal Bills of Lading Act 1916 (Pomerene Act) transfers contractual rights without the close link to the passing of property.[180]

2. Civil law

Under civil law, the transfer of a document of title does not only involve the transfer of rights to the goods but also the transfer of contractual rights. The transferee of a bill of lading therefore not only has the right to demand the goods from the

---

[178] Where property passed otherwise than upon or by reason of the consignment or endorsement of the bill of lading, either because it did not pass at all, although the buyer was at risk, or because the endorsee did not obtain full property in the goods, or because the transfer of the bill of lading was in no way instrumental in transferring property. The carrier is liable for performance of the contract of carriage to the person for whose account he performs the carriage and that is the last lawful holder of the bill of lading, i.e. the consignee. The right of action against the carrier is connected with the right to delivery of the goods. In the case of delay to, damage to or loss of the goods, it is logical that the person who is entitled to receive delivery is entitled to claim damages against the person who has made delivery. The right of the consignee, as the lawful holder of a bill of lading, to be delivered the goods includes the right to obtain compensation for damage, because it merely represents the value of the goods that the carrier failed to deliver. The consignee is entitled to claim damages against the carrier if the damage or loss occurred while the goods were in the custody of the carrier, regardless of whether he is the owner of the goods or whether he actually suffered the loss. The consignee's right of action is based on the contract of carriage and the property in the goods is not relevant for the relationship of carrier and consignee. Property may only be relevant to the relationship under the contract of sale between the consignee, as buyer, and the seller.

[179] This is achieved under Sec. 2 (1) by assignment of the right to sue the carrier to the lawful holder of a bill of lading, who may enforce the contract of carriage against the carrier irrespective of the passing of title to the goods.

[180] De Wit, *Multimodal Transport* (1995), 265.

carrier but also the right of action against the carrier for breach of contract.[181] This principle is often linked to the concept of contracting for the benefit of a third party, which is well established in all civil law systems. Under this doctrine, the parties to contract may agree that contractual rights can be transferred to a third party. The transferee of a bill of lading has not only rights, but also liabilities arising from a bill of lading, such as the obligation to pay freight. This is contrary to the nature of the contract for the benefit of a third party, which can transfer only rights and not liabilities.[182] However, the general answer for transport documents has been that no-one becomes liable merely by taking possession of a document conferring rights, but that the fulfilment of liabilities may be a condition for the exercise of the rights in the document.[183]

3. Liabilities of the transferee

a) English law

aa) The British Bills of Lading Act 1855

The remedy adopted by Sec. 1 of the Bills of Lading Act 1855 was to transfer rights of suit under the bill of lading contract to the consignee and to "every endorsee of a bill of lading to whom the property in the goods therein mentioned shall pass, upon or by reason of such consignment or endorsement". However, the Act did not impose liabilities on a holder as such. The 1855 Act was rather unclear both as to which liabilities were transferred and as to the continuing liability of a shipper after consignment or endorsement.[184]

Art. III.5[185] and IV.6[186] of the Hague/Hague-Visby Rules place specific liabilities on shippers and it might have been thought unlikely that these expressly des-

---

[181]   Pejovic, "Documents of Title in Carriage of Goods by Sea: Present Status and Possible Future Directions" [2001] *J.B.L.* 461, 464.

[182]   De Wit, *Multimodal Transport* (1995), 245.

[183]   Tiberg, "Legal Qualities of Transport Documents" [1998] 23 *Tul. Mar. L. J* 1, 3; Rabe, *Seehandelsrecht* (2000), 633 ff.

[184]   Gaskell/Asariotis/Baatz, *Bills of Lading: Law and Contracts* (2000), 131.

[185]   It was contended that the Bills of Lading Act 1855 was unaffected by the Carriage of Goods Act. In general, where the shipper's rights and liabilities in respect of the goods are transferred to a consignee under that Act, the effect will be that the word "shipper" in the Rules must be read as "consignee" since it is the contract "contained in the bill of lading" which is transferred and that contract incorporates the Rules. But Art. III.5 is carefully worded: "The shipper shall be deemed have guaranteed ..." The guarantee is hardly "contained" in the bill of lading; it is expressed rather as a recital of a (presumed) collateral agreement by the shipper and its effect is that the shipper and not the consignee is liable on it. Therefore, it seems that only the actual shipper can demand a bill of lading complying with Art. III.3, for his right to a bill of lading containing particulars furnished by him under that rule was clearly intended to be co-relative to this guarantee under Art. III.5. Colinvaux, *Carver Carriage by Sea* (1982) Vol. 1. 368; Treitel/Reynolds, *Carver on Bills of Lading* (2005), 585 f.

ignated convention liabilities would be transferred along with the bill as opposed to general contractual liabilities. However, in *The Giannis NK,* the House of Lords seems to have accepted that as the consignee or endorsee would be subject to the same liabilities as the shipper, the Hague/Hague-Visby Rules liabilities could be transferred.[187]

Sec. 2 of the Bills of Lading Act 1855 stated that nothing in the Act, i.e. the transference of liabilities under Sec. 1, should prejudice or affect ... any right to claim freight against the original shipper. It might have been inferred from the specific reference to freight that other liabilities of the shipper were prejudiced. But in *The Giannis NK* it was held that the words in Sec. 2 may have been inserted out of an abundance of caution to deal with an obvious liability and that a shipper was not divested of the obligations under Art. IV.6 by virtue of endorsement of the bill. Thus, there was no statutory innovation depriving the carrier of his rights against the shipper and substituting rights against an unknown receiver. That left open the point about whether the holder of the bill could have a concurrent liability. The Court of Appeal in *The Giannis NK* stated that the liabilities under the Bills of Lading Act 1855 Sec. 1 were concurrent, with both the shipper and consignee or endorsee sharing the same obligation in relation to them.[188] The House of Lords upheld this conclusion and emphasized that the consignee's or endorsee's liability was by way of addition, not substitution.[189] The conclusion may be correct in terms of general liabilities; however, there is still some doubt about how far it is also correct in respect of liabilities expressly designated under the Hague/Hague-Visby Rules as being on the shipper. Furthermore, as the decision in *The Giannis NK* was concerned only with the liability of the shipper, the conclusion on the liability of the subsequent holder of the bill is, strictly, obiter.

bb) The new British Carriage of Goods by Sea Act (COGSA) 1992

(1) The view of the Law Commission

The Law Commission considered carefully the arguments for and against subjecting the transferee of bill of lading rights to bill of lading obligations. Some of those consulted by the Commission thought that it might be unfair that the transferee of a bill of lading should be rendered liable for breaches of contract for

---

[186] It is argued that there is no equivalent in Art. IV.6 to the closing sentence of Rule 5 that reads "the right of the carrier ... to any person other than the carrier" nor is there anything elsewhere in the Hague-Visby Rules which implies that the liability created by Art. IV.6 is personal to the shipper and cannot be transferred to anybody else in any circumstances. Therefore, the mere use of the word "shipper" in Art. IV.6 does not mean that liability created by that Article cannot be transferred to or imposed on a third party. Mildoy/Scorey, "Liabilities of Transferees of Bills of Lading" [1992] *IJOSL* 94, 98.

[187] *The Giannis NK* [1998] 1 Lloyd's Rep. 337, 343.

[188] [1996] 1 Lloyd's Rep. 577, 586.

[189] [1998] 1 Lloyd's Rep. 337, 344 f.

which the transferee was not responsible.[190] On the other hand, there was attraction in the argument that a person who wished to take the benefit of rights under the bill should also carry the burdens. However, it was also felt that, in practice, bill of lading obligations would generally be discharged either by the shipper or the charterer and the problems with which the transferee might be faced were more apparent then real.[191]

The Law Commission came down firmly in favour of the imposition of bill of lading obligations on the transferee. They rejected the suggestion that a line either could or should be drawn between pre-and post-shipment obligations and liabilities. In particular, the Commission saw no reason why a transferee should not be subjected to the shipper's liabilities in relation to the shipment of dangerous cargo. The reasoning was explained as follows:

> It was suggested to us that special provision should be made so that the consignee or endorsee should never be liable in respect of loss or damage caused by the shipper's breach of warranty in respect of the shipment of dangerous cargo. This is said to be a particularly unfair example of a retrospective liability in respect of something for which the consignee/endorsee is not responsible. However, we have decided against such a special provision. We do not think that liability in respect of dangerous goods is necessarily more unfair than liability in respect of a range of other matters over which the holder of the bill of lading has no control and for which he is not responsible, for instance liability for loading port demurrage and dead freight.... It is unfair that the carrier should be denied redress against the endorsee of the bill who seeks to take the benefit of the contract of carriage without the corresponding burdens."[192]

(2) Transfer of liabilities under the COGSA 1992

Under the new law, where the holder of a bill of lading or any person entitled under a sea waybill or ship's delivery order takes or demands delivery of the goods, he is subject to liabilities as if he had been a party to the contract, without prejudice to the liabilities of the original contracting party.[193] The wording in the Act overcomes any privity objections by creating the fiction that the holder could be a party in certain circumstances.[194] In what sense is the holder deemed to be a party? If it had been an original party, it would presumably have been identified in the bill as a party. Thus it might be said that its liabilities would only be those that the terms of the bill directly imposed on it, as opposed to other parties to the bill, e.g. the shipper.[195] Some liabilities arising from the bill may not be directed at a

---

[190] *Rights of Suits in Respect of Carriage of Goods by Sea*, Law. Com. No. 196 para. 3.5 ff.
[191] *Ibid.* para. 3.13.
[192] *Ibid.* para. 3.22.
[193] COGSA 1992 sec.3 (1); Reynolds, "The Carriage of Goods by Sea Act 1992" [1993] *LMCLQ* 436, 443; Mildon/Scorey, "Liabilities of Transferees of Bills of Lading" [1999] *IJOSL* 94 ff. *Primetrade A.G. v. Ythlan Limited* [2005] EWHC 2399 (Comm), [2006] 1 All ER (Comm) 157 ("The Ythlan") provides guidance as to what will trigger potential liability for cargo interest under s.3(1); see also Baughen, "Sue and be Sued? Dangerous Cargo and the Claimant's Dilemma" [2006] 5(4) *S.TLI* 14, 16.
[194] Gaskell/Asariotis/Baatz, *Bills of Lading: Law and Contracts* (2000), 137.
[195] *Ibid.*

particular person and may be phrased sufficiently generally to include any relevant cargo interest that makes a claim.[196]

Liabilities which are imposed on a "shipper"[197] would not appear at first to extend to a consignee-holder or endorsee-holder. It has been held that "shipper" in Art. IV.3 and III.5 of the Hague/Hague-Visby Rules means only the shipper and not the person on whom liabilities are imposed by the 1992 Act.[198]

One difficulty with this conclusion is that the House of Lords in *The Giannis NK* seems to have concluded or assumed that Art. IV.6 obligations could be transferred by reason of the 1855 Act. It may be that the change in wording brought about by the 1992 Act means that it can still be argued that the Hague/Hague-Visby Rules obligations imposed specifically upon the shipper are not transferred, as all the 1992 Act does is remove the privity problem and not substitute "holder" for "shipper" for all purposes.[199] However, a holder might be caught by a bill which imposed obligations not on a "shipper" but on a "merchant", including holders, consignee etc.[200]

If the analysis is correct, the question may be not whether the consignee or endorsee is bound by obligations incurred after shipment or endorsement, but whether a holder is sufficiently identified in a bill so as to have liabilities imposed on it, whenever they arise. The Law Commission clearly contemplated that a lawful holder could be liable for pre-shipment liabilities.[201] They specifically declined to make a special provision exempting the lawful holder from potential liabilities

---

[196] Such as liability for freight might arise in a clause providing that "freight is payable…".

[197] By bill or by the Hague/Hague-Visby or Hamburg Rules.

[198] See *The Aegean Sea* [1998] 2 Lloyd's Rep. 39, 69. It is submitted that Art. III.5 and IV.6 seem to show an intention that the shipper only and not the consignee or endorsee should be liable under this guarantee, and the courts will probably give effect to this intention by holding a person on whom liabilities are imposed by Sec. 3 (1) of the COGSA 1992, although deemed to be a party to the contract of carriage, is not the "shipper" within this Rule. Colinvaux, *Carver Carriage by Sea* (1982) Vol. 1, 368; Boyd/Burrows/Foxton, *Scrutton on Charterparties* (1996), 434. For arguments contrary see Mildon/Scorey, "Liabilities of Transferees of Bills of Lading" [1999] *IJOSL* 94, 96 ff.

[199] Gaskell/Asariotis/Baatz, *Bills of Lading: Law and Contracts* (2000), 138.

[200] *Ibid.*

[201] However, it is said to be unfair that the final holder of the bill of lading should be liable in respect of such matters as the shipper's breach of warranty in shipping dangerous goods, demurrage incurred at the port of loading, dead freight and unpaid advance freight, as the consignee or endorsee often stands in no relation to the goods at the moment of shipment, and to make him liable in respect of pre-shipment liabilities is to make him subject to a retrospective liability for acts with which he had nothing to do. However true that might be, the Law Commission said that they did not think that a satisfactory line can be drawn at the moment of shipment with post-shipment liabilities being transferred but not pre-shipment liabilities. It seems odd to the Law Commission to say that fairness dictates that the holder should be liable for demurrage when these matters occur at the port of discharge but not at the port of loading. *Rights of Suits in Respect of Carriage of Goods by Sea* Law Com. No. 196 para.3.19 ff.

of the shipper arising from the "warranty" in respect of the shipment of dangerous cargo.[202]

(3) Liability of the original shipper

In the Court of Appeal in *The Giannis NK,* it was stated that the liabilities under the Bills of Lading Act 1855 S.1 were concurrent, with both the shipper and consignee or endorsee sharing the same obligation in relation to them.[203] The House of Lords agreed that the holder's liability was by way of addition rather than substitution.[204] This suggests that the carrier my have a choice about whether to sue the shipper or the holder.[205]

However, it is contended that the matter has not been decided conclusively against holders. The reference to "warranty" might mean the typical express clauses on dangerous cargo in bills of lading. In relation to these obligations, as mentioned above, the holder's liability may depend upon the definition clause allocating various rights and liabilities to "merchants". The better view is deemed to be that the Hague/Hague-Visby Rules Art. IV.6 allocates the responsibility solely to the original shipper.[206] Moreover, under Art. 17.1 of the Hamburg Rules, the shipper is expressly to remain liable to the carrier for inaccuracies in the bill even if it has been transferred.[207] Art. III.5 of the Hague/Hague-Visby Rules does not specifically mention the point, but it is submitted that the result is the same.[208]

On this issue, the Law Commission was of the opinion that under the Bills of Lading Act 1855 the original shipper remained liable under the bill of lading contract, despite rights acquired later by an endorsee.[209] In order to remove further doubts, the Law Commission recommended that the original liabilities of the shipper should continue.[210] The Law Commission noted that the continuing liabilities are those created by the contract, so it is still up to shippers to provide that their liabilities are to cease at a certain time, and cesser clauses are often included in charter-parties. As already mentioned, many modern combined bills of lading seek to impose wide liabilities on the "merchant" which may include an intermediate holder of a bill of lading. In some cases, the contract purports to retain their li-

---

[202] *Ibid.* para. 3.22.

[203] [1996] 1 Lloyd's Rep. 577.

[204] [1998] 1 Lloyd's Rep. 337.

[205] Gaskell/Asariotis/Baatz, *Bills of Lading: Law and Contracts* (2000), 139.

[206] *Ibid.*

[207] Hamburg Rules Art. 17.1 provides that "The shipper is deemed to have guaranteed to the carrier the accuracy of particulars relating to the general nature of the goods, their marks, number, weight and quantity as furnished by him for insertion in the bill of lading. The shipper must indemnify the carrier against the loss resulting from inaccuracies in such particulars. The shipper remains liable even if the of lading has been transferred by him…".

[208] See *The Aegean Sea* [1998] 2 Lloyd's Rep. 39, 69.

[209] *Rights of Suits in Respect of Carriage of Goods by Sea* Law Com. No. 196 para. 3.23 ff.

[210] *Ibid.*; Sec. 3 of COGSA 1992 now provides that "This section so far as it imposes liabilities under any contract on any person, shall be without prejudice to the liabilities under the contract of any person as an original party to the contract."

abilities even after the transfer of the bill of lading. The rights of an intermediate holder may be extinguished by S.2 (5) of the COGSA 1992, but the Act does not expressly deal with the extinction of liabilities. However it was noted in the Rights and Suits that on endorsement the intermediate holder loses both rights and liabilities.[211]

b) American law

The American legislation on bills of lading, the Federal Bill of Lading Act 1916 (Pomerene Act), deals only with the transfer of rights to the person to whom the bill is negotiated or transferred.[212] It does not impose liabilities on the transferee and protects the carrier only by giving him a lien for his charges.[213] Therefore, the transferee does not become liable by law. However, it might be that a consignee might be caught by the terms of the bill of lading.[214]

---

[211] *Rights of Suits in Respect of Carriage of Goods by Sea* Law Com. No. 196 para.2.40 f. It is also submitted that it is difficult to argue that the intermediate holder was in any sense an "original party" to the bill of lading within S.3 (3) of the COGSA 1992. Gaskell/Asariotis/Baatz, *Bills of Lading: Law and Contracts* (2000), 140.

[212] 49 U.S.C §§ 80105 (a)(2), 80106 (c); U.C.C. § 7-502(1)(d).

[213] 49 U.S.C. § 80109; U.C.C. s. 7-307. Treitel/Reynolds, *Carver on Bills of Lading* (2005), 174; De Wit, *Multimodal Transport* (1995), 265.

[214] *Carriage of goods by sea – Indemnity – Carrier sustaining loss as a result of hazardous cargo – Bills of lading providing for "Merchant" to indemnify carrier regardless of fault- Carrier bringing action in contract and in tort – Whether bill of lading violated COGSA – Whether carrier entitled to claim in negligence,* Case Comment [2007] 708 *LMLN* 3; "In admiralty consignees can be held to be parties to bill of lading under agency law." *United States v. Santa Clara* 887 F.Supp. 825, 836; according to 46 USCA § 40101(22) "The term 'shipper' means (A) cargo owner (B) the person for whose account the ocean transportation of cargo is provided (C) the person to whom delivery is to be made (D) a shippers' association or (E) a non-vessel-operating common carrier that accepts responsibility for payment of all charges applicable under the tariff or service contract." ; in *Senator Linie v. Sunway Line,* the Second Circuit upheld the imposition of liability for hazardous goods on a party that was named as both consignee and notify party in the bill of lading and which the District Court found to be a shipper. 291 F.3d 150, n.5; in *the APL Co. Pte. Ltd. V. UK Aerosols Ltd.,* 2006 WL 2792875 (N.D.Cal.), the plaintiff brought proceedings against the shipper and also two other parties who were endorsees of the bill of lading for damages caused by the shipment of hazardous cargo. The plaintiff alleged that all the defendants were "Merchants" within the meaning of the bill of lading which defined "Merchant" as a "Shipper, Consignee, Receiver of the Bill of Lading, Owner of the cargo or Person entitled to the possession of the cargo or having a present or future interest in the Goods and the servants and agents of any of these". The court found that endorsees are not shippers under the COGSA. However endorsees were found to as "Merchant" within the definition provided in the bill of lading; therefore were held to be obliged to indemnify plaintiff for the damage. 2007 WL 607902 (N.D.Cal.).

c) German law

Principally, in German law, the consignee is not liable by law.[215] However, the consignee might become liable if the bill of lading contains a provision for such liability. By taking the delivery of the goods vis-à-vis the carrier, he becomes subject to the provisions of the bill of lading.[216] The carrier, however, in according to the principle of good faith, is responsible for informing the consignee immediately regarding the damage arising from the cargo so that the consignee can decide if he will accept the cargo or not.[217]

H. Dangerous goods and charterparties

I. Applicability of the implied obligation not to ship dangerous goods without notice to charterparties in common law

The common law distinction between private and common carriers is not known in civil law. A common carrier is one "who exercises a public employment", "who undertakes for hire or reward to transport the goods of such who choose to employ him from place to place."[218] The implied obligation not to carry dangerous cargo without informing the carrier extends beyond the case of common carriage to the case of carriage in pursuance of a statutory duty. It was stated that this obligation is not confined to cases where the goods are tendered to a common carrier, but is capable of applying, in appropriate circumstances, to all contracts for the carriage of goods by sea.[219] Therefore even if the charterer is not the actual shipper of the cargo, the charterers will also be in breach of an undertaking implied at common law if they load dangerous cargoes without notice of their dangerous nature, unless the owners or their master knew or ought reasonably to have known of the danger they pose.[220] The implied duty is a means of risk allocation between ship-

---

[215] Abraham, *Das Seerecht in der Bundesrepublik Deutschland* (1978) 1. Teil, 527.

[216] *Ibid.;* Rabe, *Seehandelsrecht* (2000), 447. In a case where briquettes burned and damaged the ship the consignee became liable for damage by taking the delivery of the rest of the cargo t although damage was much higher than the value of the cargo. OLG Hansa 1965 S.309; MDR 1964 S.852; 1964 *VersR* 1171.

[217] *Ibid.;* Rabe, *ibid.*

[218] Chiang, "The Characterization of a Vessel as a Common or Private Carrier" [1985] 60 *Tul. L. Rev.* 299, 304 f. The distinction is considered not be useful in the law of responsibility under the contracts of carriage of goods by sea. Tetley, *Marine Cargo Claims* (1998), 35.

[219] *The Athanasia Comninos and Georges Chr. Lemos* [1990] 1 Lloyd's Rep. 277, 282.

[220] Wilford/Coghlin/Kimball, *Time Charters* (2003), 180; Gaskell, "Charterer's Liability to Shipowner, Orders, Indemnities and Vessel Damage", in Schelin (ed.) *Modern Law of Charterparties* (2003), 57. Gauci, "Risk Allocation in the Charterparty Relationship: an Analysis English Case Law relating to Cargo and Trading Restrictions" 28 *J. Mar. L. & Com.* 629. Konynhenburg, "Dangerous goods – Is the charterer liable when he is not the shipper?", <http://www.middletonpotts.co.uk/library/default.asp?p=93&c=414> (visited

per and carrier. Neither party may in fact be aware of any general or special characteristics rendering the cargo dangerous. The fact that the shipper had no opportunity to inspect the cargo is irrelevant. The duty is strict and absolute. The same logic applies to the relationship between charterer and owner, with the former bearing the loss.[221]

What if the Hague/Hague-Visby Rules are incorporated into charter-parties? For instance, in the NYPE form the common-law obligations of the charterers are supplemented by the provisions of the U.S. COGSA. § 1304(6) of COGSA, equivalent to Hague/Hague-Visby Rules Art. IV.6, falls into three parts. The first part allows the owners to land, destroy or render innocuous inflammable, explosive or dangerous goods to the shipment of which they have not consented with knowledge of their nature. The second part makes "the shipper" liable for all damage and expenses arising out of such shipment. The third part concerns shipment of such cargoes to which the owners have consented with knowledge of their nature but which become a danger to their ship or its cargo, and gives the owners the rights listed in the first part.

It is said that when properly incorporated, there seems no reason why the first and third parts of Art. IV.6 should not have full effect between owners and charterers when the Hague/Hague/Visby Rules are incorporated into charter-parties.[222] But it is not so clear that the second part should be equally effective, because this involves reading "shipper" as "charterer". However, it is contended that the second part of Sec. 4 (6) does have effect in that context.[223]

Generally, it is said that where the Hague/Hague-Visby Rules are incorporated into a charter, it is necessary to change the word "shipper" to "charter".[224] However, it is not apparent why this is necessary where there is an actual shipper.[225]

13.07.2007). However argument against that the implied duty applies to charterer is that the intermediate f.o.b. buyer chartering the vessel is unlikely to know the precise characteristics of the cargo and may not have an opportunity to inspect it prior to shipment. In contrast, the seller does have an opportunity. Since he is the party that usually enters into the contract of carriage with the carrier by procuring a bill of lading it is logical for the implied duty to bind him. Applying the same logic, the charterer who also ships the cargo as a c.i.f. seller, should also be bound; but if he does not, he should not be subject to the duty. However, it is not always the case that the party identified as the shipper on the bill of lading is the party that actually ships the cargo. The shipper may be intermediary purchasing the cargo from the producer who actually ships the cargo but is not identified as such on the bill of lading for commercial reasons. Indeed, this is generally the case where switch bills are issued. In such circumstances, the shipper is no more knowledgeable than the charterer. Nevertheless, English law still makes him liable under the implied duty. *Ibid.* As mentioned earlier in voyage charters generally cargo to be carried is agreed in the contract. By the same token it would be imprudent for a shipowner not inquiring particular state of the cargo which he agreed to carry. Therefore, strict liability of the charterer rarely arises in voyage charters.

[221] Konynbenburg, "Dangerous goods – Is the charterer liable when he is not the shipper?", <www.middletonpotts.co.uk/library/default.asp?p=93&c=414> (visited 13.07.2007).
[222] Or their enactments, such as US COGSA, U.K COGSA.
[223] Wilford/Coghlin/Kimball, *Time Charters* (2003), 181.
[224] *The Saxon Star* [1959] A.C. 133.

II. Identity of the shipper: charterer or physical shipper

In the legal framework of the carriage of goods by sea, mainly the carrier, shipper and consignee are involved. Under the Hague/Hague-Visby Rules, the carrier includes the owner or the charterer who enters into a contract of carriage with a shipper.[226] The Hague-Visby Rules do not define shipper; however, from the carrier's definition, shipper could be defined as a contractual person who enters into a contract with a carrier.[227] If the contract is a charterparty, there are carrier, charterer and consignee. Depending on the legal system, the charterer might also be the shipper.

Considering that the implied obligation not to ship dangerous goods in common law applies to charter-parties, it seems clear that if a charterer loads the cargo himself or is responsible for handing the cargo to the shipowner, he is a "physical" shipper of goods. And if he is a "physical' shipper of goods, i.e. in the sense that he loads the cargo himself or is responsible for handing the cargo to the shipowner, he can have the liabilities of a shipper at common law and under Art. IV.6 of the Hague/Hague-Visby Rules. However it is said that so long as an actual shipper can be identified, the shipper rather than the charterer should be made liable.[228]

There are number of reasons for the need to identify the shipper.[229] Considerable difficulties may arise in deciding who can actually be the shipper in the law governing carriage of goods. As a matter of principle, it would be quite understandable to have a wide definition of shipper for the purposes of dangerous goods obligations, while requiring positive evidence that a person agreed to be bound to a contract of carriage by allowing his name to be entered as "shipper" in the bill. The mere fact that a person's name appears in a bill does not mean that he is a party to a contract of carriage with a carrier, although that will be the normal in-

---

[225]   Gaskell, "Charterer's Liability to Shipowner, Orders, Indemnities and Vessel Damage", in Johan Schelin (ed.) *Modern Law of Charterparties* (2003), 57 fn. 250.

[226]   Art. 1.a.

[227]   The Rules do not really distinguish between the legal or physical shipper but the assumption of the Rules seems to be that a shipper must have a contractual relationship with the carrier. The *Pyrene v. Scindia* [1954] 2 Q.B. 402, involved a physical shipper who did not make the contract of carriage and would not have been named as shipper in the bill. The court created an implied contractual relationship between the carrier and the physical shipper, mainly as a mechanism to apply the Hague Rules regime to that shipper, e.g. in an action in tort. It was acknowledged that the physical shipper could not be sued for the freight. Does it follow that it could also not have been liable for shipping dangerous goods? It is to be noted that this case concerned a claim by the shipper against the carrier while the position in the dangerous goods scenario is the reverse.

[228]   Gaskell, "Charterer's Liability to Shipowner, Orders, Indemnities and Vessel Damage", in Schelin (ed.) *Modern Law of Charterparties* (2003), 57.

[229]   For instance, to find out who is the shipper liable for the carriage of dangerous goods, or who is the shipper liable for freight under the contract evidenced by the bill of lading.

ference.[230] It may be that the shipper box in a bill of lading is filled in merely to identify the person who happens to be physically delivering the goods to the ship for loading. This person might be a FOB seller in circumstances where the charterer has made a contract of carriage with a shipowner. The charterer cannot compel the shipper to become party to a contract of carriage by filling in his name on the bill. A party which does agree to be named as a shipper in a bill will therefore need to be aware that it might be accepting liabilities, e.g. for freight. Therefore, it may be necessary to distinguish between situations where the law is seeking to identify a person who is actually physically shipping goods and those where a person is held to be a party to the contract of carriage in a bill of lading.

In The *Athanasia Comninos*,[231] an FOB buyer (CEGB) made a contract of affreightment with a time charter but the bill of lading named the FOB seller (Devco) as "shipper" and CEGB as "consignee." It was considered that, as regards Devco, there is no room for doubt. They were shippers of the goods and were named as such in the bill of lading, without qualification.[232] As to the argument that it was CEGB that were principals to the contract of carriage, it was considered that this would not relieve Devco of liability, but would entail that both parties were principals vis-à-vis the shipowner.[233] Furthermore, the "only circumstance that would discharge Devco would be proof that they were named as participants in the contract without their consent."[234] In the light of the facts in the case, such an argument would have failed, as the bills of lading routed the goods through Devco for onward transmission to CEGB under the contract of sale and Devco did not object to their designation as shippers.[235]

---

[230] Gaskell/Asariotis/Baatz, *Bills of Lading: Law and Contracts* (2000), 42 f.
[231] [1990] 1 Lloyd's Rep. 277.
[232] *Ibid.* at 280.
[233] *Ibid.*
[234] *Ibid.*
[235] CEGB, as consignee, could not have been considered as a "shipper", as it already had the contract of carriage in the form of a voyage charter and only needed the bill of lading as a receipt. It is not clear if there were any relevant dangerous goods terms in this charter, but as it was a sub-charter, it would not have been directly relevant to the claim made by the shipowner in contract. No claims were pleaded in tort, although this may be a possibility, at least against the physical shipper. Any liability of the sub-charterer to the disponent owner, the time charter, would have been relevant to a recourse action by the latter if it had been liable to the shipowner under the time charter. In the Canadian case *The Roseline* a factual dispute as to liability for shipping dangerous goods related to effect of the spraying of sulphur, at the time of loading, with chlorinated (ordinary tap) water to reduce dust in the loading process. The buyer was a charterer and under its purchase contract the seller agreed to load and spout trim the cargo. The shipper-seller had an agreement with stevedores that the stevedores would spray the sulphur with "clean fresh sweet water" under which the stevedores accepted liability for loss or contamination of sulphur resulting from proven negligence or lack of responsible due diligence. A shipowner's charter party bill was issued which incorporated the terms of the voyage charter and named the seller as shipper. It was held that this named "shipper" was not a party to a carriage contract, even though it was named as shipper in the bill and had participated in supplying information for filling out

III. German law: shipper or actual shipper

1. Shipper and actual shipper distinguished

While three persons are involved in the legal framework of the carriage of goods by sea under the bill of lading in common law, German law knows more people: (1) the *Verfrachter* (carrier) and (2) the *Empfänger* (receiver), who cause no problems of definition; (3) the *Ablader* or person who delivers the goods to the carrier (the *Ablader* can be at the same time the receiver); (4) the *Befrachter* or person who contracts with the carrier (the *Befrachter* can be the *Ablader* or the receiver); and, in cases where the *Befrachter* does not deliver the goods to the carrier directly, (5) the *Drittablader*, who delivers the goods to the carrier instead. There is no exact English equivalent.[236] Befrachter can be said to be "shipper" and "Ablader" to be "actual shipper".

The *Ablader* is the person who acts in the name of the *Befrachter* (shipper) as an independent agent.[237] The internal relationship between *Befrachter* and *Ablader* with regard to the carriage contract is immaterial. It can be a charter, forwarding or purchase contract. The reason for the introduction of *Ablader* as a particular cargo interest in German law is based on the legal procedure involved in overseas purchasing. Under the F.O.B or F.A.S contract, the buyer concludes the contract of carriage and the seller brings the goods to the ship or on board the ship. In order for the seller to maintain right of disposal over the goods until payment is effected, the *Ablader* is solely entitled to demand the issuance of the bill of lading and to determine the receiver of the goods. In addition to German law, Turkish and Greek law recognize such a difference. However, with the exception of these laws, the difference between shipper and actual shipper is not recognized in other legal systems.

2. Liability of actual shipper

If the *Befrachter* (shipper) and *Ablader* (actual shipper) are different persons, the *Ablader* assumes the position of the *Befrachter* in the loading of the goods. In this regard without any exclusive authority he is entitled to all legal acts in relation to the shipment of goods. He has rights arising from the bill of lading, such as agreeing to deck carriage or having a claim to return of the goods before delivery as long as he has the order bill of lading in his hands.[238] He is, however, liable for the inaccuracy of the marks, number and quantity in the bill of lading according to

---

the commercial details in the bill. It had also used the bill as one of the documents to be delivered to the buyer's bank as part of the letter of credit transaction.

[236] Ashton, "A Comparative Analysis of the Legal Regulation of Carriage of Goods by Sea Under Bills of Lading in Germany, 1999 *MLAANZ* Vol. 14/2, <www.mlaanz.org/docs/99journal7a.html> (visited 9.01.2007).

[237] Rabe, *Seehandelsrecht* (2000), 303.

[238] *Ibid.* at 304.

§ 563[239] and for incorrect information with regard to the properties of goods according to § 564, under which contraband and dangerous goods fall.

§ 564b provides that "If inflammable, explosive or other dangerous goods are brought on board vessel without knowledge of the master....the *Befrachter* or *Ablader* is liable ... even if without fault." It is understood that this formulation would end "either or liability."[240] Each of these persons is strictly liable and this is considered to be joint and several liability.[241]

I. Extent of shipper's liability

I. English law

1. Liability for all damages and expenses "directly or indirectly" arising

Hague/Hague-Visby Rules Art. IV/6 provides that "... the shipper of such goods shall be liable for all damages and expenses *directly* or *indirectly* arising out of or resulting from such shipment...." Does the phrase "directly or indirectly" expose the shipper to a liability more than that which would apply in the case of an ordinary claim for damages for breach of contract or for a contractual indemnity? It is contended that the words did not affect the operation of the ordinary rule that, in the absence of a clear provision to the contrary, a person cannot enforce an indemnity where one of the effective causes of his loss is his own wrongful act.[242]

To compensate the plaintiff fully or for all that can, in some sense, be said to flow from a breach of contract would often lead to undesirable results. Therefore, the law has developed a number of rules for the purpose of limiting damages for breach of contract. Accordingly, no loss may be recovered by way of damages if it is a too remote consequence of the breach. In contracts, the test of remoteness is whether the loss in question was fairly and reasonably within the contemplation of the parties, as a probable result of the breach, as at the date of the contract.[243]

---

[239] Equivalent of Art. III.5 of Hague/Hague-Visby Rules.

[240] Gündisch, *Die Absenderhaftung im Land- und Seetransportrecht* (1999), 198 f.

[241] *Ibid.*

[242] Cooke/Young/Taylor, *Voyage Charters* (2001), 1010.

[243] The well-known statement of principle in the leading case of *Hadley v. Baxendale* (1854) 9 Ex. 341, 354, was generally considered to have embraced two rules: the first rule being concerned with the recovery of damages within the contemplation of the parties at the time the contract was made, a rule which was understood to be dependent on the knowledge of special circumstances and to require that those circumstances had been made known to the party who had broken the contract before or at the time the contract was made. Where it was the carrier who had broken the contract by failing to deliver the goods or by delaying their delivery, the application of these rules had prevented the owner of the goods from recovering as damages for breach of contract the loss of the market value of the goods. Cooke/Young/Taylor, *Voyage Charters* (2001), 564.

2. Causation or remoteness of damage?

In the *Fiona*[244] case, two submissions were made on the meaning of "directly or indirectly". The first submission was that it was sufficient to enable the claimant to hold that the shipment of fuel oil, i.e. dangerous cargo, had been a cause of loss.[245] The right to an indemnity was not limited to a situation where the shipment of that cargo was the dominant or proximate cause of the loss.[246] The contrary view was that the words "directly and indirectly" were introduced to render items of loss recoverable which might otherwise have been regarded as too remote to be recoverable.[247]

The words "directly or indirectly" are not found elsewhere in the Hague-Visby Rules.[248] A clue to the legislative purpose underlying "directly or indirectly" may lie in the fact that Art. IV.6 authorizes the carrier to land and destroy dangerous goods without incurring liability to the shipper. One possibility, therefore, was thought that those who drafted the rules were concerned with providing that a loss of profit or loss of market sustained by the carrier through the detention of his ship while the goods were landed should constitute a recoverable item of loss.[249] Prior to 1924 there had been numerous cases in the field of the carriage of goods in which a limiting principle had been applied to preclude the recovery of losses which might be considered to have resulted only remotely from a breach of contract. It may be that the legislative purpose of including the word "indirectly" in Art. IV.6 was to ensure that all losses incurred by a carrier through the detention of his ship as a result of the shipment of a dangerous cargo should be held to be recoverable. That this may have been the purpose of "directly or indirectly" receives some support from the terms of another international shipping rule, namely Rule C of the York-Antwerp Rules, 1924. Rule C provided that:

> Only such losses damages or expenses which are the direct consequence of the general average act shall be allowed as general average. Loss or damage sustained by the ship or cargo through delay, whether on the voyage or subsequently, such as demurrage, and any indirect loss whatsoever, such as loss of market, shall not be admitted as general average.

It may be significant that Rule C gives "loss of market" as a typical example of "indirect loss". In *Fiona* it was opined that it may not be right to construe Art. IV.6 by reference to English rules of remoteness of damage as they existed in 1924.[250] The words "directly and indirectly" are quite general and are just as

---

[244] [1993] 1 Lloyd's Rep. 257
[245] *Ibid.* at 286.
[246] *Ibid.*
[247] *Ibid.*
[248] Art. III.5 confers on the carrier a right of indemnity against loss arising or resulting from inaccuracies in the particulars of the goods furnished by the shipper. That indemnity does not contain the words quoted.
[249] *The Fiona* [1993] 1 Lloyd's Rep. 257, 286.
[250] *Ibid.* at 287.

applicable to causation as to remoteness.[251] Statutes enacted to give legal force to private law conventions should be construed in their international context without undue reference to the pre-existing law of any contracting state and the words used in such conventions should be interpreted "on broad principles" of general acceptance.[252]

The concept of proximate cause has played little part in the law of carriage by sea; it is rather found in the law of insurance.[253] It was said that the fact that the rule renders the shipper liable for all damage "whether directly or indirectly arising" makes it quite clear that indemnity is not limited to situations where the shipment of dangerous goods is the proximate or dominant cause of the carrier's loss.[254]

On the other hand, the wording "...directly and indirectly..." with regard to the shipper's liability in the shipment of dangerous goods had been included in the CMI/UNCITRAL draft instrument.[255] In principle, the Hague/Hague-Visby Rules govern the context of the contractual relationship between the carrier and shipper. However, if the carrier paid out a claim as a result of injury caused by the negligence of the shipper, the carrier should be able to claim compensation from the shipper. In the draft instrument, the words "directly and indirectly" were intended to mention such loss and damage.[256] However, it was suggested that the words "directly and indirectly" could interfere with issues of causation and so it was decided to delete them.[257]

II. German law

The words "directly and directly" are not included in HGB § 564b, which is equivalent to Art. IV.6 of the Hague/Hague-Visby Rules. The shipper and actual

---

[251] *Ibid.*

[252] *Ibid.*

[253] Colinvaux, *Carver Carriage by Sea* (1982) Vol. 1, 108.

[254] However, it should not be forgotten that where the facts disclose that the loss was caused by the concurrent causative effects of an excepted and non-excepted peril, the carrier remains liable. He can only escape liability to the extent that he can prove that the loss or damage was caused by the excepted peril. See *infra* p. 189 ff. In *United States v. M/V Santa Clara* 887 F.Supp.825, an action for loss overboard of containers of arsenic trioxide, as well as spill on board the vessel of magnesium phosphide, was brought by the vessel owner and operator against the shippers and consignee of those chemicals, seeking contribution and indemnification pursuant to bills of lading and under the Comprehensive Environmental Response, Compensation, and Liability Act (CERCLA). It was held that although shipper's failure to properly label magnesium phosphide as hazardous cargo breached bill of lading, that breach did not render the shipper and consignee liable for all damages associated with magnesium phosphide spill, as such damages, were not foreseeable at the time of contract, considering remoteness in time and number of intervening events. *Ibid.* at 835.

[255] A/CN.9/WG.III/WP.56, 31.

[256] A/CN.9/591, 38 f., 44.

[257] *Ibid.* at 44.

shipper are, in general, liable for all damages and expenses, such as damage to the ship, expenses for reloading, transhipment, delay, inspection and cleaning and for compensation paid to the other interested parties, e.g. crew or other cargo interests.[258]

J. The carrier's rights

As seen so far, the carrier's rights for compensation from the shipper differ according to whether the shipment of dangerous goods was performed with or without consent and knowledge as described before. The following account is founded on the assumption that the carrier has not materially broken his obligations under Article III of the Hague/Hague-Visby Rules.[259]

I. Immunity

In common law, one of the exceptions to the strict liability of the carrier is "jettison". Where goods have been intentionally and properly destroyed during the course of a voyage, in order to save the ship and the remainder of the cargo from a danger which was common to the whole, the shipper is not answerable.[260] Where the goods have been thus sacrificed, the law, except in certain cases, gives the owner a right to contribution towards his loss from those whose property was saved. But otherwise no claim for the goods or their value can be made against the carrier.

Based on this principle, irrespective of whether the carrier has consented to the carriage of the goods with knowledge of their nature and character, he has the right to "land, destroy or render the goods innocuous without compensation", i.e. in the sense of without paying compensation under Art. IV.6 of the Rules.[261] The carrier's immunity extends to claims by any party to the bill of lading contract or in common law, a bailor on terms of the bill of lading and the charterer where appropriate.

Likewise in German law, HGB § 564b provides that the master has authority to land, destroy or render innocuous at any time and any place and thus corresponds exactly to this wording of the Hague/Hague-Visby Rules.[262] Whether the carrier or master has consented or not, neither the carrier nor the master is liable to pay compensation.[263]

---

[258] Rabe, *Seehandelsrecht* (2000), 441.
[259] Corresponds to HGB §§ 559 and §§ 606.
[260] Colinvaux, *Carver Carriage by Sea* (1982) Vol. 1, 18.
[261] Cooke/Young/Taylor, *Voyage Charters* (2001), 1009.
[262] Rabe, *Seehandelsrecht* (2000), 447.
[263] HGB § 564b (2). "... Even in this case the carrier and the master is not liable to pay damages. ...".

Furthermore, it should be remembered that insofar as "fire" is involved, the carrier has the important and wide-ranging defence under Art. IV.2 (b) of the Hague/Hague-Visby Rules.[264]

It is to be noted that damage to or loss of goods caused by other reasons than exercising the right of landing, destroying or rendering the goods innocuous could amount to a distinct breach of the carrier's duties to care for goods under the Rules. Therefore, the carrier could not be exonerated from liability.[265]

II. The exercise of the right of landing, destroying or rendering the goods innocuous

The rules governing the carrier's right to render innocuous dangerous cargo which threatens the means of transport or other interests on board are not always appended to the liability rules and may have a life of their own. Art. IV.6 of the Hague/Hague-Visby Rules gives the right to the carrier to land, destroy or render the goods innocuous. The powers of neutralizing the dangers are wide-ranging to cover every possible circumstance and therefore should not present any grounds for litigation.[266]

As the main concern underlying the right of disposal is the safety of life and property at sea, it has been asserted that, even where the carrier is in breach of his obligation as to the seaworthiness and care of the cargo, the carrier's right to land, destroy, or render the goods innocuous remains unimpaired.[267] Such a right does not, however, exist when the cargo merely poses a risk of delay or detention to the ship, as it is considered that non-physical damage in the form of delay or detention is not covered by the Rules.

This right raises the question of whether the carrier is bound to act reasonably in exercising these rights. It is suggested that the carrier is so bound to both the choice between the various possible steps and to the carrying out of those steps.[268] Accordingly, if it is possible to render innocuous the cargo, that should be tried first rather than destroying it. The steps taken by the carrier should be reasonably commensurate with the danger posed to the ship and adventure and with the requirements of preserving them, but on the other hand, the standard of reasonableness should be applied with some latitude to the carrier, bearing in mind that he will normally be faced with the need to decide upon his course of action as a matter of urgency and on the basis of incomplete information.[269]

---

[264] Art. IV.2(b) "Neither carrier nor the ship shall be responsible for loss or damages arising or resulting from fire unless caused by the actual fault or privity of the carrier". HGB § 607(2).

[265] Hague/Hague-Visby Rules Art. III.1 and 2; HGB §§ 606 ff.

[266] Abdul Hamid, *Loss or Damage from the Shipment of Goods, Rights and Liabilities of the Parties to the Maritime Adventure* (Diss. Southampton 1996), 238.

[267] *Ibid.* at 239.

[268] Cooke/Young/Taylor, *Voyage Charters* (2001), 1011.

[269] *Ibid.*

1. Carrier has not consented to the shipment

The first paragraph, which applies when the carrier has not consented to the shipment, imposes no express limitation on the exercise of the rights. The carrier need not demonstrate that the goods have become a danger to ship and cargo. The reasoning is, presumably, that the carrier is not required to run a risk to which he never consented. But the absence of any express constraint makes it all the more necessary to imply some obligation, however limited, to act reasonably. Thus, if there is no reason why the goods should not be safely carried, it is contended that the carrier would not be justified in exercising his right under the first paragraph; on the other hand, if the carriage presents real risks, the carrier should not be required to wait until the danger is imminent before exercising his rights.[270]

2. Carrier has consented to the shipment

Where the carrier has consented to the shipment with the relevant knowledge, the second paragraph gives the same rights "in like manner", but subject to the qualification that the goods shall have become a danger to the ship and cargo.[271] In such a case, the carrier must clearly be under an obligation to act reasonably, given his overriding obligations as a bailee of the goods, and the fact that he or his agent had given consent to the shipment of the goods in the knowledge that their carriage gave rise to potential dangers.

3. Unreasonable acts of carrier

If the carrier acts unreasonably in his treatment of the goods, it is submitted that that would break the causative chain between the shipment and any resulting damage in accordance with the general principle could amount to a distinct breach of the carrier's duty to look after the goods, whether simply as bailee or under Art. III.2 of the Hague/Visby Rules.[272] As mentioned above, the loss of or damage to dangerous goods by reasons other than landing, destroying or rendering innocuous would also make the carrier liable.

III. Sacrifice of dangerous goods

1. In general

When the dangerous properties of a particular cargo become manifest and pose a danger to the ship or life and property on board, the master may take necessary measures, such as destroying or jettisoning dangerous goods, to abate the danger. Does the shipper of such a cargo which has been jettisoned or destroyed have a claim for a general average sacrifice against the carrier and other cargo?

---

[270] *Ibid.*
[271] *Ibid.*; Rabe, *Seehandelsrecht* (2000), 447.
[272] *Ibid.;* HGB § 606.

The principle which underlies general average is that where any property at risk in a maritime adventure is sacrificed, or where extraordinary expenditure is incurred for the common safety, the owners of any of the property at risk which completes the adventure safely should contribute to the loss or expense, in proportion to the values of their property which has survived:[273]

> There is a general average act when, and only when, any extraordinary sacrifice or expenditure is intentionally and reasonably made or incurred for the common safety for the purpose of preserving from peril the property imperiled in the common adventure.

The desire for uniformity, and the problems of identifying the correct governing law when the adventure ends at more than one discharging port, have led to the international development of a set of voluntary rules, the York-Antwerp Rules, to govern the adjustment of general average between the parties to the adventure, and these Rules are almost invariably incorporated contractually into charterparties and bills of lading and frequently into polices of marine insurance. The York-Antwerp Rules have been regularly revised, most recently in 2004.

2. Effect of fault

An exception to the rule of contribution in general average is also very closely related to dangerous cargoes: if the necessity for a general average act arose as a result of the fault of one of the parties to the adventure, the act retains its general average character and contribution is due from the parties to the adventure, subject to the important exception that the party at fault is not entitled to recover contribution from any other at whose suit the fault was actionable at the time at which the sacrifice or expenditure was made or incurred.[274] The justification for this exception has been attributed to the policy of avoiding circuity of action and to the principle that a person shall not recover from any other person in respect of the consequences of his own wrong.[275]

3. General average resulting from inherent vice or fault of the good

When a claim in general average is resisted on the ground of fault, it is usually the fault of the shipowner which is at issue, but this is not always so. A cargo which has ignited spontaneously may be jettisoned or damaged by firefighting operations and the shipowner and other cargo owners may seek to defend a claim for contribution by the owner of the sacrificed cargo on the ground that the sacrifice arose as a result of the fault of the cargo itself. Earlier it was sometimes advanced that in such a case contribution could not be recovered in respect of sacrifice of cargo if it was the inherent vice of the cargo itself which had caused the danger. However,

---

[273] Wilson/Cooke, *General Average and York-Antwerp Rules* (1997), 1.
[274] *Ibid.* at 154.
[275] *Ibid.*

this view has been said to involve a fallacy.[276] It is now settled that exactly the same principle applies in this case as in any other.[277]

In *Greenshields, Cowie & v. Stephen&Sons Ltd.*,[278] a cargo of coal ignited spontaneously as a result of inherent vice. The cargo was sacrificed and the ship-owner, in defence to a claim for general average contribution, relied on the implied exception of inherent vice in the contract of carriage and the statutory exception of fire under Sec. 502 of the Merchant Shipping Act 1894. These arguments were rejected, since the implied exception, while providing a defence to a claim for damage or non-delivery, had no relevance to a claim for contribution and the statute does not deal with general average at all.[279] The result would have been different if there had been a breach of contract in the shipping of the coal, since the loss would then have arisen as a result of the actionable fault of the cargo-owner who was claiming contribution. However, the shipowner had agreed to carry the coal with full knowledge of its nature. Thus, contribution can be recovered unless the danger arose as a result of actionable fault on the part of the owner of the cargo sacrificed. It is no defence to the claim merely to show that the damage arose from the inherent vice of the cargo.

4. General average contribution and dangerous goods

Where the carrier exercises his right to land, destroy or render innocuous the goods and the goods were loaded with his knowledge and consent, the cargo-owner's rights in general average are expressly preserved by Hague/Hague-Visby Rules Art. IV.6:

> If any such goods shipped with such knowledge and consent shall become a danger to the ship or cargo, they may in like manner be landed at any place, or destroyed or rendered innocuous by the carrier without liability on the part of the carrier, except to general average, if any.

It is therefore clear that, where the goods were shipped without the requisite knowledge and consent on the part of the carrier, the rule is intended to exclude any claim for general average contribution from the carrier.[280] The reason is that such conduct on the part of the shipper is treated as equivalent to actionable fault, which provides a defence to a claim for contribution.[281]

Whether or not the shipper could claim contribution from parties to adventure other then the carrier, such as other cargo-owners, depends upon whether the fault is actionable at their suit. There are good reasons for holding that a shipper of dangerous goods without the knowledge and consent of the carrier is in breach of his duty of care to other cargo- owners, at any rate if he knows or ought to have

---

[276] *Pirie v. Middle Dock Co.* (1881) 4 Asp.M.L.C. 388, 391.
[277] Wilson/Cooke, *General Average and York-Antwerp Rules* (1997), 163.
[278] [1908] A.C. 431.
[279] *Ibid.* at 436.
[280] Sturley, *Legislative History of the Carriage of Goods by Sea Act* (1990) Vol. 2, 511.
[281] Cooke/Young/Taylor, *Voyage Charters* (2001), 1011.

known of the danger. In such circumstances, therefore, it is contended that other cargo-owners would have a defence to the claim.[282]

5. German law

HGB regulates general average under §§ 700 etc. According to § 702, fault of the parties does not bar general average. However, the party whose fault caused the general average, e.g. the shipper or actual shipper, cannot claim contribution while he remains liable to make a contribution.

Generally speaking, although "the provision concerning distribution in case of general average remains unaffected" stands in the second paragraph of § 564b, it is also applicable in the case where dangerous cargo is carried with the consent of the carrier or master.[283] As in English law, in this case the shipper or "actual shipper" cannot claim contribution, as shipping dangerous goods without the consent of the carrier is considered equal to "fault", which deprives him of making a claim for contribution.[284]

K. Allocation of liability in concurrent causes

In the usual case where the carrier is being sued for loss of or damage to the goods once a cargo claimant has established a loss, the carrier has the burden of proving what caused the loss or damage and that such cause was an excepted peril.[285] In the absence of sufficient evidence about the cause(s) of loss, the carrier will be responsible for the (whole) loss. The carrier is therefore generally liable in cases of unexplained losses. In cases where there is combination of causes, the carrier is liable for the whole loss, unless it can prove the extent to which a quantifiable proportion of the loss was solely due to a cause for which he is not responsible.[286] In other words if the goods owner succeeds in proving unseaworthiness and that the unseaworthiness has caused the loss and if the carrier cannot prove that the unseaworthiness was not caused by want of due diligence within Art. IV.1, the carrier is unable to rely on the exceptional peril and so is liable for the whole loss or damage. This has long been the interpretation in English law.[287]

Likewise, in American Law in the case of a failure of due diligence and concurrent causation, the carrier is liable for the whole loss under the rule of the

---

[282] *Ibid.* at 1012.

[283] Abraham, *Das Seerecht in der Bundesrepublik Deutschland* (1978) 1. Teil, 527; Rabe, *Seehandelsrecht* (2000), 448.

[284] Abraham, *Das Seerecht in der Bundesrepublik Deutschland* (1978) 1. Teil, 527.

[285] Tetley, *Marine Cargo Claims* (1988), 361 ff.

[286] UNCTAD, Carrier Liability and Freedom of Contract under the UNCITRAL Draft Instrument on the Carriage of Goods [wholly or partly] [by sea], Note by the UNCTAD Secretariat, UNCTAD/SDTE/TLB/2004/2, 11.

[287] *The Fiona* [1993] 1 Lloyd's Rep. 257, 288.

Vallescura,[288] unless he can meet the burden of apportioning the loss between causes. In other words, where concurrent causes of the loss are at least possible, the burden is on the carrier to prove, if possible, the proportion of the damage attributable to his own negligence and the proportion attributable to an excepted cause. If the carrier is unable to establish the portion of the damage attributable to the excepted cause alone, the carrier must bear the entire loss.[289]

Although it has been said that the shipper is strictly liable for losses due to dangerous cargo shipped without the carrier's knowing consent, it does not necessarily follow that a shipper will in fact be liable in all such cases. It is often the case in practice, where a loss is due to a combination of causes, e.g. a breach of the carrier's obligation has contributed to the loss in question, the carrier may not be able to succeed in a claim against the shipper.[290] Thus the rule on the allocation of liability in cases of concurrent causes is significant to the risk distribution between carrier and cargo interests.

I. English law

1. Breach of Art. III.1: unseaworthiness contributing to damage

In English law, a claim by the carrier under the dangerous goods provision will be defeated if the carrier is in breach of his overriding obligation "to exercise due diligence to make the ship seaworthy and fit and safe to receive the cargo," even if this breach is only partly to blame for the damage resulting also from the shipment of the dangerous cargo.

In the *Fiona*,[291] there was an explosion in a cargo tank of the vessel when a surveyor inserted into it an unearthed temperature probe. Unknown to the surveyor, explosive gases had built up in the tank. These gases came partly from a cargo of fuel oil and partly from residues of a previous cargo of natural gas condensate which the shipowners had failed to wash away before loading the fuel oil. The shipowners claimed indemnification from the shippers of the fuel oil in respect of loss and damage caused by the explosion. It was found that the shipment of this particular fuel oil, which had emitted flammable vapours in a manner that the master had no reason to expect from this type of cargo, and the failure to wash away residues of the earlier cargo were both contributory causes of the explosion. It was held that the shipowners were therefore precluded from recovering an indemnity from the shippers under Art. IV.6, because they had failed to discharge their overriding obligation under Art. III.1.[292]

---

[288]   *The Vallescura* 293 U.S. 296.

[289]   Schoenbaum, *Admiralty and Maritime Law* (2004), Vol. 1 678 f.

[290]   Asariotis, "Main Obligations and Liabilities of the Shipper" [2004] *TranspR* 284, 285.

[291]   *The Fiona* [1994] 2 Lloyd's Rep. 506.

[292]   *Ibid.* at 519. In the first instance, it was held that it would be contrary to the scheme of the Rules and likewise inconsistent with equity and commercial common sense that a carrier should be entitled to destroy dangerous goods without compensation and without liability except to general average if the cause of the goods having to be destroyed

In other words, where unseaworthiness is at least a necessary contributory factor to the loss, the carrier will have to bear the whole responsibility in the absence of evidence identifying dangerous goods as the sole cause of part of the loss.[293] However, it should be noted that where the seaworthiness obligation is in no way related to the loss, the carrier would still be able to claim indemnity under Art. IV.6 of the Hague/Hague-Visby Rules.[294]

2. Extent of contribution of unseaworthiness

The facts of a particular case may allow a distinction to be drawn between damage to which the owner's breach of seaworthiness did contribute and damage to which their breach did not contribute, so as to preserve their indemnity in respect of the dangerous goods provision Art. IV.6.[295]

In *Northern Shipping Co. Deutsche Seerederei G.m.b.H (the Kapitan Sakharov)*,[296] containers were loaded under deck, including eight tank containers of isopentane, in spaces lacking adequate ventilation for such highly flammable liquid. The ship also loaded containers on deck. At sea there was an explosion among the containers on deck, causing a fire which first caused damage on deck and then spread below deck, into the space partly occupied by the isopentane containers. It became intense and eventually led to the sinking of the ship. It was found that the initial explosion had been within a sealed container of undeclared dangerous cargo. It was also found that the consequent fire would not have led to the sinking of the ship had her owners not negligently stowed the isopentane containers where they did, in breach of their obligation under Article III.1. It was therefore held that the owners could recover under Art. IV.6 from the shippers of the container of undeclared dangerous cargo in respect of the on-deck damage, but not in respect of damage caused and liabilities incurred in the subsequent under-deck fire and eventual sinking of the ship with her remaining cargo, to all of which their breach of Art. III.1 had contributed. The carrier's breach of Art. III.1 did not cause the initial explosion nor the loss of the surrounding containers, those losses resulting from the shipment of the dangerous cargo. The loss of vessel and other containers resulted from a combination of the shipment of the dangerous cargo and the carrier's want of due diligence, neither being a dominant cause, but that was irrelevant so long as the carrier's breach was an effective cause of loss.

Following *Fiona* and *Kapitan Sakharov,* the law can be summarized that where both a breach of the carrier's seaworthiness obligation and a breach of the shipper's obligation relating to the shipment of dangerous goods contribute to a loss, the shipowner will only be entitled to rely on the indemnity provided in Art. IV.6 to the extent that damage is clearly attributable to the dangerous nature of the

---

was a breach by the carrier of his obligations as to seaworthiness. [1993] 1 Lloyd's Rep.257, 286.

[293] Gaskell/Asariotis/Baatz, *Bills of Lading: Law and Contracts* (2000), 476.

[294] *Ibid.* at 475.

[295] Wilford/Coghlin/Kimball, *Time Charters* (2003), 182.

[296] [2000] 2 Lloyd's Rep. 255.

goods. Where the unseaworthiness was at least a necessary contributory factor to the loss, the carrier will have to bear the whole responsibility in the absence of evidence identifying dangerous goods as the sole cause for part of the loss.[297] In other words, the carrier would only be able to claim an indemnity from the shipper if and to the extent that he could prove the proportion of loss due to causes attributable to the shipper. In the absence of sufficient relevant evidence, the carrier would not be able to claim an indemnity from the shipper.

In effect, the approach is the same as the one relevant in relation to the list of excepted perils in Art. IV.2, i.e. the carrier is presumed to be at fault and the extent to which he caused such loss.[298] Consequently, where loss or damage is caused by a combination of the shipment of dangerous goods without the required knowledge and consent of the carrier and breach by the carrier of his obligations under Art. III.1, the loss will lie where it falls.[299]

3. Breach of Art. III.2: failure to properly load, handle and care for goods

No case has been decided where the shipowner's breach is only of a duty under Art. III.2 which is not an "overriding obligation" in the same way as the duties under Art. III.1.[300] Therefore, the argument to support the carrier's immunity is stronger where he has been in breach of only Art. III.2, which is expressly "subject to" Art. IV.

It is important here to distinguish those cases where the failure properly and carefully to look after the goods arises from a lack of information about their nature and character from those cases where the failure is independent of the carrier's ignorance of the true nature and character of the goods. That rule requires the carrier to act properly and carefully, which must mean with appropriate "informed" care and, if his want of informed consent is the cause of his failure to act properly and carefully, then it is asserted that he should not be deprived of his rights under Art. IV.6.[301] However, regarding damage resulting from his own independent or informed failure properly and carefully to care for goods,[302] he

---

[297]   Gaskell/Asariotis/Baatz, *Bills of Lading: Law and Contracts* (2000), 476.

[298]   *Ibid*.

[299]   Cooke/Taylor/Young, *Voyage Charters* (2001), 1014. This reflects the principle accepted in *Government of Ceylon v. Chandris* [1965] 2 Lloyd's Rep.204, 216, where it was held that "if part of the damage is shown due to a breach of contract by the claimants, then the general rule that the burden of proof lies on the person claiming relief applies and the claimant must show how much of the damage was caused otherwise than by his breach of contract failing which he can recover nominal damages only."

[300]   Cooke/Young/Taylor, *Voyage Charters* (2001), 1013.

[301]   *Ibid.*

[302]   Note that in certain cases the duty requires the carrier to disregard instructions from the shippers as to the stowage or carriage of the goods.

would be deprived of his right to indemnity on the grounds that his own wrongful act was an effective cause of his loss.[303]

For instance, a seaworthy vessel is carrying two reactive chemicals whose nature and character are not disclosed and, by an inadvertent pumping error during discharge, i.e. a breach of Art. III.2, the two are brought together and cause an explosion. The ensuing damage would, in one sense, have resulted from the shipment of the dangerous goods and in another from the interaction of that shipment and the breach of Art. III.2 in failing simply to keep two distinct parcels separate, but it may be that, if the carrier had had knowledge of the true nature and character of the goods, he would have taken special precautions to prevent a mixing of the two. If special precautions had been taken which would probably have averted the mixing, it is asserted that the breach of Art. III.2 should not preclude recovery under Rule IV.6, because it could fairly be said that the operative cause was the failure to inform, not the consequential loss of care.[304] However, where the failure to inform has no causative impact on the carrier's conduct, it is asserted that the carrier would fail in his claim because he would be seeking indemnity in respect of his own negligence.[305]

4. The Law Reform (Contributory Negligence Act) 1945

At common law, the court was not entitled to apportion damages claimed for breach of contract based upon the plaintiff's contribution to his or her own losses. Thus, even if the plaintiff was partially responsible for the loss, the defendant was held totally liable if found to have caused a portion of the loss. Legislation in the field of tort law was enacted to permit the court to apportion damages reflecting the "contributory negligence" of the plaintiff.[306]

In England, the Law Reform (Contributory Negligence) Act 1945 removed the defence but permitted the tribunal determining a claim where the claimant was partly to blame for his loss to reduce the damages recoverable to reflect his blameworthiness.[307] The Court of Appeal decisions above preclude the possibility of an apportionment of damages where the carrier has committed a causative breach of Art. III.1, but that possibility is not necessarily extinguished by those decisions in the case of a breach of Art. III.2. It is however, difficult to utilize the jurisdiction of the court or arbitrators to apportion damages in the above cases under Sec. 1(1) of the Act that provides that:

> Where a person suffers damage as the result partly of his own fault and partly of another person or persons, a claim in respect of that damage shall not be defeated by reason of the fault of the person suffering the damage, but the damages recoverable in respect

---

[303] Cooke/Young/Taylor, *Voyage Charters* (2001), 1013.

[304] *Ibid.* at 1014.

[305] *Ibid.*

[306] Polat, "Contributory Negligence in Contract Law: Toward a Principled Approach" [1994] 28 *U.B.C.L.R*, 142-169, 143; Bristow, "Contributory Fault in Construction Contracts" [1986] 2(4) *Const. L.J.* 252-261, 252.

[307] Cooke/Young/Taylor, *Voyage Charters* (2001), 565.

thereof shall be reduced to such extent as the court thinks just and equitable having regard to the claimants share in the responsibility for the damage:

Provided that:

(a) this subsection shall not operate to defeat any defence arising under a contract

(b) where any contract or enactment providing for the limitation of liability is applicable to the claim, the amount of damages recoverable by the claimant by virtue of this subsection shall not exceed the maximum amount so applicable.

According to Sec. (4)

... fault means negligence, breach of statutory duty or other act or omission which gives right to a liability in tort or would, apart from this Act, give rise to the defence of contributory negligence.

Accordingly, the facts must be such that a liability in tort would exist, whether or not there is also a breach of contract.[308] Whilst the breach by the carrier of his obligation under Art. III Rules 1 and 2 might well qualify as "fault" within this definition, it does not follow that the shipper's liability is absolute and is not necessarily based upon negligence or principles of "fault" and where a claim is based on the breach of an absolute duty without negligence, the Act does not apply.[309] It may, however, be that in the context of the present rule, the nature of the breach rather than the nature of the duty should be the guiding principle and that the machinery of the Act can be utilized, but only where the circumstances of the shipment of the dangerous cargo are such that it can fairly be said that the shipper is at "fault", in the sense of being in some way negligent or lacking proper care, skill or prudence as a matter of general law.[310]

II. German law

In German law, where the loss is partly caused by the plaintiff's fault, the defendant's obligation to make compensation and the extent of such compensation will depend on the circumstances, especially upon how far the loss has been caused predominantly by the one or the other party.[311] This principle is laid down in BGB § 254 (I).[312] In § 254(II), it is expressly stated as applying to two special situations: firstly, when the fault of the plaintiff consisted only in an omission to warn the defendant of the danger of unusually high which the defendant neither knew nor ought to have known or secondly, when the plaintiff has omitted to avert or mitigate the loss. The wording of BGB § 254, as well as its position in the general part of the law of obligations, show that it applies to contracts and torts alike.[313]

---

[308] Because the legislation was designed to deal with negligence and other tort actions, the courts were reluctant to apply it to contract claims.

[309] Cooke/Young/Taylor, *Voyage Charters* (2001), 1015.

[310] *Ibid.*

[311] Markesinis/Lorenz/Dannemann, *The German Law of Obligations* (1997), 661 ff.

[312] German Civil Code (bürgerliches Gesetzbuch).

[313] Markesinis/Lorenz/Dannemann, *The German Law of Obligations* (1997), 661.

HGB § 559 provides that a carrier must provide a ship that is both seaworthy and cargoworthy, and is liable "for the damage" caused by a lack of seaworthiness. The carrier is not liable where the defect could not be detected by the "due diligence of an ordinary carrier before the commencement of the voyage." Due diligence as defined by common law courts – "reasonable skill, care and competence in light of the circumstances reasonably apparent at the time" – is the same under German law.

As in English law, the carrier is not protected by the exemptions of § 606, § 607 and § 608, which are the equivalents of Art. IV.2 of the Hague/Hague-Visby Rules. Namely, if a lack of seaworthiness or cargoworthiness caused the damage to the cargo, the carrier cannot rely on the exception provisions.[314] However, the carrier might be protected by § 564(b) (1), i.e. dangerous cargoes being carried without the knowing consent of the carrier or master. If both the unseaworthiness and the special quality of the cargo that has not been notified by the shipper have contributed to the damage, this is the typical situation of "contributory fault" in German law.[315] Accordingly, the extent of liability will be assessed based on the extent of the contribution to the causation of the damage and/or of negligence.

III. American law

In the United States, the contributory negligence of the plaintiff was also originally a bar to any recovery whatsoever from the defendant who was also at fault.[316] In the context of modern justice, this inflexible rule is harsh and inadequate, but it was attenuated by such doctrines as the "last clear chance".[317] Modern law has abolished the rule. Accordingly, negligent plaintiff is not barred from recovery if both the plaintiff and the defendant are at fault; the plaintiff can still recover under the comparative fault doctrine.[318] The doctrine limits the plaintiff's recovery based on his or her negligent conduct. In maritime law, the Supreme Court in *United States v. Reliable Transfer Co.* Inc.[319] adopted a rule of comparative fault.[320]

However, under such a law, negligence which is attributable to the plaintiff is considered only in mitigation of damages in tort action.[321] U.S. Courts have not

---

[314] Ashton, "A Comparative Analysis of the Legal Regulation of Carriage of Goods by Sea Under Bills of Lading in Germany, [1999] *MLAANZ* Vol. 14/2, <www.mlaanz.org/docs/99journal7a.html>.

[315] Gündisch, *Die Absenderhaftung im Land- und Seetransportrecht* (1999), 198.

[316] Woods, *Comparative Fault* (1978), 1 ff.

[317] *Ibid.* at 13 ff. Here the court would conclude that even if the plaintiff was at fault, he was entitled to full damages if the defendant had the last chance of avoiding the accident but negligently failed.

[318] CJS Negligence § 291.

[319] 421 U.S. 397, 95 S.Ct. 1708.

[320] Schoenbaum, *Admiralty and Maritime Law* (2004) Vol. 1, 193 ff; Tetley, *Marine Cargo Claims* (1988), 630.

[321] CJS Negligence § 291.

considered a case where both dangerous cargo and unseaworthiness contributed to the damage. However, the obligation to provide a seaworthy ship is an overriding obligation in American law as in English law and comparative fault is applicable in tort cases, thus possible outcome might be similar to that in English law as held in *The Fiona* and *The Kapitan Sakharov.*

L. Liability towards third parties

I. Shipper's liability to the other cargo owners

The shipper has no contractual relationship with other cargo-owners. Therefore, the responsibility of the shipper is in delict or tort and does fall neither under the contract of carriage nor Art. IV.6 of the Hague/Hague-Visby Rules.[322] German law, however, as a code-based system, expressly provides for the shipper's and actual shipper's liability vis-à-vis other parties interested in the particular voyage, e.g. shipper, consignee, passengers and crew.[323]

II. Carrier's liability to the other cargo owners

1. Hague/Hague-Visby Rules

As mentioned in Chapter 2, Classes 1 to 5 of the IMDG Code generate explosion and/or fire hazards. Two designated types of danger in Art. IV.6 of the Hague/ Hague-Visby are also explosivity and inflammability. Under Art. IV.2 (b)[324] of the Rules, the carrier is excused from liability for loss or damage resulting from fire.[325] In this regard, as far as fire is involved in the damage to cargoes, the carrier is liable for neither dangerous goods nor other goods on board.

Moreover, Art. IV.2(q) provides that "Neither the carrier nor the ship shall be responsible for loss or damage arising or resulting from … any other cause arising without fault or privity of the carrier, or without the fault or neglect of the agents or servants of the carrier… ." This catch-all exemption was included to take the place of the various other exemptions traditionally used in bills of lading.[326] This

---

[322] Tetley, *Marine Cargo Claims* (1988), 462; Colinvaux, *Carver Carriage by Sea* (1982), 843 fn. 67; Du Pontavice, "The Victims of Damage Caused by the Ship's Cargo" in Grönfors (ed.) *Damage from Goods* (1978), 29, 36 f.

[323] HGB §§ 564b, 564, 563 and 512.

[324] U.S. COGSA § 1304(2)(q), HGB § 607(2).

[325] Colinvaux, *Carver Carriage by Sea* (1982) Vol. 1 180, 378; Tetley, *Marine Cargo Claims* (1988), 411 ff.; Treitel/Reynolds, *Carver on Bills of Lading* (2005), 607 f. ; Schoenbaum, *Admiralty and Maritime Law* (2004) Vol. 1, 692 ff.; Karan, *The Carriers Liability under the International Maritime Conventions the Hague, Hague-Visby, and Hamburg Rules* (2004), 294 ff.

[326] For a list of exemptions see Boyd/Burrows/Foxton, *Scrutton on Charterparties* (1996), 208 ff.

exemption is broad in comparison to other exemptions and can cover virtually any cause for loss of cargo, including those in clauses IV.2 (a)-(p). However, what it gives the carrier in terms of substantive breadth it takes away in terms of the burdens of proof and persuasion that are placed upon the carrier.

As long as the carrier does not know the dangerous nature of the cargo and the necessary precautions to be taken, he can rely on Art. 4.2 (q). Proof of sole cause of the damage to be the dangerous cargo unknown to the carrier, master or agent vis-à-vis other cargo-owners is also proof of "that neither the actual fault nor privity of the carrier nor the fault or neglect of his agents or servants contributed to the damage."[327]

However, there may be situations where the carrier might be found liable, for instance by failing to inspect carefully the cargo brought on board, or both the carrier and shipper might be found to have contributed to the damage of the other cargo.[328] In such cases the carrier can recourse on the shipper under Art. IV.3 or Art. IV.6 of the Rules.[329]

2. German law

While the carrier is exempted from liability under HGB § 607b for damage arising from fire as in the Hague/Hague Visby Rules Art. IV.2(b), there is no equivalent to Art. IV.2(q) in the HGB. The carrier is liable to other cargo-owners according to §§ 559 and 606 ff., which are equivalent to the Hague/Hague-Visby Rules Art. III.1 and 2.[330] Although HGB does not comprise Art. IV(q) of Hague/Hague-Visby Rules, proof of the cause of damage to other goods to be dangerous cargo would be also proof of faultless of the carrier.

On the other hand, the carrier may be found negligent to other cargo owners for taking onboard and not exploring true nature or for wrong handling of dangerous cargo.[331] He can nevertheless recourse on the shipper under § 564.

If both carrier and shipper or actual shipper are found to have contributed to damage to other cargo, their respective liability will be estimated according to BGB § 254.[332]

---

[327] Mustill, "Carriers' Liabilities and Insurance" in Grönfors (ed.) *Damage from* Good (1978), 69, 79; *In Goodwin, Ferreira & Co. v. Lamport & Holt, Ltd.* (1929) 34 Ll. L. Rep. 192, 196, insufficiency of packing damaged other cargo. The shipowner has discharged onus of proof showing that insufficiency of packing cargo caused damage to another cargo. It was held that accident arose without fault or privity of the carriers or without fault or neglect of the agents or servants of the carriers.

[328] Du Pontavice, "The Victims of Damage Caused by the Ship's Cargo" in Grönfors (ed.) *Damage from Goods* (1978), 29, 50.

[329] *Ibid.* at 51 f. ; Mustill, "Carriers' Liabilities and Insurance" in Grönfors (ed.) *Damage from* Good (1978), 69, 84 f.

[330] Rabe, *Seehandelsrecht* (2000), 448.

[331] Rabe, *Seehandelsrecht* (2000), 448; Feldhaus, *Zur Entwicklung der Haftung beim Seetransport gefährlicher Güter* (1985), 40.

[332] See *supra* p. 194 f.

M. Shipper's liability in the CMI/UNCITRAL draft instrument

I. In general

Under the draft instrument as contained in A/CN.9/WG.III/WP.32, Art. 30 complemented Art. 25 with regard to liability. It dealt with the liability of the shipper vis-à-vis the carrier for losses other than those due to failure of the shipper to comply with his obligations under Art. 27 and 28. The Draft had three different variants and all proposed variants established fault-based liability of the shipper. According to all three variants, the shipper would be liable unless he could rebut negligence or prove that the loss was due to a force majeure type of event which a diligent shipper could not have avoided or prevented. Thus, irrespective of which of the approaches reflected in the different variants might finally be adopted, it is clear that a shipper would be liable to the carrier for all injury, loss or damage caused by the goods, including third-party liability, unless the shipper could disprove fault. The shipper's liability would thus not be strict in the technical sense, but as the burden of proof is reversed, the shipper would be liable unless he could discharge the burden of proof.[333]

As mentioned before, in the first Draft, in contrast to the Hague/Hague-Visby Rules, no specific reference was made to dangerous goods in any of the different variations of Art. 30; thus problems of defining dangerous goods would not have arisen. In contrast to the Hague/Hague-Visby Rules, the liability of the shipper would have, also in principle, arisen where dangerous goods shipped with the carrier's knowing consent were to give rise to a loss.

Liability resulting from Art. 27 and 28 is regulated by Art. 29. Accordingly, the shipper and carrier are liable to each other, the consignee and controlling party for loss or damage caused by either party's failure to comply with its obligations under Art. 26-28.

II. Inclusion of strict liability with regard to dangerous goods

This liability scheme set out in the Draft was re-considered at the 13th Session.[334] A proposal was made for the replacement of draft Articles 29 and 30 by a provision along the following lines:

1. subject to Articles 25, 27 and 28, the shipper is liable for damage or loss sustained by the carrier or a sub-carrier that the shipper has caused intentionally or by his fault or neglect.
2. if the shipper has delivered dangerous goods to the carrier or the sub-carrier without informing the carrier or sub-carrier of the dangerous nature of the

---

[333] Asariotis, "Main Obligations and Liabilities of the Shipper" [2004] *TranspR* 284, 288.
[334] Relevant articles were considered as contained in A/CN/WG.III/WP.32.

goods and of necessary safety measures, and if the carrier did not otherwise have knowledge of the dangerous nature of the goods and the necessary safety measures to be taken, the shipper is responsible for the damage or loss sustained by the carrier.

By way of explanation it was stated that the shipper should be liable for damage it had caused to the carrier through fault or negligence.[335] The proposed text was said to introduce a balance between the carrier's and the shipper's liabilities. Paragraph 2 of the proposed text was intended to establish strict liability of the shipper for not informing the carrier of the dangerous nature of certain goods.

Consequently, it was decided that draft Articles 29 and 30 should be re-drafted entirely to reflect the general principle that the liability of the shipper should be based on fault, but a specific provision should be inserted at an appropriate place in the draft instrument to deal with the issue of dangerous goods, based on the principle of strict liability of the shipper for insufficient or deficient information regarding the nature of the goods.

III. Extensive duties of shipper

Subsequently, the provision on the shipper's liability became Art. 31 and re-drafted as follows: "(1) The shipper is liable for loss, damage [, delay] or injury caused by the goods, and for breach of its obligations under Article 28 and Para. 30(a), [unless][unless and to the extent that][except to the extent that] the shipper proves that neither its fault nor the fault of any person ...caused or contributed to the loss, damage [, delay] or injury."[336] Moreover, a special provision dealing with dangerous goods was drafted.[337]

There was agreement with the general observation that draft Art. 31 was of particular concern with respect to the inclusion of more extensive shipper's obligations in comparison with existing maritime transport regimes. It was thought that the introduction in this provision of a fairly extensive strict liability regime on the shipper, without any right to limit his liability, was quite problematic.[338] There was particular concern with regard to presumed fault on the part of the shipper. It was observed that presumed fault amounted to a reversal of burden of proof onto the shipper that had no parallel in existing maritime transport regimes.

IV. Presumed fault and burden of proof

Generally, the carrier had the burden of proof in proving that the loss or damage was caused by a breach of obligation or negligence of the shipper, such as failure to provide necessary information. Once the carrier established the cause of the loss

---

[335] A/CN.9/552 para.138 ff.
[336] A/CN. 9/WG.III/WP.56, 30. There are also two variants of this article. See *ibid.*
[337] A/CN.9/WG.III/WP.56, 31.
[338] A/CN.9/591 para.137.

or damage, it was open to the shipper to prove that the loss or damage did not arise as a result of his fault. This general regime was thought to reflect the fact that the carrier was usually in a better position to establish what had occurred during the carriage, since he was in possession of the goods. There was support for the view that the traditional approach to fault-based liability as set out in Art. 12 of the Hamburg Rules and Art. IV.3 of the Hague/Hague-Visby Rules should be preserved as the general regime, with strict liability only in certain situations.[339]

An alternative view said that draft Art. 31(1) was appropriate and that the approach taken in the Hamburg Rules was not necessarily fair to the carrier, since most containers in modern transport were packed by shippers, thus making it difficult for the carrier to prove the cause of the loss.

V. Loss, damage or injury

It was suggested that the word "injury" in draft Art. 31 should be deleted in order to clarify that it did not intend to create a claim for third parties.[340] It was further suggested that "damage" should also be deleted and that reference should be made only to "loss". However, it was considered that the draft instrument should ensure that if a carrier paid out a claim as a result of injury caused by the negligence of the shipper, the carrier should be able to claim compensation from the shipper as a loss suffered by the carrier.[341] And the loss here was understood to include damage to the ship as well.[342]

VI. Delay

Liability for delay contained in Art. 31 was particularly problematic as a basis for the shipper's liability, since it could expose the shipper to enormous and potentially uninsurable liability. For instance, a shipper who failed to provide a necessary customs document could cause the ship to be delayed and could be liable not only for the loss payable to the carrier, which could include enormous consequential damages, but also for the losses of all the other shippers with containers on the ship. It was, therefore, suggested that the shipper's liability for "delay" should be deleted from the draft text.[343] Alternatively, if delay was retained in the text, a reasonable limitation should be placed on the liability of the shipper.[344]

A contrary view with regard to the deletion of delay stated that the liability of the shipper and of the carrier for delay was an important aspect of the draft con-

---

[339] *Ibid.* para.138.
[340] *Ibid.* para.141.
[341] *Ibid.* It was suggested that this could be achieved by referring to "loss sustained by the carrier."
[342] *Ibid.*
[343] *Ibid.* para.143.
[344] *Ibid.*

vention.[345] It was observed that deleting delay called into question the rationale for creating strict liability for submitting incorrect information, since inaccurate information was the most common cause for delay.

Support was also expressed for the view that delay should not be easily discarded as a basis of liability, but rather considered as a separate basis for liability, whether caused by the shipper or the carrier.[346]

Furthermore, it was considered that the basis for the liability of the carrier in the draft instrument also included "delay." Thus, if delay was removed as a basis for the shipper's liability, a corresponding change should be made to the carrier's liability. It was explained that this was not simply a matter of balancing the overall rights and obligations of the shipper and the carrier in the draft convention, but that it would not be fair to hold the carrier liable for a delay for which he may not be responsible.[347]

VII. Strict liability

With regard to the consideration of which of the shipper's obligations should be subject to a strict liability regime, there was general support for the view that the shipper should be held strictly liable for the accuracy of information provided by the shipper to the carrier under Art. 30(c) unless inaccuracy was caused by the carrier.[348] It was also suggested that a separate provision could be created for such a strict liability obligation along the lines of the special treatment given to dangerous goods in draft Art. 33.[349] It was also suggested that in addition to the provision of inaccurate information to the carrier and with respect to dangerous goods, there was a third category of obligations for which there should be strict liability on the part of the shipper.[350] This was security-related obligations and should apply to those goods that are prohibited due to their potential relationship with weapons of mass destruction or similar uses. It was said that in these situations, the carrier could be subject to major losses and penalties as a result of the shipper's breach and the shipper's liability in these circumstances should be strict.[351]

VIII. Conclusion with regard to liability of the shipper: general regime fault-based strict liability exception

After discussion, the Working Group decided that a fault-based regime should be adopted as the general regime for the basis for shipper's liability for breach of his

---

[345] *Ibid.*
[346] *Ibid.* para.145.
[347] *Ibid.* para.146.
[348] *Ibid.* para.148.
[349] *Ibid.* para.150.
[350] *Ibid.*
[351] *Ibid.*

obligations under draft Articles 28 and 30, while strict liability should be the basis of shipper's liability in respect of dangerous goods under draft Art. 33 and for providing inaccurate information under Art. 30(c).[352]

IX. Strict liability for dangerous goods

Draft Art. 33 deals with dangerous goods. In terms of shipper's liability it provides that: [353]

> (2) The shipper must mark or label dangerous goods in accordance with any rules, regulations or other requirements of authorities that apply during any stage of intended carriage of goods. If the shipper fails to do so, it is liable to the carrier ... for all loss, damages, delay, and expenses ... arising out of or resulting from such failure.
> (3) The shipper must inform the carrier of the dangerous nature or character of the goods in a timely manner before the consignor delivers them to the carrier or a performing party. If the shipper fails to do so and the carrier or performing party does not otherwise have knowledge of their dangerous nature or character, the shipper is liable to the carrier and any performing party for all loss, damages, and delay and expenses directly or indirectly[354] arising out of or resulting from such shipment.

It was indicated that these provisions established strict liability.[355] The view was expressed that given the harsh burden of strict liability, this provision should be refined to cover only those cases in which the shipper failed to comply with mandatory regulations. It was also proposed that packing should be added to the shipper's obligations.

Furthermore, it was suggested that draft Art. 33 (2) should not impose strict liability on the shipper when the carrier was aware of the dangerous nature of the goods.[356] There was support for the proposal that appropriate language inspired by Art. 13(3) of the Hamburg Rules should be inserted in draft Art. 33(2) to refer to the carrier's lack of knowledge.

---

[352] *Ibid.* para.153.
[353] A/CN.9/WG.III/WP.56, 31.
[354] The words "directly and indirectly" were decided to be deleted as they could interfere with the issues of causation. A/CN.9/591 para.166. See *supra* p. 181 ff.
[355] A/CN.9/591 para.162.
[356] *Ibid.*

Part 5: Limitation of liability and insurance

A. Shipper's or charterer's exposure to liability in general

As seen so far, the charterer or shipper may be exposed to liability vis-à-vis the shipowner or carrier due to the shipment of dangerous goods, whether strict or fault-based liability, for resulting damage to the vessel or other cargo onboard, injury to the crew, damage, the costs of disposing of the cargo and cleaning the vessel, and even for damage to the environment. Moreover, not only dangerous goods but any good which damages the ship's other cargo due to the defective nature or packing or lack of information with regard to handling conditions may result in liability. Likewise, a shipper or charterer may be responsible for loading and unloading goods and thus liable for resulting damage.

Furthermore, under a charterparty, where the charterer has some control over the ship and thus more rights than a shipper of a general cargo under a bill of lading, the charterer is exposed to even more liabilities.[1] The liability exposures of a voyage or time charterer mainly arise out of the terms of the charterparty. The charterparty typically permits the charterer to order the chartered vessel to a particular port or berth for loading. However, most charter-parties also contain a safe port/safe berth clause stating that the selected port or berth must be safe or must allow the vessel to lie "safely afloat" at all times. If the vessel is damaged by some unsafe port condition the charterer did not warn the shipowner about, the charterer can be held liable for the resulting loss, which could include both physical damage to the vessel and resulting loss of use. Some charterparties require the charterer to pay for and provide fuel for the vessel. If the quality of the fuel is substandard, the charterer can be liable for any consequences, such as damage to the vessel's engine.

Consequently, charterers' or shippers' liability arises from different occurrences and charterers or shippers may be exposed to high liabilities. This part will discuss whether the charterer or shipper can benefit from a feature of maritime law that is limitation of liability, and the insurance options available for charterer as well as shipper.

---

[1] Nunes, *"Charterer's Liabilities under the Ship Time Charterer"* [2004] 26 *Hous. J. Int'l. L.*561.

B. Limitation of liability

Maritime law recognizes two different limitations: one is package or kilo limitation of carriers for loss of or damage to goods and the other is limitation of liability in a broader context for maritime claims, which is an important theme of maritime law.

C. Package or kilo limitation for cargo loss or damage

I. Hague/Hague-Visby Rules

For over 100 years, ocean carriers have attempted to avoid their heavy responsibility as common carriers by inserting non-responsibility clauses into bills of lading. The Hague Rules of 1924 were a compromise whereby non-responsibility clauses were disallowed by Art. III.8 but, on the other hand, the carrier was not responsible for more than £100 per package or unit as stipulated in Art. IV.5. The Visby Rules of 1968 adopted a limitation of 10,000 Poincare gold francs per package or unit, or 30 Poincare gold francs per kilo. In 1979, a second protocol to the Hague Rules was adopted, replacing the 10,000 Poincare gold francs per package and 30 Poincare gold francs per kilo by 666.67 Special Drawing Rights (SDR.) per package and 2 SDR per kilo.[2]

1. History of package or kilo limitation of Hague Rules

The net effect of Art. IV.5 is to normalize and standardize the liability of carriers. Excessive cargo claims are reduced to more appropriate amounts. Particularly in connection with the carriage of valuable goods, the consequence is that the cargo-owner or his insurer may have to bear a substantial part of the loss or damage in question. In maritime law in general, a provision of this character is by no means unique. When adopted, Art. IV.5 did not represent any innovation in the field of maritime law. In most countries, contractual provisions of similar content had long been known and frequently inserted in bills of lading.[3] The validity of these clauses was recognized by courts in most countries but they were probably most extensively applied in American bills of lading.[4] Under the choice of rates doctrine, American courts held that provisions fixing a maximum amount for the liability of the carrier were not invalid under the Harter Act. Influenced by this widespread practice, the Canadian Water Carriage of Goods Act, 1910, § 8 established a similar statutory limitation of liability. This Canadian provision was sub-

---

[2]   Under Sec.1304 (5) of the US COGSA, the limitation amount is $500 per package or, in the case of goods not shipped in packages, per customary freight unit.

[3]   Selving, *Unit Limitation of Carrier's Liability* (1961), 24.

[4]   *Ibid.*

sequently used as a model for Art. IV.5 of the Hague Rules. The draft Art. IV.5 met severe criticism from members of the Hague Conference, particularly from representatives of cargo-owners in 1921.[5] It was said that there was no use in opposing the traditional exemption clauses if the carriers could persistently carry on their practice of inserting bills of lading provisions limiting their liability to ridiculously low amounts. Such stipulations had been enforced by courts in most countries, and in effect they frequently allowed recovery of nominal damages only. It was therefore made a condition for approval of the Hague Rules by cargo-owners that Hague Rules Art. IV.5 should establish a liberal per package limitation of liability, which at the same time should constitute a mandatory minimum liability on the part of the carrier.[6] The shipper may avoid the package or kilo limitation by making a declaration as to "the nature and value" of the goods "before shipment and inserting it in the bill of lading".[7]

In this connection, Art. IV.5 was in fact adopted for the protection of cargo-owners. Principally, its purpose was to invalidate limitation of liability clauses which imposed upon the carrier a liability inferior to that of Art. IV.5. In the course of time, however, attention has been directed more towards its concrete application in cases where the parties have not inserted in the contract provisions concerning the extent of any liability on the part of the carrier. Hence, Art. IV.5 mainly grants a privilege to the carrier.[8]

2. Legislative considerations

The liability provisions of the Hague Rules were intended to create an incentive on the part of the carrier to prevent loss or damage to the cargo. The position of the representatives of the cargo-owners argued that it would be completely futile to invalidate exemption clauses unless the conference simultaneously adopted provisions for the purpose of prohibiting clauses limiting the liability of the carrier. On the other hand, shipowner representatives at the Hague Conference emphasized that carriers should be protected against excessive and quite unanticipated cargo claims.[9] There should be some proportionality between the amount of freight charged and the extent of liability of the carrier.

Art. IV.5 was originally adopted for the purpose of limiting the liability of the carrier for loss or damage to valuable goods shipped in packages. If the true nature and value of such goods had not been disclosed, the carrier could have unexpectedly incurred quite exceptional liabilities, without having been in the position to give the cargo any special protection which its value might warrant.[10] With this limitation system, the carriers are given protection against excessive cargo claims

---

[5] *Ibid.*
[6] *Ibid.* at 26.
[7] Hague Visby Rules Art. IV.5(a)
[8] Selving, *Unit Limitation of Carrier's Liability* (1961), 26.
[9] *Ibid.* at 28.
[10] *Ibid.* at 29.

by the fixing of a general limit for any liability irrespective of the specific value of the cargo.

Art. IV.5 was a compromise between the interest of shipowners in avoiding any liability and the cargo-owners' interest in obtaining full compensation for any loss or damage to the cargo. The significance of Art. IV.5 frequently appears only in the settlement between the carrier's liability insurer, the P&I insurer, and the cargo insurer. It is therefore contended that this provision, like several other rules of maritime law, should be considered primarily from a marine insurance point of view.

Unfortunately, both shippers and carriers have hoped to obtain terms favourable to them by avoiding the positive steps provided by the Rules, i.e. proper description of the number of packages on the bill of lading, the declaration of value, the charging of appropriate freight. In effect, shippers and carriers seek a commercial advantage over their competitors without wishing to accept their responsibilities under the Hague/Hague-Visby Rules. As a result, the package limitation for goods shipped in containers is a dilemma, often giving rise to ridiculous results and overwhelmingly serving the interests of carriers.[11]

3. Limitation for delay?

With regard to delay, the Hague/Hague-Visby Rules do not provide specifically any limitation. One possibility for a claim for delay to be *prima facie* a claim in respect of which limitation can apply to that is a claim for "loss ... in connection with the goods."[12] However, a difficulty arises expressly since the limit is to be calculated with reference to particulars of "the goods lost or damaged." It would appear to strain language to say that "delayed goods" are "lost" or "damaged" goods and if it is not possible to calculate the limit, it does not seem possible in fact to limit liability for delay.[13]

As continental contracting states incorporate the expression "loss or damage" in Art. IV.1, 2 and 4 into their domestic law legislation as "loss of or damage to goods" and interpret it as "physical" loss or damage as distinct from practice in common law, the carrier is held liable for non-physical damage arising from late delivery under general provisions in obligation law.[14] Therefore, these claims for delay are not subject to limitation under the Rules.

---

[11]   Tetley, *Marine Cargo Claims* (1988), 640.

[12]   Under English law, it is possible to claim damages if the delivery of the cargo has been delayed due to the carrier's breach of contract. Griggs/Williams/Farr, *Limitation of Liability for Maritime Claims* (2005), 145; Tetley, *Marine Cargo Claims* (1988), 309; Gaskell/Asariotis/Baatz, *Bills of Lading and Contracts* (2000), 342.

[13]   *Ibid.* In *The Breydon Merchant"*[1992] 1 Lloyd's Rep. 373, it was held in the context of Art. 2(1)(a) of the 1976 Limitation Convention that goods which suffered a diminution in value as a result of "actionable delay" were to be treated as damaged.

[14]   Karan, *The Carriers Liability under the International Maritime Conventions the Hague, Hague-Visby, and Hamburg Rules* (2004), 217 f.

II. Hamburg Rules

1. Limitation of liability for loss of or damage to goods

Limitation of liability was designed to encourage investment in shipping and its survival is largely due to the stronger bargaining position of the carrier.[15] In a period when the economic climate indicates a need for a reduction in shipping capacity rather than for increased investment, the justification for the retention of the principle in the Hamburg Rules was less evident. Such retention was, however, favoured on the grounds that it was of value in enabling the carrier to calculate risks and so establish uniform and cheaper freight rates, while at the same time protecting him from the risks associated with cargo of a high, undisclosed value. Nevertheless, the agreed limits must be fixed at a sufficiently high level to encourage the carrier to look after the cargo and, in this respect, the figures established by the Hague Rules had been severely eroded. The Hamburg Rules had considered two urgent problems: quantitative unit of calculation and the amount of the monetary unit of account. Art. 6 is entitled "Limits of Liability" and provides that:

> 1.(a) The liability of the carrier for loss resulting from loss of or damage to goods according to the provisions of Art. 5 is limited to an amount equivalent to 835 units of account per package or other shipping unit or 2,5 units of account per kilogram of gross weight of the goods lost or damage, whichever is higher.

Accordingly, the limits of liability are based on the amount per package or shipping unit or alternatively per weight, whichever is the higher. Where the weight is unknown, the package or unit will be the only applicable test.[16] In the case of bulk cargo, the weight test will clearly be beneficial to the shipper. The adoption of this system led to the appearance of Para.2 on containerization:

> 2. For the purpose of calculating, pallet or similar article of transport is used to consolidate goods, the package or other shipping units enumerated in the bill of lading, if issued, or otherwise in any other document evidencing the contract of carriage by sea, as packed in such article of transport are deemed packages or shipping units. Except as aforesaid the goods in such article of transport are deemed one shipping unit.

The cargo-owner's right to declare the full value of the cargo under the Hague Rules, which was little used, was deleted from Art. 6 of the Hamburg Rules. However, it is still open to parties to agree on a higher level of liability than that provided in the Article.[17]

---

[15]   Wilson, "Basic Carrier Liability and the Right of Limitation", in Mankabady (ed.) *The Hamburg Rules on the Carriage of Goods by Sea* (1978), 138, 146.

[16]   Mankabady, "Comments on Hamburg Rules", in Mankabady (ed.) *The Hamburg Rules on the Carriage of Goods by Sea* (1978), 27, 62.

[17]   Hamburg Rules Art. 6.4: "By agreement between the carrier and the shipper, limits of liability exceeding those provided for in paragraph 1 may be fixed".

2. Limitation of liability for delay

As the Hague/Hague Visby Rules contain no specific provisions for the recovery of loss caused by delay in delivery of the cargo, the position is not clear. In order to remove all doubts and to bring carriage by sea into line with carriage by other modes of international transport, the Hamburg Rules expressly provide for liability for delay but subject to limitation:

> 6. 1. (b) The liability of the carrier for delay in delivery according to the provisions of article 5 is limited to an amount equivalent to two and a half times the freight payable for the goods delayed, but not exceeding the total freight payable under the contract of carriage of goods by sea.

D. Unlimited liability of shippers

In contrast to the limited liability of carriers for loss of or damage to goods and in some countries for delay, the liability of the shipper or transferee of the bill of lading is unlimited. It is not unknown for dangerous cargoes to have led to the constructive or total loss of both ship and other cargoes.[18] Unsafe stowage can have similar consequences. In a case where a bill of lading expressly or by incorporation imposes safe-port obligations on the shipper, the potential liabilities can again be substantial. The above-mentioned monetary limits only apply to protecting the ship. Open-ended liabilities which a shipper may incur call in question the justification for limitation of liability for loss of or damage to or delay of goods.

Although it may be contended, likewise justifiably, that the shipper could invalidate limitation of liability by declaring the nature and value of the goods, the same could not be said in respect of delay where limitation is also applied in cases of delay.

Moreover, it is interesting to note that when the carrier fails to exercise his duty of care in respect of the goods and to provide a seaworthy ship at the commencement of the voyage, he can still benefit from the limited liability under Art. IV.5; however, when goods are damaged by other goods on board, the shipper of the damaging goods, whether dangerous or not, is not able to limit his liability.[19]

---

[18]   In *In Re M/V DG Harmony* calcium-hypochlorite hydrated caused fire lasted three weeks. As a result the US $16 million vessel was declared a constructive total loss and virtually all her cargo was destroyed.

[19]   German law expressly states in HGB §§ 564b, 564, 563, 512 the shipper's direct liability to other persons having an interest in the voyage, such as other shippers, consignee. In common law or where the liability of the shipper is not expressly provided in a code, the shipper may be liable to other parties in tort. *Supra* p. 196 f.

E. The CMI/UNCITRAL draft instrument

I. Consideration of limitation of the shipper's liability in the draft instrument

The draft instrument imposed extensive obligations on the shipper and provided basis for liability of carrier and shipper for delay. As a result of these, it was suggested that a limit should be placed on the shipper's liability, if "delay" was retained as a basis for the shipper's liability, given the large and potentially uninsurable liability that could be covered.[20] Moreover, it was suggested that such a limitation on the liability of the shipper for consequential losses should exist in any event, as the shipper could be held responsible for broad, but probably insurable, liability for damage to the ship.[21] Concern was particularly expressed with respect to delay. In order to maintain a fair balance in the draft convention, it was essential to include a mirror provision establishing liability for a shipper who caused the delay and exposed a carrier to losses resulting from delay claims against him by other shippers, and that because carrier liability for damage due to delay would be limited, such shipper liability should also be subject to a reasonable limitation.

However, determining a reasonable limitation is associated with extreme difficulties. There was a general agreement that such a limitation should be at a high enough level so as to provide a strong enough incentive for the shipper to provide accurate information to the carrier, but that it should be foreseeable and low enough so that the potential liability would be insurable.[22] Accordingly, a limitation based on the freight paid by the offending shipper was deemed to be unreasonably low by the carriers' interests, while the shippers' interests found other formulations, such as full responsibility for damage due to delay to all other shippers on the vessel, unreasonably high.[23] It was also indicated that a carrier should be fairly protected against any losses he incurred for delay damage caused by a shipper, albeit resultant liability on one shipper could be significant.

It was concluded that the only equitable solution to this dilemma would be to remove the concept of liability for damage due to delay from the draft for both shipper and carrier unless agreed in the contract.

II. Submission of European Shippers' Council with regard to limitation of shipper's liability

As mentioned in Part 4, shippers have extensive duties under the CMI/UNCITRAL draft instrument. From the shipper's point of view, the use of specific

---

[20]   A/CN.9/591 para. 147.
[21]   Ibid.
[22]   Ibid.
[23]   A/CN.9/594 para. 202.

provisions to underline the shipper's obligations is consistent with the trend towards holding all transport actors accountable. However, shipping interests are concerned about unlimited liability, which they stated in their submission to the Working Group, stressing that the Council is concerned not about shippers being held liable but about the fact that shipper's liability appears to be unlimited.[24] In this situation, unilateralism accentuates the lack of balance in a draft convention that is already unfavourable to shippers. Therefore, it was said that to achieve a better balance, it would be desirable for shippers to be subject to a liability regime equivalent to that envisaged for carriers, with a limitation of liability.[25] It was also said that the grounds for treating the shipper and the carrier differently are questionable and can be explained only by the existence of a long tradition of imbalance between shipowners and maritime transport users, in both economic and legal terms.[26]

F. Global limitation of liability for maritime claims

The privilege of the global limitation of liability, that is, the ability of a shipowner to discharge all liabilities by surrendering his interest in a vessel that has been involved in a maritime catastrophe, is an idea conceived to serve the needs of commerce. The limitation of a shipowner's liability has been justified as a commercially practicable device by which the effects of a maritime disaster are reasonably apportioned.[27] The original rationale for enabling shipowners to limit their liability was the unusually risky nature of shipping and the need to encourage people to go into the business.[28] It was clear at the very beginning of maritime commerce in Italy in the eleventh century that a shipowner stood to lose a great deal more than his investment as a result of liabilities which might arise as part of the same incident which ruined the voyage.[29]

---

[24]   A/CN.9/WG.III/WP.64 para. 43.

[25]   *Ibid.*

[26]   *Ibid.*

[27]   The preferential treatment afforded vessel owners under a limitation system has been justified on many grounds, including that of national defense. However, it seems that the economic rationale is the more persuasive. As waterborne commerce began to expand in the Middle Ages, it became more difficult for shipowners to accompany their vessels on the increasingly longer voyages. In addition, it was recognized that sea carriers incurred greater financial risks than land carriers as sea adventures are peculiarly liable to mishaps of appalling extent since the owner must entrust this ship to servants who, no matter how carefully selected, may by a moment's inattention or carelessness, cause a disaster. Donovan, "The Origins and Development of Limitation of Shipowners' Liability" [1979] 53 *TLR* 999, 1002.

[28]   Seward, "The Insurance Viewpoint", in *Limitations of Shipowners' Liability: The New Law* (1986), 161.

[29]   Donovan, "The origins and development of limitation of shipowners' liability" [1979] 53 *TLR* 999, 1001.

By the middle of the nineteenth century, three distinct systems of limitation had developed: (1) the German system, which placed no liability on the owners but subjected the ship to an *in rem* action; (2) the French system of abandon, under which the owner, although personally liable, could discharge his obligations by releasing the ship and freight to the claimants; and (3) the English system of limiting the shipowner's liability to the value of the ship measured before the infliction of the damage.[30]

It was not until 1734 that limitation of liability became a feature of English admiralty law. At the time, stimulated by liability burdens not shared by their counterparts in continental Europe, English shipowners succeeded in importuning Parliament to pass a statute that relieved shipowners from liability for acts of their master and crew done without the "privity or knowledge" of the owner, to the extent of the value of the ship, its equipment and the freight to be earned on its particular voyage.[31]

The full extent of this potential liability went unrecognized by the American shipping industry until the Lexington decision[32] in 1848, which held a shipowner fully liable for the loss of a shipment of money in specie, despite a contractual provision limiting liability. Although unaware of the nature of his cargo, he was fully liable for the loss of $18,000 in gold coin. At that time, the hazards of maritime commerce were substantially greater than they are today and the owner had little or no control over the conduct of the vessel once it left port. Neither the corporate form of limited liability nor marine insurance was generally available to protect an owner from potentially ruinous liability for acts of the master and crew.[33] American shipowners, feeling the full force of their potential liability and their competitive disadvantage with respect to English shipowners, asked Congress to find a solution. Congress responded by enacting the Limitation of Shipowners' Liability Act in 1851. The Act has been amended several times since adopted.

I. Importance of limitation of liability

The principle of the limitation of liability remains vital to those involved in the shipping industry and even to those who seek its services. It provides a mechanism for sharing costs between all those involved in, and benefiting from, the marine

---

[30] Eyer, "Shipowners' Limitation of Liability-New Directions for Old Doctrine" [1964] 16 *Stan. L. Rev.* 370, 372.

[31] Thomas, "British Concept of Limitation of Liability," 53 *TLR* 1205, 1205 ff. Shortly after this development, the privilege of limited shipowners' liability made its first statutory appearance in the United States, in the states of Massachusetts (1819) and Maine (1821).

[32] *New Jersey Steam Nav. Co. v. Merchant's Bank*, 47 U. (6 How.) 344 (1848).

[33] Eyer, "Shipowners' Limitation of Liability-New Directions for Old Doctrine" [1964] 16 *Stan. L. Rev.* 370, 372.

adventure.[34] A certain number of accidents are inevitable and there would seem to be a very strong case for sharing the cost of these rather than imposing all the losses on the shipowner, particularly in cases where there was no real fault on his part.

The limitation of liability has another highly important function under a modern insurance-based system, which is that it makes the shipowners' liabilities insurable. The structure of shipowners' liability insurance is founded on the right of the shipowner to limit his liability and thus to bring a degree of comparability between different ships trading in different areas and exposed to different risks.[35] The fact of the matter is that the marine insurance industry relies on it; marine liability policies are written on the general premise that shipowners have the right to limit their liability and that such limitation will indirectly benefit underwriters.[36] The more important reason for maintaining the principle of the limitation of liability is that the commercial marine insurance market has only so much capacity at realistic rates.[37] The P&I Clubs, which provide the bulk of shipowner's liability coverage, often furnish unlimited coverage to their members. However, they can only do so because the laws of maritime nations stipulate a reasonable amount to which shipowners can usually limit their liabilities.[38] Furthermore, P&I Clubs are dependent in turn on their reinsurance costs. The concept of unlimited liability ignores the problem of realistic insurable limits. The insurable limit is the maximum amount of overall coverage available at a realistic cost in respect of any one catastrophe; however, it may be divided among primary underwriters, excess underwriters and reinsurers. That amount in turn determines the cost to the assured and, ultimately, the cost to the consumer of the goods being carried by the shipowner.

Thus, if the limitation of liability were not provided, both claimants and shipowners would suffer, because the P&I insurance market for the shipping industry is limited and it would be technically impossible to insure to the maximum extent of potential liability. Even if it were technically possible, the cost would be unbearably high. Without limitation, the P&I insurers would charge higher premiums, because they would have to reimburse the shipowner the full amount of proven claims. This would eventually hurt consumers in the form of higher prices. Moreover, the very important justification for the limitation of liability is the need for international uniformity.

Consequently, the existence of the right to limit makes it possible for a shipowner to obtain cover for his total exposure and not just some proportion,

---

[34]   Seward, "The Insurance Viewpoint," in *Limitations of Shipowners' Liability: The New Law* (1986), 161, 163.

[35]   *Ibid.*

[36]   Buglass, "Limitation of Liability Form a Marine Insurance Viewpoint" [1979] 53 *TLR* 1364, 1364. Long ago it was estimated that liability insurance premiums might increase twenty-five to thirty percent if shipowners (and therefore underwriters) were deprived of this shield. Affidavit of a marine underwriter in *In re Independent Towing Co.*242 F.Supp.950, (E.D. La. 1965).

[37]   *Ibid.*

[38]   *Ibid.*

limited instead by reference to a figure in his contract of insurance. This means that the shipowner is not "exposed to ruin" because he can insure the whole and not just part of his exposure.

II. Convention on Limitation of Liability for Maritime Claims

1. Development

The international character of shipping suggests that substantial differences between the national laws on shipowners' liability are undesirable. Accordingly, at the beginning of the twentieth century, the limitation of shipowners' liability was taken up by the Committee Maritime International ("CMI") as a subject appropriate for international unification. In 1924, this work was concluded by the adoption of an international Convention for the Unification of Certain Rules relating to Limitation of Owners of Seagoing Vessels, which was a most unhappy compromise between the existing systems. Although the 1924 Convention was once implemented in about 15 countries, it never achieved its objective, notably because the United Kingdom did not accede.

After Word War II the efforts to achieve international uniformity were resumed. The result was the 1957 Convention relating to the Limitation of Owners of Seagoing Ships. With this Convention, the English system of limitation of liability received full international recognition. Only a few elements were added to refine the system.[39]

2. 1976 Convention on Limitation of Liability for Maritime Claims (LLMC 1976)

Over the years it had been generally accepted that the rules relating to the limitation of liability for maritime claims enshrined in the 1924 and 1957 Limitation Conventions required updating. The International Conference on the Limitation of Liability for Maritime Claims took place in 1976. It was agreed at that Conference that the limitation figures should be accompanied by a mechanism in which the right to limit should be forfeit needed reviewing. It was recognized that the previous system of limitation had given rise to excessive litigation and there was a desire that this should be avoided in future. There was agreement that a balance needed to be struck between the desire to ensure on the one hand that a successful claimant should be suitably compensated for any loss or injury which he had suffered and the need on the other hand to allow shipowners, for public policy reasons, to limit their liability to an amount which was a reasonable premium.

The solution which was finally adopted to resolve the competing requirements of claimant and defendant was: the establishment of a much higher limitation fund, which was as high as a shipowner could cover by insurance at a reasonable

---

[39] Selvig, "An Introduction to the 1976 Convention," in *Limitation of Shipowners Liability: The New Law* (1986) 3, 5.

cost, and the creation of a virtually unbreakable right to limit liability. Therefore, the 1976 Convention represented a compromise. In exchange for the establishment of a much higher limitation fund, claimants would have to accept the extremely limited opportunities to break the right to limit liability.[40]

a) Persons entitled to limit liability

Art. 1 of the 1976 Convention provides that following persons are entitled to limit their liability:

1. Shipowners and salvors, as hereinafter defined, may limit their liability in accordance with the rules of this Convention for claims set out in Article 2.
2. The term "shipowner" shall mean the owner, charterer, manager and operator of a seagoing ship.
3. Salvor shall mean any person rendering services in direct connection with salvage operations. Salvage operations shall also include operations referred to in Article 2, paragraph 1(d), (e) and (f).
4. If any claims set out in Article 2 are made against any person for whose act, neglect or default the shipowner or salvor is responsible, such person shall be entitled to avail himself of the limitation of liability provided for in this Convention.
5. In this Convention the liability of a shipowner shall include liability in an action brought against the vessel herself.
6. An insurer of liability for claims subject to limitation in accordance with the rules of this Convention shall be entitled to the benefits of this Convention to the same extent as the assured himself.

The Act of invoking limitation of liability shall not constitute an admission of liability.

aa) Owner

It is well established since early times that the owner of a ship is entitled to limit his liability. It is a right enjoyed by both legal and equitable owners. The wording in Article 1 includes a part owner.[41]

bb) Charterer

The definition of shipowner includes the charterer of a ship. The word "charterer" is meant to include all charters, e.g. demise, time and voyage.

The question arises whether the slot charterer, a party who has the right to use a vessel on a particular voyage and who often issues his own bills of lading, can limit his liability. It is submitted that all the other parties referred to in Article 1(2)

---

[40] Germany is a contracting State to the LLMC 1976. The Protocol increasing the financial limits of liability came into force as of 13 May 2004.

[41] Griggs/Williams/Farr, *Limitation of Liability for Maritime Claims* (2005), 11; Hodges/ Hill, *Principles of Maritime Law* (2001), 528.

have an interest in the whole ship and the limitation on their liability is calculated by reference to the tonnage of the whole ship.[42] There is no provision which allows a slot charterer to limit his liability proportionately to the space which he has chartered, and accordingly the decision, and hence the choice, would seem to be between allowing the slot charterer to limit his liability according to the full tonnage of the vessel, despite the fact that his contract allows him to use merely part of it, or not allowing him to limit at all on the basis that he is not a charterer or operator of a ship.[43]

cc) Other persons

Art. 1 (4) of the 1976 Convention extends the right to limit to "any person for whose act, neglect or default the shipowner or salvor is responsible." This wording is apparently wide enough to encompass agents and independent contractors such as stevedores, provided that the shipowner is responsible for their actions as a matter of law.

dd) Liability insurer

Art. 1 (6) of the 1976 Convention introduces an innovation. Where a person entitled to limit liability is insured, the insurer is entitled to the benefits of the Convention "to the same extent" as the assured himself. The intention of this provision appears to be to place the liability insurer in no worse a position than the assured, insofar as limitation is concerned, should a direct action be brought against the insurer. It follows that the liability insurer should not be able to limit his liability if the assured cannot.[44]

b) Claims subject to limitation

Art. 2 of the 1976 Convention provides that

1. Subject to Articles 3 and 4 the following claims, whatever the basis of liability may be, shall be subject to limitation of liability:
   (a) Claims in respect of loss of life or personal injury or loss of or damage to property (including damage to harbour works, basins and waterways and aids to navigation) occurring on board or in direct connection with the operation of the ship or with salvage operations, and consequential loss resulting therefrom;
   (b) Claims in respect of loss resulting from delay in the carriage by sea of cargo, passengers or their luggage;

---

[42] *Ibid.* at 7 ff.
[43] *Ibid.* at 11. By contrast, it is argued that the wording of Art. 1 is unqualified and so it must refer to all charterers; Hodges/Hill, *Principles of Maritime Law* (2001), 529.
[44] *Ibid.* at 15.

(c) Claims in respect of other loss resulting from infringement of rights other than contractual right, occurring in direct connection with the operation of the ship or salvage operations;

(d) Claims in respect of the raising, removal, destruction or the rendering harmless of a ship which is sunk, wrecked, stranded or abandoned, including anything that is or has been on board such ship;

(e) Claims in respect of the removal, destruction or the rendering harmless of the cargo of the ship;

(f) Claims of a person other than the person liable in respect of measures taken in order to avert or minimize loss for which the person liable may limit his liability in accordance with this Convention, and further loss caused by such measures.

aa) Claims in respect of loss of life or personal injury or loss of or damage to property

Under the previous limitation regime, the benefit of limitation was available in respect of injury or damage to a person or to property on board, or, if they were not on board, only if such injury or damage was caused by a person on board or by a person not on board in the course of specific activities which were laid down by Art. 1 of the 1957 Convention. Therefore, limitation was restricted to acts or omissions done by a person on board or in the navigation or management of the ship, or in the loading, carriage or discharge of its cargo, or in the embarkation, carriage or disembarkation of its passengers.[45] These restrictions have resulted in unfortunate decisions.[46]

The 1976 Convention seeks to deal with this problem by replacing the list with a wider definition of claims which are subject to limitation by referring to events occurring "on board or in direct connection with the operation of the ship, or with salvage operations, and consequential loss resulting therefrom". Hence, as long as there is the necessary link between any of the loss or damage mentioned in Art. 2 (1) (a) and the ship against which the claim is made, it would appear that the right to limit in the case of such claims is established.[47]

bb) Claims in respect of loss resulting from delay

An apparent innovation introduced by this subsection is the extension of the benefit of limitation to claims resulting from delay in the carriage of cargo, passengers or their luggage. Such a claim would have fallen under the notion of damage in previous legislation. Although there is nothing novel about Art. 2 (1) (b), it is perhaps necessary to refer briefly to the specialized conventions on the subject, namely the Hague-Visby Rules, Hamburg Rules and Athens Convention 1974, for

---

[45] *Ibid.* at 18.

[46] Such as in the *Tojo Maru* [1971] 1 Lloyd's Rep. 341, where the House of Lords held that the salvors were not entitled to limit their liability, since the negligent act of the diver was not an act done either in the management of or on board the tug.

[47] Griggs/Williams/Farr, *Limitation of Liability for Maritime Claims* (2005), 19.

the purpose of noting that no provision has been made in any of these conventions as regards limitation of liability for delay. These conventions, however, have made express provisions allowing for consultation to be made with any international conventions or national law relating to the limitation of liability of owners.[48]

cc) Claims in respect of infringement of rights other than contractual rights

Art. 2 (1) (c) covers all claims resulting from infringement of rights provided that they do not arise in contract. Examples of claims which may fall under this article are those for loss of use and loss of profits made by fishing-boat owners, yacht owners, fish and shellfish farm owners, shellfish harvesters, fishing-net and fishing-pot owners, shop owners, local municipalities, local government and the state itself consequent upon pollution.[49]

dd) Claims in respect of the raising, removal, destruction or the rendering harmless of a ship including anything on board such ship

The raising, removal, destruction and the rendering harmless of a ship are services generally performed by a contractor, a salver or harbour or conservancy authority.[50] Though the opening line of Art. 2 (1) declares that claims stipulated in the Article shall be subject to limitation "whatever the basis of liability", this is in fact not always the case.

In relation to a claim made by a contractor for payment for such services rendered under contract, Art. 2(2) states that claims under Art. 2(1)(d), (e), and (f) shall not be subject to limitation of liability to the extent that they relate to "remuneration under a contract with the person liable". The phrase "person liable "must necessarily mean the person who is liable under the contract. The purpose of this exception is to preclude a party, for example a shipowner or a cargo-owner, who has entered into a contract with a contractor for the rendering of such services from pleading limitation when a claim is brought against him for payment for the services rendered. The reasoning is that parties to a contract must be held to their agreement.

ee) Claims in respect of the removal of the cargo of the ship

By virtue of Art. 2(1)(e) of the 1976 Convention, limitation is expressly available in respect of certain claims relating to the removal, destruction or rendering harmless of cargo except where such claims relate to remuneration under a con-

---

[48] See Art. VIII of the Hague-Visby Rules, Art. 25(1) of the Hamburg Rules and Art. 19 of the Athens Convention 1974.
[49] Griggs/Williams/Farr, *Limitation of Liability for Maritime Claims* (2005), 22.
[50] The U.K has made a reservation to exclude application of 2(1) (d) of the 1976 Convention. See Sec. 185 (2) of the MSA 1985.

tract with the person liable.[51] Conflicts may arise between the provisions of Art. 2(1) (d) and 2(1) (e). Under 2(1) (d), the expression "anything that is or has been on board such ship" could include cargo.[52]

ff) Claims in respect of measures taken to avert or minimize loss

To recover under Art. 2 (f) for claims in respect of measures taken to avert or minimize loss, the first requirement is that the claimant must be a person "other than the person liable."[53] On the other hand, for a claim to be within the scope of Art. 2(1) (f), it is sufficient if it is in respect of measures taken, as opposed to a claim "for measures to be taken", to avert or minimize a loss. Therefore, the claimant does not necessarily have to be the person who has incurred expenses in order to avert or minimize the loss.[54] Moreover, in order for a claim to fall within the scope of this article, it must relate to measures taken to avert or minimize a loss which would be subject to limitation.[55]

gg) Claims brought by way of recourse or for indemnity

Art. 2(2) in the first part of the 1976 Convention provides that "claims set out in Paragraph 1 shall be subject to limitation of liability even if brought by way of recourse or for indemnity under a contract or otherwise." The purpose of this article is to ensure that the claims listed in Art. 2 (1) maintain their character at all times. In *The CMA Djakarta*,[56] it was held that a charterer would be entitled to limit his liability for an indemnity claim brought by the owners of the ship in relation to payments made by the owners in settlement of cargo claims brought against them as carriers under bills of lading.

III. Can a charterer limit his liability against the shipowner?

The original privilege of limitation was granted to shipowners, not charterers. But the whole history of the limitation of liability displays a natural tendency for claimants to try to avoid the statutory limits, e.g. by finding someone to sue who is not a shipowner, or by framing the claim in such a way that the claim does not fall within the precise wording of the appropriate Convention or national statute. Once the loopholes become apparent, there is a natural tendency for the legislature to try to close the gaps in order to preserve the original rationale for limitation.[57] The

---

[51] LLMC 1976 Art. 2(2)
[52] Hodges/Hill, *Principles of Maritime Law* (2001) 540 ff.; Griggs/Williams/Farr, *Limitation of Liability for Maritime Claims* (2005), 24.
[53] *Ibid.* at 551 f.
[54] *Ibid.*
[55] *Ibid.*
[56] [2004] 1 Lloyd's Rep.460, 469.
[57] Mustil, "Ships are different- or are they?" [1993] *LMCLQ* 490. Demise charterers were granted the right to limit under the Merchant Shipping Act 1894 by the Merchant

1957 Convention extended the limitation provisions to the "charterer, manager and operator" of the ship "as they apply to an owner himself." In other words, if the shipowner is entitled to limit his liability, e.g. towards cargo-owners, a time charterer should also be entitled to limit, e.g. where he was the contracting carrier.[58] This phrase indicated that the protection envisaged for the charterer was against liabilities of the same kind as are expected by a shipowner when sued by third parties.[59] The same phrase, however, did not appear in the 1976 Convention.

1. The Aegean Sea case

In the *Aegean Sea Traders Corp. v. Repsol Petroleo SA, (The "Aegean Sea")*,[60] the *Aegean Sea* went aground at La Coruña in Spain on 3 December 1992, after a voyage from the U.K with a cargo of 80,000 tones of crude oil. The accident itself caused enormous pollution, gave rise to many compensation claims and raised important questions of maritime legal principle. As a result of grounding, the vessel broke up and then exploded. The vessel, together with most of her cargo, was lost and there was large-scale pollution. This led to massive claims against the vessel's owner for pollution damage and also a hefty claim for salvage services.

The owners sought for pollution damage, through arbitration, to recover from Repsol Oil International Ltd., the vessel's charterer, some US$65m in respect of the vessel, her bunkers, lost freight, an indemnity for pollution claims, liability to CRISTAL and salvage services.[61] The owners claimed that the charterers nominated an unsafe port and/or that the charterers were responsible under an implied indemnity.[62] The charterers denied that the nominated port was unsafe and attributed the accident to the negligence of the master. They further contended that in the event they were found liable to the owners, they were entitled as charterers of the *Aegean Sea* to limit their liability under the LLMC 1976.[63] If so, the cap of the charterers' liability in this case would be about US$12m as opposed to the full claim of US$65m. However, the answer was "no". In other words, the charterers were not entitled to the benefit of limitation when sued by the owners. In coming to this conclusion, the court looked very closely at the wording of the 1976 Convention and the history of how it had evolved.

Firstly, the 1957 Convention was examined.[64] The papers in preparation for the 1957 Convention do not say anything about the question of whether charterers could limit liability as against owners, although the fact that the Convention provides for only one fund makes the question virtually superfluous. If one thinks

---

Shipping Act 1906, presumably because they were thought to deserve the same protection as shipowners sued by third party claims.

[58] Gaskell, "Pollution, Limitation and Carriage in the Aegean Sea", in Rose (ed.) *Lex Mercatoria*, (2000), 71, 81.

[59] Ibid.

[60] [1998] 2 Lloyd's Rep. 39.

[61] *Ibid.* at 40.

[62] *Ibid.*

[63] *Ibid.*

[64] *Ibid.* at 45.

about it, for there to be such a limit available, two separate funds would have to be established in respect of the liabilities of charterers and owners respectively. The concept of a single fund pointed to the total liabilities of the shipowners and charterers being treated as one, with nothing in the travaux préparatoires pointing clearly one way or the other.[65]

The intention was that charterers should benefit from limitation in situations where they had their "owner's hats" on and incurred "owner-type" liabilities, e.g. through issuing bills of lading, not when sued by owners.[66] The 1976 Convention made a number of changes to limitation law, but there was nothing to indicate that there was an intention to change the position of the charterer, and there was nothing in the travaux préparatoires either. It was found that policy considerations were balanced, i.e. it might be unfair that a shipowner to be able to limit his liability against a charterer but not vice versa, although charterers were able to protect themselves by the terms of the charterparty.[67] In the absence of a clear guide in the words of the 1976 Convention, which approach should be adopted? The fact that the words "as they apply to an owner himself" did not appear in the 1976 Convention was held to be irrelevant as the same means were achieved by different drafting techniques in the 1976 Convention.[68] More significant was Art. 9, which provided for aggregation of all claims against the shipowner, including those persons categorized as shipowner in Art. 1.[69] If there was one fund for all claims against shipowners and charterers, how would that work when the charterer himself brought a claim against the shipowner in his capacity as operator of the ship? Of course, this would mean that the other third-party claimants would suffer, as the limitation fund itself would be reduced by the charterer's claim.[70] Consequently it was held that that the shipowner's claim against the charterer was not subject to limitation under the 1976 Convention.

---

[65]   *Ibid.*

[66]   *Ibid.* at 47.

[67]   *Ibid.*

[68]   *Ibid.* at 48.

[69]   LLMC 1976 Art. 9 provides:

1. The limits of liability determined in accordance with Article 6 shall apply to the aggregate of all the claims which arise on any distinc occasion:

(a) against the person or persons mentioned in paragrapf 2 of Article 1 and any person for whose act, neglect or default he or they are responsible; or

(b) against the shipowner of a ship rendering salvage services from that ship and the salvor or salvors operating from such ship and any person for whose act, neglect or default he or they are responsible; or

(c) against the salvor or salvors who are not operating from a ship or who are operating solely on those ship to, or in respect of which, the salvage services rendered and any person for whose act, neglect, or default he or they are responsible.

2. Limits of liability determined in accordance with Article 7 shall apply to the aggregate of all claims subject to thereto which may arise on any distinct occasion against the person or persons mentioned in paragraph 2 of Article 1 in respect of the ship referred to in Article 7 and any person for whose act, neglect or default he or they are responsible.

[70]   *Aegean Sea* [1998] 2 Lloyd's Rep.39, 50.

The logic of the decision is that even a ship manager would not be entitled to limit liability if the shipowner sued him for damage caused to the ship itself, for instance where the manager sent the ship to sea in a unseaworthy state.[71] If a third-party claimant sued the manager, the manager would be able to limit in respect of that claim, but would not have protection against any claim by the shipowner. It could be said that the manager and the charterer are in a different position when sued by the shipowner, compared with being sued directly by third-party claimants, as they can control their liability to the shipowner through contract.[72]

2. The CMA Djakarta case

a) Facts of the case

In *CMA CGM S.A v. Classica Shipping Co. Ltd. ("The CMA Djakarta")*,[73] the container ship *CMA Djakarta* was chartered under the terms of the NYPE. Bills of lading on receipt of containerized cargo were issued by the time charterers. An explosion occurred on board the chartered vessel, causing substantial damage to the ship and other cargo and resulting in salvage being performed and general average declared. The dispute was referred to arbitration. In the arbitration procedure, the charterers pleaded an entitlement to limit their liability pursuant to the provisions of the 1976. The arbitrators concluded that the explosion had taken place in two containers containing bleaching powder and that the incident amounted to a breach of the obligation of the charterer under the NYPE to employ the vessel for carrying lawful containerized merchandise "excluding any goods of a dangerous, injurious, flammable or corrosive nature". The arbitrators found in favour of the owners and held that the charterers were liable to the owners for the cost of repairs to the vessel amounting to US$26,638,032, including sums paid for the salvage services rendered to the vessel. The charterers appealed.

b) Issues

The appeal raised the question of whether the charterers were able to limit their liability in respect of the claims made against them by the shipowners under the provisions of the LLMC 1976. The Court of Appeal held that the word "charterer" in Article 1(2) was to be given its ordinary and not a restricted meaning.[74] Having regard to the proper interpretation of Art. 2 (1) (a), the only claim the charterers had a right to limit was the claim by the shipowners for an indemnity for liabilities incurred by them to the cargo-owners.[75] All other claims were not limitable.

---

[71] Gaskell, "Pollution, Limitation and Carriage in the Aegean Sea", in Rose (ed.) *Lex Mercatoria* (2000), 71, 82 f.

[72] *Ibid.*

[73] [2004] 1 Lloyd's Rep. 460.

[74] *Ibid.* at 467.

[75] *Ibid.* at 469.

c) Interpretation of the convention

The crucial question in that case is the proper interpretation of the word "charterer" as used in Article 1(2). This raised the question of statutory interpretation of a particular kind, for the word derived from an international convention. The approach to be adopted in this situation has been considered on many occasions. The general approach is that the convention is to be interpreted having regard to its own language and structure, free of English-law perceptions and principles.[76] The Convention is to be interpreted by reference to broad and general principles of interpretation which may not always be easy to identify, but assistance may be given by such international instruments as the Vienna Convention on the Law of Treaties 1969,[77] Art. 31 and 32.

The Court of Appeal considered that the upshot of this approach was that the court was obliged to ascertain the ordinary meaning of the words used in their context and also take into account the object and purpose of the Convention. It was open to the court to have regard to the *travaux préparatoires* and the circumstances of the conclusion of the convention. The interpretation given to the same word in a previous international convention can only be referred to once the ordinary meaning has been ascertained.[78] Such a reference may affirm the court in its conclusion as to the ordinary meaning of the word or cause it to reassess its conclusion in cases where the adoption of the ordinary meaning renders the convention ambiguous or uncertain or leads to a manifestly absurd or unreasonable result.[79]

d) Ordinary meaning of "charterer" in the 1976 Convention

Looking at the convention as a whole, the court concluded that the word "charterer" was to be given its ordinary meaning, i.e. a charterer acting in his capacity as such.[80] There was no justification for placing a gloss on the concept of charterer and restricting its interpretation to circumstances when the charterer acted *qua* owner that is as if he were the charterer.[81] The mere fact that the reference to charterer appeared in the definition of "shipowner" did not of itself justify such an approach. Moreover, the distinction would often be difficult to make for certain responsibilities, such as the loading of cargo, which are allocated variously by the terms of the charterparty.[82] One method of ensuring certainty would be to confine the meaning of "charterer" to demise or bareboat charterer, but there was abso-

---

[76]   *Ibid.* at 463.
[77]   It was ratified by the United Kingdom on 25 June 1971 and came into force on 17 January 1980.
[78]   [2004] 1 Lloyd's Rep.460, 464.
[79]   Thomas, "Limitation of Liability – London Convention 1976 – Definition of Charterer – Right to Limit – Limitable Claims – Articles I(2) and 2(1)(a), CMA CGM SA v. Classica Co. Ltd" [2004] (10) *JML* 122, 123.
[80]   [2004] 1 Lloyd's Rep.460, 464.
[81]   *Ibid.* at 465.
[82]   *Ibid.*

lutely no justification for adopting this approach.[83] Limitation of liability had been extended for the first time to charterers by the Limitation Convention 1957, but there was nothing in this convention which suggested a departure from the adoption of the ordinary meaning of the word "charterer" as used in the 1976 Convention.[84]

e) The right to limit only extended to claims which were limitable under Art. 2

To conclude the investigation of charterers, it was necessary to ascertain whether a claim for damage to the ship by reference to which a charterer seeks to limit his liability is a claim which falls within Art. 2. There is the further question of whether the charterers can limit their liability for any of the other claims brought by the shipowner, but the shipowner's main claim is for the cost of the extensive repairs required for the vessel. So the question here is whether a claim for loss or damage to the vessel by reference to which a charterer seeks to limit his liability is a claim which falls within Art. 2.[85]

If properly interpreted, Art. 2 (1) (a) allows for several different categories of claim in respect of loss of life or personal injury and for claims in respect of "loss of or damage to property occurring on board." The Court found that this latter phrase does not include loss of or damage to the ship itself, since neither the loss of a ship nor damage to the ship can be said to be loss or damage to property on board.[86] Property on board means something on the ship and not the ship itself.[87] Furthermore, the third category of claims is a claim in respect of loss of or damage to property "occurring … in direct connection with the operation of the ship." The most obvious reason for including this third category of claim is to cater for cases of collision with another ship. Loss or damage to that other ship or its cargo is not "loss of or damage to property … occurring on board" but is "loss of or damage to property … occurring … in direct connection with the operation of the ship." Therefore, the wording of this category of claim did not envisage including loss or damage to the ship itself, that it is the ship the tonnage of which is to be used to calculate the limitation fund, but loss to something other than that ship herself.[88]

f) Limitation fund and aggregation of claims

It follows that the restriction on which claims are limitable applies equally to charterers and further provides a justification for the procedural processes by which limitation funds are established under Art. 9 and 11 of the Convention. Art. 9.1 provides for claims against (a) persons mentioned in Art. 1.2 to be aggregated if they arose on distinct occasions; likewise for claims against (b) the owner

---

83 *Ibid.*
84 *Ibid.*
85 *Ibid.* at 467.
86 *Ibid.*
87 *Ibid.*
88 *Ibid.*

of a ship rendering salvage services and a salvor operating from that ship and (c) a salvor not operating from a ship. Art. 9 (2) then deals with passenger claims. Art. 10 provides that liability can be limited without the creation of a fund. Art. 11.3 then provides for the constitution of a limitation fund when that is, in fact, done; it provides for separate funds for the "shipowner" category of those entitled to limit and the "salvor" categories (and for passenger claims) by providing:

> A fund constituted by one of the persons mentioned in paragraph 1(a), (b), (c) or paragraph 2 respectively.

Accordingly, through the references to Art. 9.1 (a), all those persons designated as shipowners in Art. 1.2 of the Convention are brought as a single unit for the constitution of the fund.[89] Thus, the fact of the restriction removes the potential anomaly of charterers claiming limitation under a fund established by the shipowner in respect of a claim brought by the shipowner against the charterers.[90]

Accordingly, the claim by the shipowners for the cost of repairing the ship as a result of damage caused by the charterers' breach of contract was not limitable under Art. 2(1)(a), nor were the claims to be indemnified for salvage, which was held to be included in the claim for damage repair and general average contribution.

g) Claims limitable

The only claim that was limitable was the shipowner's claim to be indemnified for liabilities incurred to other cargo interests.[91] The bills of lading had been issued by the charterers and the shipowners had been sued in tort. The primarily liability was that of the charterers and had they been sued directly, they would have been able to limit liability for the claim under Art. 2 (1) (a), the claim falling within the words "loss of or damage to property … occurring … on board the ship."[92] The fact that the claim was being brought against the charterers indirectly, through the shipowner's claim for indemnity, did not change the situation. The claim continued to be one within Art. 2 (1)(a) and the charterers were entitled to limit.

h) Conclusion

The approach taken by the Court of Appeal means that the position of charterers under the 1976 Convention is not governed by the question of status with reference to the claim made.[93] Accordingly, whether a claim is limitable raises the question of interpretation and although difficult and complicated questions may

---

[89]   *Ibid.* at 468.

[90]   *Ibid.*

[91]   *Ibid.* at 469.

[92]   *Ibid.*

[93]   Thomas, "Limitation of Liability – London Convention 1976 – Definition of Charterer – Right to Limit – Limitable Claims – Articles I(2) and 2(1)(a), CMA CGM SA v. Classica Co. Ltd" [2004] (10) *JML* 122, 124.

arise, the general approach to the question of interpretation is well appreciated. The interpretation given to Art. 2 (1) (a) will also has the effect of limiting the occasions when charterers are able to claim limitation of liability when sued by shipowners. While final words of the House of Lords on this were awaited the respective parties have settled the case, leaving the law reverse in charterer's favour.

IV. United States law

Under 46 U.S.C. § 183, only an owner of a vessel, whether American or foreign, may limit his liability. The term "owner" is defined to include the charterer who actually "mans, victuals and navigates the vessel", which is interpreted to mean demise and bareboat charterers.[94] However, the Limitation Act does not extend these liability limitations to time charterers, voyage charterers, insurers or other parties as distinct from the LLMC 1976.[95] The Limitation Act provides that a shipowner may be entitled to limit his liability to the post-casualty value of the ship plus any pending freight. In order for a shipowner to avail himself of the Act's liability limitations, a shipowner must demonstrate that the relevant accident or tort occurred without his "privity or knowledge."[96] In the context of a corporation, "privity and knowledge" means the privity and knowledge of a managing agent, officer or supervising employee, including shoreside personnel.[97]

Obviously, in American law the question of whether a time or voyage charterer can limit his liability to a shipowner does not arise, as a time or voyage charterer is not one of those persons who are entitled to limit.

V. Considerations regarding charterer's liability against shipowner and limitation of liability

However seldom, the above cases well reveal that charterer's liability may arise towards a shipowner in different cases. Probably at the time the Convention drafted no one had envisaged shipowner's claims against charterer of a size sufficient to warrant consideration.[98]

With respect to pollution damage, the international community has produced a finely balanced regime for the compensation of pollution incidents, in which the limitation of liability is central. One reason for this was that the insurance market said that it needed the certainty of limits in order to be able to cover the potentially massive liabilities. Of course, CLC is designed to protect claimants and, to a great

---

94   46 U.S.C. § 186.
95   46 U.S.C. § 183(a).
96   46 U.S.C. § 183(a).
97   46 U.S.C. 3 183(e).
98   Andrewartha/Hayhurst, "English Maritime Law Update: 1998" [1999] 30 *J. Mar. L. & Com.* 457, 486.

extent, does so; and the system of channeling liability only to the registered ship-owner means that there should be no need for double insurance. It is therefore a little ironic that the one consequence of the decision in *The Aegean Sea* is that charterers may well need to take out extra insurance cover for their potential pollution liabilities if sued by shipowners. Yet their liability insurers may well be the same, in the sense that they may be members of the International Group of P & I Clubs, who have always been anxious to avoid double insurance, in order to make full use of the capacity of the market to provide the largest possible amount to claimants. With respect to damages to the ship and shipowner, these cases raise a new issue and highlight the necessity of insurance.

As a matter of policy, the question is whether the regime of limitation of liability is improved or undermined by these decisions on this point. The history of limitation of liability shows that the ingenuity of lawyers and courts has always created gaps. These stimulate litigation on narrow technical points and can cause uncertainty and imbalances in protection. It could be said that the decisions in *The Aegean Sea* and *CMA Djakarta* create unnecessary distinctions between types of claim and complicates what should ideally be a clear overall picture of liabilities and their limitation.[99] Experience shows that litigation is encouraged where in the same adventure there are some persons who are, and some who are not, subject to limitation of liability. If one of the aims of a system of limitation of liability is to encourage settlement between parties who have access to insurance, the conclusions on this issue may not be entirely welcome.

In a recourse action, as in *The Aegean Sea* for pollution and in *CMA Djakarta* for cargo claims, the charterer would not have to pay more than the shipowner himself paid. However, it demonstrates that the charterer may have to indemnify the shipowner without limit and would therefore need high insurance cover in the cases where the shipowner is deprived of the right to limit or limitation is not allowed for claims in respect of damage to the ship. It is obvious that charterers and managers need to be advised in the future to seek to obtain protection. In terms of pollution damage, this protection could be obtained through the terms of a charterparty. There could either be straight exclusion of liability, e.g. for pollution consequences of port nominations, or provisions which by contract give the charterer the same protection against the shipowner as if there was a right to limit under LLMC 1976.[100] However, there is no guarantee that such clauses would always be effective and enforceable in all parts of the world, so more consideration will need to be given to charterers' insurance cover and its extent. With respect to damage to the ship by the charterer, it seems more difficult to make to make contractual exclusions. Therefore, an insurance arrangement is an inevitable object of the charterer's liability for damage to the ship.

On the other hand, it is worth noting that the Maritime Code of the Nordic Countries incorporates a legislative commentary which reveals that under the

---

[99]   Gaskell, "Pollution, Limitation and Carriage in the Aegean Sea", in Rose (ed.) *Lex Mercatoria*, (2000) 71, 85.

[100]   *Ibid.* at 86.

Code not only can a "part charterer" limit but also a shipper.[101] However, it is evident from an examination of the travaux préparatoires to the LLMC 1976 that the drafters were restricting the right to limit to those controlling the whole of the ship.[102] In the interest of achieving international uniformity on this issue and in the light of recent cases, it may be that that the definition of shipowners in the Convention should be reviewed. If part charterers, and even maybe shippers, are to be allowed to limit under the Convention, it will be necessary to prescribe whether the fund should be restricted by reference to the proportion of the ship's space which the part charterer is using at the time in question.[103]

VI. Criticisms on the limitation of liability

It is clear that the privilege to limit liability originally was granted to shipowners. Under the LLMC 1976, charterers have the right to limit their liability for certain claims but not against the shipowner. The idea of limiting the charterer's liability against the shipowner raises also a technical difficulty, i.e. how the single fund would cover both charterers and owners and how charterers can claim the benefit of limitation through that fund. The privilege to limit liability, however, has been already subject to criticism. These recent cases might bring up the issue of a justification for such limitation.

It is said that the need to encourage shipowners to invest is no longer a valid reason for limitation, despite the currently parlous financial position of shipping and the potentially vast liability exposure as evidenced by enormously expensive accidents or incidents at sea.[104] The original purpose of building up the shipping industry seems to have lost much of its force with the availability of insurance, bills of lading statutes that put substantive limits on liability for cargo loss and the ability to limit claims by contract.[105]

The modern theory is that insurance and corporate structure are available and these remove the danger of disaster from the shipowner.[106] Most shipowners are shareholders in shipping companies. A shipowner is normally liable merely up to

---

[101] Swedish Maritime Code chap.9.1.

[102] Griggs/Williams/Farr, *Limitation of Liability for Maritime Claims* (2005), 11.

[103] *Ibid.*

[104] Six motives and three situations which have impelled the enactment of limitation statutes. The motives have been pointed out. The motives are the idea of joint venture, high cargo values, limits on share capital, ruin without fault, the attraction of local venture capital and general benefit to users. Situations in which a right to limit may be regarded as desirable are various, but there are three very broad ones: "closed" situations, "partly closed", "open" situations. Mustill, "Ships Are Different, or Are They?" [1993] *LMCLQ* 490, 491 ff.

[105] Schoenbaum, Admiralty and Maritime Law (2004), 759.

[106] Kim, "Shipowners' Limitation of Liability: Comparative Utility and Growth in the United States, Japan and South Korea" 6 *U.S.F. Mar. L.J.* 357, 363; Seward, "The Insurance Viewpoint", in *Limitations of Shipowners' Liability: The New Law* (1986), 161, 163.

his percentage of ownership in the company. Even if the losses were to exceed the total assets of the company, the shipowner would not be liable beyond this investment. Moreover, there are also basic insurance systems available to the shipowner, such as liability insurance, covering the shipowner against claims of third parties, and property insurance, covering the shipowner from damage to his vessel.[107]

In response to the above arguments, it is said in favour of the limitation of liability that such a limitation makes P&I insurance feasible. However, this can be answered by pointing out that such arguments cannot be supported because economic forces will create an adequate market for potential liability once the need exists.[108] Moreover, it is also contended that exceptionally high insurance premiums are normally imposed only upon large passenger ships and so limitation can be repealed with respect to cargo ships and small passenger ships.[109] Since insurance premiums constitute a substantial portion of the shipowner's expenses, limitation is merely a means of restricting business costs rather than a device to limit the shipowner's loss or to protect his investment. Consequently, it is asserted that limitation is merely a subsidy to the maritime industry which is unusually granted, not by the treasury but rather by fortuitously selected shippers, passengers and crew members.[110]

It is said that the only modern proposition that it is in the interest of society at large is that shipowners should be permitted and indeed encouraged to remain in their traditional businesses, performing it in the traditional way.[111] This proposition may be right, but there are an increasing number of contrary opinions. It has been stated that the principle of limited liability is that full indemnity, a natural legal right, should be curtailed for political reasons.[112]

The world is changing rapidly. In the light of recent situations where shippers are subject to more duties and unlimited liability and where the charterers have no right to limit their liability against the shipowner, it would be interesting to see developments on the limitation of liability.

G. Insurance

I. Marine insurance in general

Activity often combines the opportunity of advantage with the risk of loss. Thus, the financial rewards of international trade have to be offset against the numerous risks involved in transporting goods, particularly in the potentially dangerous environment of the sea. A person engaged in international commerce could bear

---

[107] *Ibid.*
[108] *Ibid.* at 364.
[109] *Ibid.*
[110] *Ibid.* at .390.
[111] Mustill, "Ships Are Different,or Are They?" [1993] *LMCLQ* 490, 500.
[112] *The Amalia* (1863) Br. & L. 151.

the risks himself, with legal recourse for compensation against anyone who has caused him a loss by breach of contract or tort, but otherwise relying on his own resources. However, the adverse risks of a particular adventure or type of trade are a disincentive to participation in commerce. This is not only against the interests of individual traders but also against the public interest in promoting and benefiting from trade.

In the past, one could identify no fewer than four categories of marine coverage: cargo, hull, freight and builder's risk. Later on, three great sectors in the practice of marine insurance were defined: cargo, hull and freight, with protection and indemnity (P&I) insurance as virtually a footnote to the hull business. Today, it is impossible to envisage a marine industry without P&I insurance; it would be unthinkable to omit it from any description of major marine departments. While cargo, hull and P&I may be viewed as the true modern division of marine insurance, the increased sophistication of the business and the diverse interests of international commerce and finance have created insuring arrangements that ultimately defy the neat compartmentalization of coverage. For this reason, it is perhaps best to divide marine insurance into the same basic categories applicable to all commercial insurance: property and liability.

Given the potential risks, shippers and charterers seek to protect themselves against these risks by obtaining insurance. Charterers and shippers obtain insurance policies to cover the risks associated with liabilities. A charterer usually obtains legal liability insurance for charterers, which will cover the liability risks associated with a charterer's legal liability. In fact, there is a number of P&I clubs which are available solely to their members, who are charterers that will insure such risks. General cargo shippers, on the other hand, could take out general liability insurance suiting their needs.

II. Protection and indemnity (P&I) insurance

Generally speaking, P&I insurance is third-party liability insurance covering vessel owners for specifically named risks. P&I policies do not provide comprehensive general liability coverage, nor do they cover all possible risks or liabilities to which a vessel owner may be exposed. Instead, a P&I policy only insures the vessel owner against those risks specifically set forth in the policy and no others. P&I insurance was developed to provide protection to shipowners for losses arising in connection with the operation of their vessels. This relatively modern coverage provides protection not usually obtained in other standard marine policies.

A great portion of the P&I market is composed of the P&I "clubs", which are akin to mutual, non-profit insurance companies. Their development stemmed from the shipowners' need to obtain protection from the increased liability exposure for the losses of passengers and crew. It can certainly be said that, along with developments in shipping and the substantially increased regard for life and property, the range of solvency-threatening liabilities faced by shipowners and operators has increased enormously. Modern P&I insurance covers two broad categories of exposure: shipowner third-party and contractual liabilities. Although P&I clubs generally operate in a manner similar to other mutual insurers, their formality of operation differs.[113] Clubs mutually do not issue policies *per se*, and the assured are referred to as members.

1. P&I club cover

There are two principal forms of insurance in respect of liabilities to third parties. One form is know as "liability" insurance, which places an obligation on the insurer to pay any damages which the assured is likely to pay as a result of occurrences which are defined in his insurance cover.[114] The insurer's duty to indemnify arises on the occurrence of the insurance event. The other form of third-party insurance is "indemnity" insurance, which places an obligation upon the insurer to reimburse or indemnify an assured only to the extent that the assured has incurred and discharged his liability.[115] The insurer's duty to indemnify does not arise until the assured has paid damages to the third party. The type of insurance cover provided by P&I clubs has traditionally been one of indemnity.

2. Marine risks insured by P&I clubs

The insurance cover available from P&I clubs has been described as a somewhat miscellaneous group of leftovers.[116] New types of cover have emerged as social changes and technological and legislative developments in the past century and a half have demanded. A shipowner with full protection and indemnity cover will be insured against liabilities in respect of everything from catastrophic oil spills, dramatic and costly collision and extreme loss of life to feeding stowaways, compensating passengers for loss of luggage and bailing out drunken crew members.

A shipowner's hull and machinery insurance is designed primarily to protect the assured against losses to his vessel, whereas protection and indemnity insurance seeks to indemnify an owner in respect of the discharge of legal liabilities he has incurred in operating his vessel.[117]

---

[113] Ships are "entered", i.e. become insured, if the club accepts the vessel. If acceptance is granted, a certificate evidencing this fact is issued. Although no policy is issued, coverage is detailed in the association's "Rules", which effectively function as policy language. The Rules pertain not only to matters of coverage, but also extend to such considerations as reserves, premium (calls) meeting and disputes.

[114] Hazelwood, *P&I Clubs Law and Practice* (2000), 141.

[115] *Ibid.*

[116] *Ibid.* at 152.

[117] The old hull clubs developed into "protecting" or "protection" clubs and the cover in respect of "indemnity" was added later, in most cases as a separate class. The protection class was regarded as having reference to liabilities and accidents arising from the ownership of a vessel, whereas the "indemnity" class of cover was regarded as having reference to the liabilities which arise from the employment of the vessel. This meant that insurance cover in respect of cargo liabilities and trading was included in the indemnity class, whereas all other types of liability relating to the management and navigation of the vessel were covered in the protection class. Nevertheless, there was

A typical P&I club rule-book provides indemnity insurance in respect of a member's liability arising from the following events: collisions and non-contract damage; damage to fixed and floating objects; cargo claims; property on board; loss of life, personal injury and illness; passengers; crew liabilities; supernumeraries and others on board; fines; inquiries and criminal proceedings; quarantine expenses; stowaways, etc.; diversion expenses; life salvage; unrecoverable general average; ship's proportion of general average; liabilities relating to the wreck of the entered vessel; pollution; towage contracts; expenses incurred pursuant to directions of the club; and the "Omnibus Rule."[118]

3. Special risks insured by the P&I clubs

In addition to the usual risks insured for the benefit of ordinary members, clubs are also prepared to agree special terms for owners engaged in trades which involve special risks peculiar to the trade or type of vessel operated by the owner. Such special cover usually involves a separate agreement with the club and payment of an additional premium. Some examples are charterers, salvors, off-shore operations and containers.

4. Club membership

Club Rules which deal with the eligibility for insurance are usually widely framed. Such rules provide that the club may enter into a contract of insurance in respect of any vessel with any of the following persons: the registered owner, owners in partnership, owners holding separate shares in severalty, part-owners, mortgagees, trustees, charterers, operators, managers or builders of the vessel, any other person whatsoever interested in or in possession of the vessel or any other person who in the opinion of the managers should be regarded and treated as any of the persons previously mentioned. The categories of person which the clubs will admit as "club members" are in fact more restricted than this all-embracing catalogue of persons.[119] The most usual members of a P&I club are shipowners or ship operators. Although mortgagees and charterers can acquire the protection of a shipowner's P&I club, they are rarely "members" in the full sense of the term.

5. Insurable interest

In order to gain entry to a P&I club, the prospective member must, as with any insured person, have an insurable interest. Not all persons with an insurable inter-

---

never any clear definition of protection and indemnity and no clear borderline between the two.

[118] This rule provides that the committee shall have absolute discretion in paying a member any amount in respect of losses, liabilities, costs or expenses incidental to the businesses of owning, operating or managing vessels which in the opinion of the committee of directors falls broadly within the scope of the club's cover.

[119] Hazelwood, *P&I Clubs Law and Practice* (2000), 81.

est, however, are entitled to join a P&I club. A shipowner has an insurable interest in liabilities which he may incur by owning or operating his vessel. In order to become a P&I club member, a person must be subject to the potential liabilities in respect of which he seeks protection and indemnity.

6. Members

a) Shipowners

From a study of club rule-books and other documentation, it could appear that it is the named vessel which is insured in the P&I club rather than the owner of the vessel. Entry is effected by registering the name of the vessel with the club and the club agrees to insure losses arising in connection with it. It is not the vessel itself, however, that is entered or insured as such. The registration of a vessel in a club's register of members is simply a convenient device for defining the extent of the club's liability. Clubs generally provide in their rules that their protection extends to members only in respect of "entered vessels" and only in their capacity as ship-owners or otherwise.

b) Chartered vessels

There are clubs designed specially for charterers, offering defence cover in respect of legal costs of pursuing and defending claims and P&I cover providing insurance for charterers' liabilities. Additionally, charterers continue to seek protection from the traditional shipowners' P&I clubs.

In respect of charterers, clubs have separate conditions for their insurance and do not give a general consent to insure time charterer's interests under owner's insurance. Charterers who are not operators or demise charterers, but merely time or voyage charterers may become members of P&I clubs. The normal practice is for the operator or shipowner to enter his vessel in respect of his interests and for the charterer to enter as a member in respect of his interest so that both become members of the club.[120]

In the case of demise or bareboat charters, the ordinary P&I insurance for the vessel is normally taken out by the operator, that is, the demise charterer. Nevertheless, the owner of the demise chartered vessel remains in need of protection and indemnity insurance, particularly in respect of those claims which give rise to maritime liens and those other claims for which he will be liable as a shipowner. Such an owner will normally obtain protection and indemnity cover under the insurance of the demise charterer or operator.

c) Charterers as members

Members of shipowners' P&I clubs who are charterers are usually charterers by demise, time or consecutive voyage charterers, but there is nothing in the rules of

---

[120]   *Ibid.* at 98.

shipowners ' clubs which would prevent a mere voyage charterer from effecting entry in respect of a voyage.[121]

Charterers may become separate members of a club which is not the owner's club, or separate members of the owner's club, or may join the owner's club under a family arrangement whereby both shipowner and charterer enter the same club as co-assured parties and co-members.

A charterer's insurance cover is usually more limited than that required by a shipowner. Where both owner and charterer of the vessel are entered in the same P&I club, occasions may arise where there will be a conflict of interest between the two members.[122] Charterers often arrange for a fixed premium entry rather than enter under the usual mutual underwriting system.

The most likely type of charterer seeking P&I coverage will be time charterers, but the clubs will also consider covering so-called "space" and "slot" charterers. Some traditional P&I risks will be applicable to charterers whereas others will not. Charterers will be concerned with traditional owner-type risks such as liabilities for loss of or damage to cargo, for loss of life and personal injury, oil pollution[123] and any other liabilities to third parties to which charterers are subject to by virtue of their being equated with shipowners. Also, charterers will require indemnities in respect of liabilities, losses and expenses which are peculiar to charterers as opposed to owners.[124]

The charterer's liability to the owner will depend, of course, upon the wording of the applicable charterparty and charterers will seek insurance against liabilities which they might have to the owner. Most shipowners' P&I clubs offer full ownership type P&I cover to charterers together with specially agreed terms.

---

[121] *Ibid.* at 99.

[122] Such instances may arise particularly where cargo interests, in respect of lost or damaged cargo, proceed against owner and charterer as carrier. In such a case, it would appear that although the claims-handling department of the club will deal with both of its members, it is contrary to the club's practice for claims against members with conflicting interests to be handled by the same individuals within the claims department. Where the conflict develops into a dispute which requires litigation, the particular club will, of course, instruct separate firms of lawyers to deal with the separate conflicting interests.

[123] Most states besides the USA are party to the CLC under which pollution claims are channelled solely to registered owner, if the cause of the pollution is something for which the charterer is responsible then the charterer may be obliged to indemnify the owner. On the other hand, in certain jurisdictions, Calirofnia, Alaska, Japan, a charterer can incur direct and even strict liability for pollution caused by the ships.

[124] Wu, "What are the key charterers' risks?", *Club Cover* CLC01/07, 2 f. <www.ukpandi. com/ukpandi/resource.nsf/Files/charterers%20brochurejan2007/$FILE/charterers+broc hurejan2007.pdf> (visited 3.3.2007).

7. Charterers' liability for damage to hull

At one time, shipowners' P&I clubs refused to provide indemnities to charterers in respect of their liabilities for loss or damage to the chartered vessel.[125] The view was that the clubs do not cover property damage, the insurance of which should be obtained on the proprietary market as with hull and machinery insurance.[126] The clubs would, as agents, arrange such cover for charterers as separate insurance on the proprietary market

The charterers' special club did offer this cover within its rules and now most shipowners' P&I clubs have extended their cover for charterers to include this type of insurance. This cover is in respect of a charterer's liability to the owner in respect of loss or damage to the vessel itself and any equipment, including navigational equipment, stores or bunkers, which is owned or leased by the owner or any associate of the owner. Where any such items are owned not by the owner but by the charterer, these would not be insured under the charterer's P&I entry, as this would be the charterer's own property and should be insured as such elsewhere. This cover is only in respect of charterer's liability to the owner and not in respect of losses to the charterer's own property.

8. Conditions of cover and dangerous goods

Since P&I cover is based on mutuality, it is expected that each Member should fulfil certain common conditions and maintain minimum standards. These are general requirements, not directly or necessarily linked to the specific dangers of dangerous cargoes. One basic principle is that a member should not knowingly assume a risk or unreasonably ignore a risk which other members are taking proper steps to avoid.[127] Mutual insurance is based upon the idea that, whilst errors, mistakes and the inevitable risks of shipowning are covered, deliberate wrong-doing or a cynical disregard for basic precautionary measures should not be insured.

Hence, no claim shall be recoverable from the Club if it arises out of or loss is consequent upon an entered vessel carrying contraband, running blockades or being employed in a trade unlawful in the countries of shipment, destination or transit or performance.[128] Moreover, if a member is in breach of contract of carriage and that breach of contract arises from, or is constituted by, the wilful or reckless act or default of the Member, there shall be no recovery in respect of the Member's liability or expenses arising out of that breach of contract. So far as dangerous cargoes are concerned, a charterer is expected to take particular care to

---

[125] Hazelwood, *P&I Clubs Law and Practice* (2000), 231.

[126] *Ibid.*

[127] Martin, "HNS: A P&I Club Perspective", in *The Transportation of Hazardous Cargoes by Sea: Managing Your Risks and Undertaking the Consequences of the Law* (1993), 16 f. This exception in club rules is the warranty implied by the Marine Insurance Act 1906 that the adventure insured is a lawful one and will be carried out in a lawful manner. Sec. 41.

[128] *Ibid.*; Hazelwood, *P&I Clubs Law and Practice* (2000), 176.

stow and handle the goods properly. He cannot knowingly and recklessly ignore the risks associated with the cargoes and refrain from taking the necessary precautions. Therefore, if a member loads dangerous goods and ignores dangerous goods regulations, thus knowing that the risks of carriage will be increased as a result, he cannot expect to be covered by the club if the worst happens and a casualty ensues.[129] However, there may be cases in which even strict compliance would not be sufficient to prevent the damage, i.e. unknown dangers, for which charterers need protection.

III. Charterer's liability insurance

Alternatively, the charterer may choose liability insurance in the open market using the standard wording. Under a usual clause, the policy would cover the liability of the charterer for damage to the hull and the liability of the charterer for damage to the cargo.[130]

Like the charter contract that forbids the carriage of certain goods, a clause can occasionally be found in the insurance contract in which the insurer grants no coverage for liability claims which are due to the transportation of cargo that is contrary to the terms of the agreement.[131] As we have already seen, many charterparties exclude the shipment of dangerous cargo unless notice is given of its nature. Consequently, if such an exclusion clause is contained in the insurance contract, the charterer has no insurance protection for the whole area of liability for damage due to the transportation of dangerous goods contrary to the agreement.

Furthermore, like the P&I exclusion with regard to the violation of dangerous goods regulations, there might be provisions excluding the damage caused by such a violation. For instance, German DTV Hull Clauses Art. 14.1 which replaced the Art. 60 of the German General Rules of Marine Insurance[132] provides that "Underwriters do not cover loss or damage resulting from goods which are precluded from carriage or which are only admitted for carriage in accordance with the German regulations for the carriage of dangerous goods if these regulations were violated in connection with the carriage and the loss or damage is due to such violation, unless the assured proves that he has complied with the regulations and has performed everything necessary to secure the observance thereof in connection with the carriage or neither knew or ought to have known of such carriage." As a result there is no coverage if the GGVSee[133] is violated; but the liability of the insurer still remains when the charterer is not at fault as regards the violation

---

[129] Martin, "HNS: A P&I Club Perspective", in *The Transportation of Hazardous Cargoes by Sea: Managing Your Risks and Undertaking the Consequences of the Law* (1993), 19.

[130] For the coverage of charterers' liability insurance, see Schwampe, *Charterers' Liability Insurance* (1988), 24 ff.

[131] *Ibid.* at 107.

[132] Allgemeine Deutsche See-Versicherungsbedingungen (ADS).

[133] Or local rules where the shipment is from a foreign port. DTV Hull Clauses Art.14.2. See *supra* Part 1 and 2.

of the regulations or the loading with dangerous goods.[134] This, however, only affects the scope of liability of the charterer's liability insurance insofar as the insured is liable regardless of fault, i.e. under the dangerous goods provision HGB § 564b.[135]

English policies generally contain no general exclusion of cover for liabilities caused by the transportation of goods contrary to the terms of the contract.[136] The insurer's liability for special types of cargo which appear particularly likely to cause damage is excluded or restricted on other ways: Either condition is included in the policy, which states that the charterer is expressly permitted in the charter contract to transport the goods. In this case because it does not come first to a liability of the charterer for damage caused by the cargo, there is therefore no obligation for the insurer to provide cover.[137] Or a list of "non approved cargoes" is appended to the policy, for which a considerably higher franchise is agreed.[138]

It is obvious from the wording of general conditions that failure to conform to applicable rules and regulations prevent insured's recovery.[139] Thus, any violation of dangerous goods regulations bars recovery.

IV. Insurance for shippers

1. Public liability insurance

Public liability insurance is a broad residual class of insurance intended to cover claims that are not met by specific forms of cover. Claimants are businesses or members of the public who suffer injury to their property or bodily injury through the negligent conduct of business activities. For the vast majority of businesses, there is no primary legislation that requires them to buy liability insurance. That is the case in the carriage of dangerous goods by sea. Neither German nor English

---

[134] Schwampe, *Charterers' Liability Insurance* (1988), 108.

[135] Because where liability is dependent on fault, in conjunction with the regulations of the GGVSee and the positive violation of contract, the lack of fault in the matter of the contravention of the regulations or loading of dangerous goods nullifies the liability of the charterer, so that to this extent there is no question of an insurer's obligation to provide cover.

[136] Schwampe, *Charterers' Liability Insurance* (1988), 116.

[137] As it is accepted that carrier has agreed to carry risk. See *supra* p. 159 ff.

[138] *Ibid.*

[139] Charterer's Liability Insurance Conditions 2002 LSW 1122 Art. 2.A.9 titled "Carriage of Cargo" provides that:
Cover under this insurance is only in respect of cargoes listed as Approved Cargoes in the Policy. The assured shall exercise due diligence so far as it is within The Assured's control to ensure that cargo:
  (a) conforms in type, quality and quantity to that permit in the Charterparty; and
  (b) is carried and stowed with the approval and consent of the Owner and/or Master of the Insured Ship; and
  (c) is carried and stowed in conformity with all relevant international, national and local conventions and regulations.

nor American legislation on the transport of dangerous goods by sea requires liability insurance. In practice, however, it is essential for majority of businesses and particularly essential for businesses dealing with the shipment of dangerous goods, where the shipper is subject to strict liability, to take out insurance protection.

There is no standard liability for public liability insurance. The policies may differ somewhat from one insurer to another. Moreover, many marine insurers can arrange tailor-made insurance coverage suiting the customer's needs as far as possible. Public liability insurance policies generally exclude pollution damages. Additional coverage for pollution damage may be taken out. Furthermore, in most countries damage caused by products manufactured or sold is covered by separate products' liability insurance. In the case of carriage of goods the product liability cover may be relevant to the manufacturer and shipper of end products who sold these.[140]

2. Shipper may run hazard of hazard

In spite of the possible tailor-made marine insurance coverage, shippers' marine insurance policies are commonly written to exclude coverage for damage caused by hazardous materials. In an action in *Zen Continental Co. v. Intercargo Insurance Co.*,[141] factually related to the *Senator Linie*,[142] where the shipper was found strictly liable for the shipment of dangerous cargo, one of the NVOCC,[143] which was shippers in the given case, sued its liability insurer to recover clean up costs and fines imposed and the costs of defending a lawsuit for damage arising out of the carriage of shipments of toxic chemicals, including the TDO carried on the *Tokyo Senator*.[144]

The defendant insurer refused to defend or indemnify its insured on the grounds that the policy of insurance excluded toxic pollutants from coverage and had a one-year statute of limitations. The policy styled in "International Transit Liability Insurance Policy" covered certain risks attendant on the insured's businesses as an NVOCC. As is typical of most insurance policies, the policy at issue contained numerous exclusions which had the effect of denying coverage for certain risks.[145]

---

[140] Schultsz, "Insurance aspects of shippers' liability" in in Grönfors (ed.), *Damage from Goods* (1978), 60, 63.
[141] 151 F.Supp.2d 250.
[142] 291 F.3d 145; see *supra* Part 4.
[143] Non-vessel-operating common carrier.
[144] 151 F.Supp.2d 250.
[145] The relevant exclusion entitled "Hazardous Materials/Pollution/Contamination" provided that:

"We will not cover any claims for environmental damage, pollution, or contamination of any kind however caused, including but not limited to: claims arising out of accidental, sudden or gradual, foreseeable, intentional or unintentional occurrences.

We will not cover any claims arising out of any activity, transaction, incident or occurrence involving any explosives; pressurized gases; nuclear parts, fuels, materials

On motion and cross motion for summary judgment, it was held that the policy was not intended to provide the insured with coverage given the broadly-worded hazardous materials/pollution exclusion, which provides that "any claims arising out of any activity, transaction, incidence or occurrence involving materials that are "hazardous" or "toxic" will not be covered.[146]

Imposing strict liability on the shippers for the shipment of hazardous goods without the knowing consent of the carrier but having no coverage for this liability is somehow questionable. In *Senator Linie v. Sunway Line*,[147] the court noted that imposing strict liability under the COGSA § 1304(6) should not alter insurance industry as shippers currently know it, but instead will establish a clear allocation of risk that should assist maritime actors and insurers in preparing for the marginal possibility of an unknown hazard.[148] In the case in question, the shipper did not show that additional insurance coverage for unknown inherent danger is unavailable or prohibitively priced.[149] Moreover, the court said that to the extent that federal laws already regulate the transportation of known hazardous and dangerous materials[150] and impose civil and criminal penalties for knowing violations of these laws, the interpretation of § 1304(6) in imposing strict liability should not render shipper liability incoherent and uncertain, but should instead fill a gap with respect to materials whose inherent hazards are unknown to regulators.[151]

V. Coverage for liability arising out of dangerous goods

Liability insurance for hazardous materials is available in some parts of the world. Except a few countries[152] insurance for hazardous materials is not compulsory. This type of insurance is a scarce industry commodity and one of the most expen-

---

or devices; hazardous, radioactive, toxic; ... or flammable materials; any weapons or armaments; or any means of biological or chemical warfare.

Further, we will not cover claims arising out of the actual, alleged or threatened discharge, disposal, release or escape of pollutants in any stage or storage, handling or transportation; ... whether accidental, sudden or gradual, foreseeable or unforeseeable, intentional or unintentional.

Pollutants mean but are not limited to; any solid, liquid, gaseous, thermal, radioactive, sonic, magnetic, electric or organic irritant; contaminant; or anything which causes or contributes to damage, injury adulteration, or disease. This includes, but is not limited to smoke, vapour, soot, fumes, acid, chemicals and waste."

[146] 151 F.Supp. 2d 250, 265.
[147] 291 F.3d 145
[148] *Ibid.* at 169 (fn. 38).
[149] *Ibid.*
[150] *Supra* Part 1.
[151] 291 F.3d 145, 169, (fn. 38).
[152] U.S. 49 CFR 387.9 requires compulsory liability insurance for road vehicles with coverage between $ 1,000,000 and 5,000,000. Giermanski/Neipert, "The Regulations of Freight Forwarders in the USA and its Impact on the USA-Mexico Border" [2000] 9 *WTR Currrents: Int'l Trade L.J.* 11, 16. In Turkey there is compulsory liability insurance on dangerous substances.

sive insurances.[153] Thus, it seems unlikely that many shippers have liability insurance for dangerous goods.[154]

Some years ago insurers have curtailed insurance coverage for many hazardous materials and for firms disposing waste.[155] In order to provide coverage for any industrial activity which has the potential for a very high "maximum credible loss" an insurer must put together a team of co-insurers and re-insurers on the risk.[156] Without re-insurers, direct insurers would be unable to provide third party coverage for many industrial firms. Thus availability of particular insurance depends on the capacity of insurance market.

As mentioned in Part 1 Convention on Civil Liability for Cause during Carriage of Dangerous Goods by Road, Rail and Inland Navigation Vessels (CRTD) was done on 18 October 1989; however so far only two states signed the convention. In order to investigate what difficulties would prevent accession to the CRTD a questionnaire was prepared. Replies to questionnaire show that one of the main concerns was high limit of liability and thus difficulty in finding insurers to cover this limit.[157] The fact that dangerous materials insurance is one of the most expensive type of insurances and shippers and charterers have no right of limitation against carrier or shipowner respectively, make debatable the shipper's liability insurance for dangerous goods.

As seen below in Part 6 shipper interests have always hesitated being liable for the shipment of hazardous substances. Indeed, in the light of cases involving dangerous goods, it is the fact that shippers may be held liable for great amounts under the contract of carriage. With the increase of such occasions shipping and marine insurance industry would adjust this field.

---

[153] In the U.S. it is estimated that depending on materials handled, minimum premium for this insurance would be $15,000 per year. *Ibid.*

[154] Webster, "Managing Risk" [2004] 18(9) *M.R.I.* 15, 16.

[155] Baram, "Insurability of Hazardous Materials Activities" [1988] (3.) *J. Statist. Sci.* 339, 340.

[156] *Ibid.* at 342.

[157] UNECE Trans/WP.15/2001/17/Add.2, Trans/WP.15/2001/17/Add.4.

Part 6: Third-party liability for damage arising from the carriage of HNS

A. In general

So far we have seen liability issues in terms of contractual relation between shipper and carrier. In this part third party liability arising from the shipment of hazardous substances will be considered.

The volume of hazardous and noxious substances (HNS) other than oil carried by sea has constantly increased. However, the international community did not initially perceive a strong need for an international liability scheme for HNS accidents.[1] It generally lacked experience in dealing with HNS pollution damage and the IMO doubted the likelihood of high clean-up costs. As world opinion focused on the problem of oil pollution from tankers during the 1970s, the international marine community became concerned about damage caused by hazardous and noxious substances released into the marine environment after accidents to the vessels carrying them. Since dangerous chemicals are more easily dispersed in the marine environment and their levels of toxicity are likely to be higher than that of oil, there is a greater premium on quick, efficient and consequently expensive clean-up operations of HNS.[2] A number of incidents affecting European coasts in the early 1980s have evidenced the magnitude of the problems associated with chemical spills. The international maritime community was also concerned with the risk of chemical explosions resulting from accidents with HNS-carrying vessels. Aside from international legislation prohibiting the intentional dumping of wastes and providing compensation for nuclear incidents, no effort had been made to fashion an international regime to address this problem. Thus, it was almost inevitable that international attention would turn to this issue.

In 1984, the IMO convened a conference to consider a new instrument for dealing with compensation for accidents involving hazardous and noxious sub-

---

[1] Schuda, "The International Maritime Organization and the Draft Convention on Liability and Compensation in Connection with the Carriage of Hazardous and Noxious Substances by Sea: An Update on Recent Activity" [1992] 46 *U. Miami L.Rev.*1009, 1017; De Bièvre, "Liability and Compensation for Damage in Connection with the Carriage of Hazardous and Noxious Substances by Sea" [1986] 17 *J.Mar.L.&Com.* 61, 66; Goñi, "Watching from Coastal Shore the Passing through of Vessels Loaded with Crude Oil or Dangerous Merchandises" [1991] 26 *E.T.L* 143, 144.

[2] Sasamura, "Development of the HNS Convention," in *13th International Symposium on the Transport of Dangerous Goods by Sea and Inland Waterways* (1998), 491, 492.

stances, but the issue proved to be so complex that the attempt had to be abandoned. It was not until 1996 that the matter could be considered again, but this time the attempt was successful. The HNS Convention was adopted by a diplomatic conference held in May 1996 under the auspices of the IMO.[3] The Convention aims to ensure adequate, prompt and effective compensation for damage to persons and property, the costs of clean-up and reinstatement measures and economic losses caused by the maritime transport of hazardous and noxious substances.

B. Need for a HNS Convention

I. Potential damage

HNS incidents pose significant risks due to the immense and growing amount of international trade in hazardous and noxious substances, and potential damage resulting from pollution, fire or explosion can be enormous.[4]

Numerous factors influence the magnitude of the damage resulting from an HNS incident. These factors include the type and character of the substance released, the particular characteristics of the accident, the quantity of the substance released, the physical and meteorological conditions at the time of the incident, the ecological sensitivity of the body of water and coastline to pollution, the technical difficulties arising from the incident, the use of the body of water for navigation, recreation or fishing and the population and type of property located near the incident.[5]

II. Inadequacy of existing liability schemes

In general, existing conventions dealing with the transportation of hazardous and noxious substances emphasize prevention and safety. For instance, the MARPOL 73 regulates the control and handling of bulk and packaged noxious and harmful

---

[3]   The liability and compensation regime in relation to hazardous substances releases from vessels in U.S. waters is governed by the Comprehensive Environmental Response, Compensation and Liability Act (CERCLA) and the Federal Water Pollution Control Act (FWPCA). For more information see Edgcomb, "Hazardous Substance Releases from Vessels: Current U.S. Law, the HNS Convention and Its Potential Impact if Ratified" [1997] 10 *U.S.F. Mar. L.J.*73 ff.

[4]   For incidents involving hazardous and noxious substances see *Monitoring Implimentation of the Hazardous and Noxious Substances Convention, Report on Incidents Involving HNS,* IMO LEG 85/INF.2; McKinley, *The International Convention on Liability and Compensation for the Carriage of Hazardous and Noxious Substances by Sea: Implications for State Parties, the Shipping, Cargo and Insurance Industries* (Diss. Cape Town 2005), 7 ff.

[5]   Pawlow, "Liability for Shipments by Sea of Hazardous and Noxious Substances" [1985] 17 *Law & Pol'y Int'l Bus.* 455, 458.

substances, oil, sewage and garbage in an attempt to eliminate intentional pollution and reduce accidental pollution of the marine environment. The existing maritime shipping liability and compensation treaties include CLC 1992, the Fund Convention 1992 and the LLMC 1976 Convention. The CLC and Fund Convention provide limited compensation for pollution damage and some funding for measures to prevent the discharge of oil from ships, but do not provide compensation for damage caused by hazardous substances. The LLMC 1976 Convention and its 1996 Protocol provide compensation for HNS-related damage, but it applies to a wide range of claims and the limit of liability is inadequate.

C. Development of HNS Convention

I. In the wake of the 1969 Brussels Conference

The governments represented at the Brussels Conference, which in the aftermath of the *Torrey Canyon* disaster adopted the 1969 CLC and the International Convention Relating to Intervention on the High Seas in Cases of Oil Pollution Casualties ("The Intervention Conference"), urged that the IMO should intensify its work on all aspects of pollution by agents other than oil. The issue was also raised by the Legal Committee of the IMO at its 17th session, recognizing the need to solve the problems raised by marine pollution by agents other than oil as a matter of urgency. However, the general consensus was that any further work could not be undertaken until the necessary technical information concerning the types of pollutants other than oil and their polluting potentialities became available.

As an international conference on marine pollution was scheduled to be held in 1974 and a convention for the control of vessel-source pollution was envisaged, the Committee delayed its decision until a later date. As the preparation of this conference proceeded, including the conduct of studies designed to provide and analyse further technical information on pollutants, it was felt that the Committee should consider the extension of both the private and public law conventions to cover pollution damage caused by hazardous and noxious substances other than oil, at least in the initial stages.[6] Relevant technical data were made available to the Committee's 13th session in 1972, which involved a list of about 250 substances by GESAMP which were considered to be capable of causing pollution damage and presenting appreciable degrees of environmental hazard[7] because of their inherent nature or because of the quantities in which they were carried.

Although the technical information was available, the general opinion of the Committee was that "immediate progress in the extension of civil liability and compensation principles and rules to pollutants other than oil remained handi-

---

[6]   De Bièvre, "Liability and Compensation for Damage in Connection with the Carriage of Hazardous and Noxious Substances by Sea" [1986] 17 *J. Mar. L. & Com.* 61, 63.

[7]   Such as human health hazards, hazards likely to affect the quality of the marine environment or marine resources and hazards merely diminishing or deteriorating amenities.

capped by lack of experience." The main argument was that the international community was not in need of international regulations in this respect. To put it another way, most government delegations pointed out that they had lack of experience with maritime accidents involving substances other than oil and causing significant damage.

Despite the serious concern of the international maritime community with the increased dimensions of the risks posed by the maritime transportation of dangerous cargoes, regulatory efforts remained limited to the formulation and adoption of technical safety rules and standards aimed at damage prevention. Civil liability and compensation matters were clearly not of immediate concern for there to be special regulations to provide recoveries for third parties suffering serious damage, including property damage, loss of life, and damage to health, as long as there was a lack of experience in the sense of recurrent maritime accidents of a harmful kind.

II. Consideration of a limited extension of the CLC 1969 and the elaboration of a new convention for damage by other pollutants

To obtain further information and comments as to the need for an extension of 1969 CLC, the Committee prepared and approved a questionnaire for submission to governments and interested organizations at its 18th and 19th sessions in 1973. The result appeared that although there were not enough data available on a comprehensive basis concerning all the possible types of substances capable of causing pollution damage of a specific and serious nature, "it would be undesirable and dangerous to wait for a major catastrophe before taking action" with respect to the development of a legal regime covering liability and compensation.[8] However, in the subsequent discussion of the Committee it appeared that doubts continued to exist on the urgency of proceeding with further work on the subject as a matter of priority. Due to many practical difficulties neither immediate international action on the drafting of a protocol to the 1969 CLC nor a new independent convention was feasible. Some delegations said that they could not see a real need for the creation of a new liability and compensation regime. They argued that the nature of most types of marine pollution by chemicals and other hazards and noxious substances was such that clean-up costs were unlikely to be high and the higher limitation amounts most likely to be adopted by the LLMC 1976 Convention should solve the problems that could arise in connection with damage other than pollution damage resulting from the transportation of dangerous goods by sea.

Nonetheless, the Committee decided to devote time at its next session to further discussion of the possibility of beginning work immediately on a CLC protocol of limited scope of application, involving only an extension to those oils not yet covered by the CLC 1969, i.e. non-persistent oil. Apparently the Committee intended to deal with the whole subject of civil liability and compensation for oil-pollution damage arising from the carriage of hazardous and noxious substances

---

[8]   *Ibid.* at 66.

by adopting a step-by-step approach. Moreover, there was a consensus in the Committee that it would be desirable to cover in an entirely separate convention all hazards and liabilities relating to substances other than oil as comprehensively as possible and to elaborate such an instrument as the requisite technical information became available.[9] Furthermore, it was suggested that the large gap between the 1969 CLC and the "nuclear" conventions should be filled by a single, comprehensive convention.

A new questionnaire was prepared to obtain further information and comments for the purpose of determining how to proceed in possibly preparing a limited extension of the CLC 1969 to cover non-persistent oil and the elaboration of a new international scheme regulating liability and compensation for damage arising from all remaining uncovered risks. The responses showed some difficulty in providing definite scientific criteria for the development of a sufficiently precise or manageable list of substances to form the basis of an international liability and compensation system.[10] Eventually, at its 32nd session in 1977, the Committee expressed the general view that an extension of the CLC 1969 could well be desirable and should be envisaged in terms of a protocol only covering the kind of damage dealt with by that Convention, i.e. pollution damage. Moreover, it was decided to proceed to the preliminary discussion of a new and comprehensive convention to provide a civil liability system for damage from all substances other than oil considered to be particularly harmful, as the existing conventions did not satisfactorily deal with such damage.

III. The preparatory work for the draft on the HNS Convention

1. Proposed systems of liability

At the 33rd session in 1977, the highest priority was given to work on a new comprehensive convention to provide a liability and compensation system for hazardous and noxious substances other than oil.[11] The essential issue was the question

---

[9]   The reasons for proceeding in this way were that the risk and nature of damage which can be caused by other noxious and hazardous substances differed considerably from the risk and nature of damage from persistent and other oils. The system, therefore, contained in the CLC 1969, which involved placing primary liability on the shipowner, might prove to be inappropriate in respect of damage from substances other than oil, since in the case of those substances the risk would arise more from the inherent quality of the substance than from the act of transportation. It might be more appropriate for primary liability to fall on a party other than the carrier, such as the manufacturer, the consignor, the owner of the substance or any other party interested in the cargo. And the system contained in the CLC 1969 might be appropriate only in respect of cargo carried in bulk.

[10] It was particularly so, if the nature of damage to be covered by such a list was not confined to pollution in the strictest sense, but was to extended to cover other hazards, such as toxicity, fire and explosion. LEG XXXII/3 para.8.

[11] LEG.XXXIII/5 para.52.

of party liable, before the remaining basic issues could be properly resolved. Different systems of party liability were considered:[12]

- joint and several liability of shipper and shipowner
- a two-tier system of liability providing for a primary liability of the shipowner and an excess ("residual") liability of the shipper
- exclusive liability of the shipper
- shipowner liability alone
- cargo interests' exclusive liability as product liability.

Of all these systems, considerable support was given to some kind of shipper's liability.[13] The reason for that was the fact that pollution and other types of damage arising from maritime accidents involving hazardous and noxious substances result not only from inadequacies in the ship's conduct but also from the inherent harmful characteristics of the cargoes carried, as well as from substandard shipping practices, such as unsafe packing and inadequate or inexact dangerous goods declarations.[14] It must be pointed out that the concept of shipper's liability was a relatively new development in maritime law. Traditionally, maritime law places legal liability on the shipowner as he is the one having the highest degree of operational control over the way in which cargoes are handled during their carriage by sea. However, it does not necessarily mean that the shipowner has no means to protect himself against higher claims which might arise from incidents involving damage caused by hazardous cargo. The long-established practice in the case of the transportation of dangerous cargo is that the cost of additional insurance protection taken out by the shipowner is passed on to the cargo interests.[15] However, such an arrangement is very different from the concept of shared or exclusive shipper's liability, where the cargo interests are personally and directly liable for loss or damage, in spite of the fact that they have no custody over the goods during their carriage.[16]

---

[12]   *Ibid.* at 69.

[13]   Pawlow, "Liability for Shipments by Sea of Hazardous and Noxious Substances" [1985] 17 *Law & Pol'y Int't Bus.* 455, 465.

[14]   Shippers opposed, arguing that traditional maritime accidents, such as grounding, collision or fire, result from operator negligence and shipowners control crew training, vessel operation and navigation. Therefore, shippers could do little to reduce those risks of carriage. Schuda, "The International Maritime Organization and the Draft Convention on Liability and Compensation in Connection with the Carriage of Hazardous and Noxious Substances by Sea: An Update on Recent Activity" [1992] 46 *U. Miami L.Rev.*1009, 1026.

[15]   De Bièvre, "Liability and Compensation for Damage in Connection with the Carriage of Hazardous and Noxious Substances by Sea" [1986] 17 *J. Mar. L. & Com.* 61, 70.

[16]   The only existing approach involving personal and direct liability of the shipper is in the field of liability for nuclear damage. The 1963 Vienna Convention on Civil Liability for Nuclear Damage and the 1971 Brussels Convention Relating to Civil Liability in the Field of Maritime Carriage of Nuclear Materials extend the liability of the operator of a nuclear installation to damage caused by a nuclear incident during the transportation of nuclear substances by sea.

It must be noted that the primary reason for supporting shipper's strict liability was to establish a legal regime providing full and adequate recoveries for victims, particularly in those cases where catastrophic damage arose from the maritime carriage of highly hazardous and noxious substances other than oil, including not only pollution damage but also damage by fire, explosion and toxicity.[17] As the dimensions of the latter type of damage could be particularly massive, involving extensive property damage and/or wrongful death or major health hazards, it would be extremely difficult from the victim's point of view to disregard these types of damage.

2. A new source of insurance: imposition liability on shipper

The decision to proceed with the elaboration of a liability regime restricted to compensation for extensive damage caused by ultra-hazardous substances led to the majority suggesting that it was more appropriate to hold the shipper of particularly harmful cargoes strictly liable. Particularly, it was thought that internationally agreed global limits of liability for maritime claims might well be insufficient to provide satisfactory compensation for the exceptional hazards envisaged to be covered under the new convention. Therefore, the amount of additional insurance required might be found beyond the shipowner's ability to provide. This would require the imposition of liability upon another party and the search for a new source of insurance.[18] It would mean that the capacity and the cost-effectiveness of the existing cover made available by the insurance market for both the shipowner's general liabilities by LLMC 1976 and his oil pollution liability would be protected.[19] It was clear that the principle of shipper's liability offered a new and necessary means of safeguarding and promoting the growth and economic efficiency of the insurance market through the application of the principle of "risk spreading."

IV. The 1984 Conference on HNS

The debate for an HNS Convention comprised three discrete issue areas: defining the scope of the regime, allocating and limiting the liability, and establishing an institution for maintaining the system. The successful resolution of each of these issues was essential for the favourable outcome of the negotiations. As in any regime that aspires to allocate risk and reward for an economic activity, these issues tended to be both interconnected and competing, but had to be accommodated.

---

[17] LEG. XXX.IV/7 para.20.
[18] LEG. XXX.IV/7 para.22, 49, 61.
[19] De Bièvre, "Liability and Compensation for Damage in Connection with the Carriage of Hazardous and Noxious Substances by Sea" [1986] 17 *J. Mar.L &Com.* 61, 71.

1. Cargoes covered by the Draft HNS Convention

In creating a new legal regime, it is essential to determine what will be covered by that regime. The draft HNS Convention submitted to the conference a relatively short list of substances to be covered by the new Convention, limited to bulk cargoes. The draft excluded chemicals carried in packaged form, because only a small number of very specialized and easily identifiable vessels carried HNS in bulk. That would limit the scope of the Convention and make it more acceptable. If packaged goods had been included, the Convention would apply to many more ships because most commercial vessels can carry packaged chemicals.

a) Bulk HNS only

Favouring bulk cargoes only resulted particularly from difficulties encountered with the formulation of a practicable definition of the term "shipper." In other words, it was thought that identifying the shipper would be easier in the case of bulk transport, which normally involves only a very limited number of consignments carried by each vessel.[20] As a result, the exclusion of packaged HNS substances was justified on the basis that the new convention was aimed primarily at providing a liability and compensation system for catastrophic damage and for such damage caused by great quantities of hazardous substances, i.e. quantities carried in bulk. The controversy over whether to add packaged chemicals to those carried in bulk also illustrated how misleading it could be to translate the experience of drafting CLC 1969.[21] As a result, the list included only 45 noxious liquid commodities when carried in bulk.[22] The criterion of bulk as opposed to packaged form was not based solely on the magnitude of potential damage but was considered as the only one available to produce a reasonable list.

b) Problems in the inclusion of packaged HNS

The list of HNS to be covered by the new convention had been prepared by an "Informal Working Group of Technical Experts." The Group emphasized that it was practically impossible to produce a list of substances carried in packaged form, although in principle some substances carried in packaged form might in some circumstances cause catastrophic damage. Particular reference was made to the carriage of explosives listed in Class1 of the UN Transport of Dangerous Goods Code[23] which, although carried in packaged form, could cause extensive damage. Due to time and technical considerations, the Informal Working Group of

---

[20] This would highly facilitate the identification of the party liable in excess of the shipowner's liability in the event of an accident, as well as the enforcement of compulsory insurance requirements.

[21] Bederman, "Dead in the Water: International Law, Diplomacy, and Compensation for Chemical Pollution at Sea" [1986] 26 *Va.J.Int. L.* 485, 495.

[22] LEG/CONF.6/3, Annex. They were taken from the Bulk Chemicals Code, the Gas Carrier Code and the 1973 MARPOL Convention.

[23] See *supra* Part 2.

Technical Experts did not produce another list. It was also explained that the elaboration of a list of HNS carried in packaged form would entail a number of nearly insurmountable technical problems which would render practically impossible the retention of the bulk criterion which constituted the premise of the whole work.

Another significant issue was the identification of those substances on the bulk list which can also be carried in packaged form. In this respect it was pointed out that if the new convention were to cover also those cases in which the substances appearing in the bulk list could be carried in packages, the bulk list of 45 HNS would be rendered inappropriate because some substances carried only in packaged form are far more intrinsically dangerous than some on the bulk list.[24] The final recommendation of the group of experts assisting the Legal Committee was that the inclusion of the packaged substances capable of causing devastating damage would be inconsistent with the rationale of a convention in which the risks are to be associated with bulk carriage.[25]

Furthermore, it was impossible to state whether the damage likely to result from the explosion of such packaged explosives would be less or more extensive than the pollution capabilities of some pollutants which had been excluded from the list solely because they are carried only in packaged form. Although the Working Group of Technical Experts acknowledged the unique property of some packaged explosives to cause devastating damage, it was thought that the inclusion of any packaged cargo would ruin the whole scheme.[26]

Consequently, it appeared that it was impossible to propose a list of hazardous substances for the very purpose of civil liability. There were many basic uncertainties in scientific knowledge about the precise characteristics of a great number of hazardous substances as well their potential damage. A substantial number of governments, some of whom had already experienced costly maritime accidents involving such substances, decided that it was not possible to justify the exclusion of packaged HNS.[27] Some other governments recognized that the new liability and compensation adopted might well lack the necessary political credibility to become widely enforced within a reasonable period of time unless all the relevant risks were covered, including those in packaged form.[28]

It was pointed out that ships carrying HNS in bulk, especially LNG carriers, but also chemical tankers, are not only purpose-built and therefore generally safe ships, but are operated with the utmost care by specially trained shipboard and

---

[24] LEG XXXIX/WP.1.

[25] Any fears on the part of potential victims in relation to this question were disregarded in view of the utmost care normally associated with the transport and handling of explosives and the high safety standards existing both in the manufacturing and packaging procedures and in the stowage rules for transportation on board ships.

[26] Bederman, "Dead in the Water: International Law, Diplomacy, and Compensation for Chemical Pollution at Sea" [1986] 26 *Va.J.Int. L.* 485, 496.

[27] *Ibid.* at 486, fn. 7.

[28] Opponents of the inclusion of packaged cargo preferred to address packaged cargo in a separate convention. Pawlow, "Liability for Shipments by Sea of Hazardous and Noxious Substances" [1985] 17 *Law & Pol'y Int't Bus.* 455, 474.

terminal personnel. In comparison, packaged HNS, except packaged explosives, are frequently carried at random on general ships. Incidents involving the loss overboard of hazardous packaged chemicals have been increasingly reported. Likewise, experience has shown that the transport of packaged HNS in containers is very often done without care or vigilance, especially with regard to safe stowage standards. Furthermore, HNS bulk carriage by sea is far more limited in scope than the transportation of packaged dangerous products in ordinary cargo vessels or container ships, which has been steadily growing. Consequently, there was fear that the limitation to bulk carriage could have the effect that the packaged mode of transport would become favoured by HNS carriers and shippers. This would be highly undesirable from the viewpoint of safety, especially if the "shift" were to tank containers involving sufficiently large quantities to cause massive damage in the event of an accident.[29]

2. Allocation of liability between shipowner and shipper

The Draft HNS Convention incorporated a two-tier system of strict liability, whereby primary liability was channelled to the shipowner and excess liability was placed on the shipper. The imposition of excess liability on the shipper was aimed at ensuring the availability of an additional source of compensation to provide recovery and above the amounts involved in the existing global limitation of liability for maritime claims, i.e. the limits set by the LLMC 1976.[30] A considerable number of delegates favoured this type of liability because of the conservation of the 1976 limits and the perception that the extent and nature of the damage caused during maritime carriage depends considerably on the inherent nature of the cargoes carried. The struggle to allocate the liability for this compensation between shipowners and shippers became the chief stumbling block of the negotiations.

Critics of the draft HNS convention methodology attacked its ambiguous treatment of "shipper" and "shipowner" as representations of commercial reality. It was said that the term "shipper" primarily involved a commercial concept. Therefore, this was highly difficult to define in the legal sense for the purpose of a treaty instrument which aimed at channelling liability for third-party damage caused by harmful cargoes as far as possible towards a person or persons having the highest degree of control over such cargoes when carried by sea and over accidents which might occur at sea. The shipowner has the exclusive responsibility for the safe operation of the vessel and the goods stowed, but cargo interests put HNS into circulation with all their inherent dangers.[31]

---

[29]   Bederman, "Dead in the Water: International Law, Diplomacy, and Compensation for Chemical Pollution at Sea" [1986] 26 *Va.J.Int. L.* 485, 495.

[30]   De Bièvre, "Liability and Compensation for Damage in Connection with the Carriage of Hazardous and Noxious Substances by Sea" [1986] 17 *J. Mar. L.&Com.* 61, 74.

[31]   Bederman, "Dead in the Water: International Law, Diplomacy, and Compensation for Chemical Pollution at Sea" [1986] 26 *Va.J.Int. L.* 485, 500; Pawlow, "Liability for Shipments by Sea of Hazardous and Noxious Substances" [1985] 17 *Law & Pol'y Int't*

For the purpose of allocating liability, the IMO Legal Committee decided that the shipper should be clearly identifiable and should also have access to an insurance market.[32] In contrast to the shipowner, however, it would be difficult to identify the appropriate HNS shipper in order to hold him liable for pollution damage. Again, dual allocation of liability in the oil pollution regime has worked because the oil industry is vertically integrated, with the companies handling the drilling, transport, refining and distribution of the product. Thus, identifying the shipper is easy. In the case of HNS, however, various parties might have an interest in the cargo: the manufacturer, the consignee or buyer, the freight forwarder, the reseller or the trading company. The draft defined the "shipper" merely as the person on whose behalf or by whom as a principal the hazardous substances are delivered for carriage.[33] It seemed that the Committee intended to pinpoint the seller of the goods, irrespective of whether or not he was a party to the contract of carriage.[34] Nevertheless, the emphasis on the "principal" entirely undermined the goals of certainty and insurance access. Since this formulation had a self-defining character, a number of different parties might disclaim their principal role in the transaction. The alternative of identifying the shipper as the party who bears the risk of loss would have the virtues of being more concrete than the draft definition and relevant for insurance purposes, but it would be confusing and dependent on the terms of the individual carriage contracts.

Consequently, it appeared that the principle of shipper's liability raised practical questions affecting commercial and insurance practices.

3. Insurance of shipper liability

Insurance of shipper liability was also one of the controversial issues. Insurance interests[35] had pointed out that there existed no traditional insurance market capacity to cover the civil liability of the shipper, and that the cost of creating the required insurance cover might be expensive.[36] In addition to the concern relating to the fact that the draft HNS convention prescribed a system of compulsory insurance in connection with the liability of both the shipowner and the shipper, there

---

Bus. 455, 471; S Schuda, "The International Maritime Organization and the Draft Convention on Liability and Compensation in Connection with the Carriage of Hazardous and Noxious Substances by Sea: An Update on Recent Activity" [1992] 46 U.Miami L.Rev.1009, 1026.

[32] LEG.44/7.

[33] Draft HNS Convention Art. 1 para.4.

[34] LEG.47/7.

[35] Represented by the International Union of Marine Insurers (IUMI) and the International Association of Producers of Insurance and Re-Insurance (BIPAR)

[36] De Bièvre, "Liability and Compensation for Damage in Connection with the Carriage of Hazardous and Noxious Substances by Sea" [1986] 17 J.Mar.L.&Com 61, 74. Shipowning interests argued that the marine insurance market cannot support increased coverage of shipowners, while shippers have access to other insurance markets since they usually have deeper pockets. Pawlow, "Liability for Shipments by Sea of Hazardous and Noxious Substances" [1985] 17 Law & Pol'y Int't Bus. 455, 470.

was also concern about the implications of the concept of shipper's liability for the long-term capacity and cost-effectiveness of the insurance market.[37] Another important factor was mentioned in connection with the alleged need for predictability and economic efficiency, related to the clarity of the definition of "shipper", which, if sufficiently precise, would reduce the possibility of duplicated insurance, which can only lead to reduced capacity and increased cost.

In the insurance market, the question of sustainable insurance market capacity to absorb a risk of a potentially catastrophic level is dependent on the amount of premiums available. This, then, depends on how much premium the assured can afford and is willing to pay. The question of long-term availability of cost-effective insurance relates not only to the frequency of loss but also to the actual amount lost in relation to each risk covered.[38] A two-tier system of mixed liability of the shipowner and the shipper indicated a general awareness of the main principles of insurance based on limitation of liability and risk- spreading. However, the conference was unable to agree on what specific limitation amounts should be adopted for the respective liabilities of the shipowner and shipper. The genuine search for an improved, victim-oriented liability and compensation regime providing an additional, cost-effective source of compensation was thus abandoned as a result of conflicting views relating to the distribution of the financial cost of implementing the new regime between ship-owning and cargo interests.[39]

4. Alternative: cargo liability insurance

When it became clear that agreement on the definition of the term "shipper" could not be reached, a modified text of the draft HNS convention was introduced which replaced the personal liability of the shipper: a system of compulsory cargo liability insurance.[40] Accordingly, the principle of compulsory insurance, which is essential for guaranteeing payment when civil liability has been imposed, was preserved; but the need to identify the shipper was removed by permitting claimants to proceed directly against the cargo insurance compensation fund.[41] However, the system of direct access was rejected by the representatives of the insurance interests. The argument was that if there was to be international regulation of shippers' liability, it would have to be based on the system of personal liability. It seemed that it was not so much the principle of shippers' liability *per se* which worried insurance interests, but rather the fact that the removal of personal liability threatened an insurer's right of recourse against the shipper, particularly in the case of fault. Likewise, the absence of direct accountability of the shipper in the event of an incident causing damage was perceived to seriously erode the ability

---

[37]   LEG/CONF.6/22.

[38]   The "per vessel, per incident" level.

[39]   De Bièvre, "Liability and Compensation for Damage in Connection with the Carriage of Hazardous and Noxious Substances by Sea" [1986] 17 *J.Mar.L.&Com.* 61, 78.

[40]   LEG CONF.6/C.1/WP.22, Art. 13.

[41]   De Bièvre, "Liability and Compensation for Damage in Connection with the Carriage of Hazardous and Noxious Substances by Sea" [1986] 17 *J.Mar.L.&Com.* 61, 77.

of insurers to charge premiums on the basis of client experience, as well as to discourage the exercise of care and loss-prevention practices on the part of the assured.[42]

5. Outcome of the 1984 Conference on HNS

The Draft HNS Convention was given serious consideration at the Conference. However, lack of agreement on basic issues triggered a lack of consensus on most of the interrelated HNS provisions. As a result, the draft instrument on HNS was refused.[43] This outcome was inevitable given the very cramped timetable of the diplomatic conference and the lack of confidence in the level of elaboration of the draft treaty and preparatory work of the IMO Legal Committee. The Conference concluded that it would not be feasible, in the time available, to resolve the many complex issues in the draft convention and reach broad agreement on a treaty instrument which would receive wide acceptability.[44] However, it was decided to send the draft convention back, so that the Organization could arrange to prepare a new and more widely acceptable draft for submission to a diplomatic conference which might be convened in the future. It was also recommended that the Organization assign priority to the preparation of such a new draft and arrange for a diplomatic conference to examine the new draft at the earliest possible time.

Most significantly, the Conference made it clear that although there were differing points of views about the many complex issues raised by the draft and their implications, there was also awareness of the risk posed by the carriage of hazardous and noxious substances by sea and the need to adopt uniform international rules to deal with issues of liability and compensation in respect of damage caused by such substances. Hence, despite the divergence of opinion on the subject, it was obvious that there was need for an international convention.

V. Developments since 1987

Despite the 1984 Convention's inability to reach an agreement, the need for international HNS liability was recognized. In 1987, the nations jointly submitted an outline of options to resume work, suggesting two alternative liability schemes: primary shipowner liability with secondary compulsory shipper cargo insurance and shipowner sole liability with increased liability limits of the LLMC 1976 up to a level providing sufficient compensation for all damage, including HNS, or alternatively creating a special layer available exclusively for HNS damage above and

---

[42] LEG CONF.6/C.1/WP.24 and WP.25.
[43] It was the first time that a diplomatic conference rejected outright a convention proposed by the IMO. Bederman, "Dead in the Water: International Law, Diplomacy, and Compensation for Chemical Pollution at Sea" [1986] 26 *Va.J.Int. L.* 485, 492.
[44] LEG./CONF.6/C.1/5.

beyond the general liability limits.[45] This proposal suggested also that any new system of HNS liability should include packaged HNS, not only bulk; liability should rest with an easily identifiable party; as far as practicable, strict liability should be imposed; and any limit of liability should be sufficiently high to adequately compensate for damage.[46]

The IMO Legal Committee prepared a list of questions for use as a focal point of future work:[47]

1. What should be the geographic scope of application?
2. What specific substances should be covered by the convention and by what method should a list of substances be incorporated into the convention and amended?
3. Should packaged substances be covered?
4. Should residual products and waste material transported for dumping or incineration at sea be included in the scope of application of the new convention?
5. How should "damage" be defined in the new convention?
6. Should costs for prevention and for clean-up be compensated under the new convention?
7. Should liability rest solely with the shipowner or should it be shared with cargo interests?
8. Should liability be strict?
9. What limitation amount should be established?
10. Who should be able to take advantage of the limitation of liability limits?
11. How should the new convention relate to the 1976 LLMC?
12. What are the insurance implications of each of the above questions?

At the 60th session, there was support for a regime that involved sharing the costs of compensation between the shipowner and cargo interests. The Legal Committee also agreed that the new convention should be based on strict liability. Attention focused on the lack of specific information regarding the availability, total capacity and cost of insurance. The Group of Shipowners' Protection and Indemnity Associations submitted a paper asserting that the volatile state of the insurance market made speculation as to whether market capacity could cover increased liability under the limits of an HNS convention impossible.[48]

A proposal for a second tier was introduced to provide supplementary compensation when the limits of the first-tier shipowner's liability were exceeded.[49] Cargo insurers would fund this second tier of compensation by collecting a levy from

---

[45] Schuda, "The International Maritime Organization and the Draft Convention on Liability and Compensation in Connection with the Carriage of Hazardous and Noxious Substances by Sea: An Update on Recent Activity" [1992] 46 *U.Miami L.Rev.*1009, 1032.

[46] *Ibid.*

[47] "Consideration of a Possible Convention on Liability and Compensation in Connection with the Carriage of Hazardous and Noxious Substances by Sea (HNS)" *Note by Secretariat, IMO Legal Comm.*, LEG 62/4, Annex I, 15 December 1989.

[48] LEG 60/3./3.

[49] LEG 60/3/4.

shipping interests in order to finance the purchase of a pool of comprehensive insurance. Cargo interests would prepay levies on export cargoes of substances listed in the IMDG Code, based on the value of the cargo.[50] A potential problem with this proposal is that a substance's potential risk might bear little or no relation to its value. However, the advantage of this scheme is its certainty. Cargo insurers would be responsible for collecting the levies and applying them to insurance. The definition of damage was also firmly determined.

Regarding the issue of substances to be covered, whilst some delegations continued to advocate covering only bulk cargo because of the ease of administering a bulk-cargo system, others advocated covering only HNS which could cause catastrophic damage.[51] The majority of the delegations, however, supported a wide scope of application to include all HNS incidents caused by either bulk or packaged cargo. The Legal Committee focused on all levels of HNS damage, not only on catastrophic incidents.[52] Furthermore, the Committee agreed to proceed with discussions on the assumption that the new convention would apply to packaged HNS.[53]

At the 62nd session in 1990, HNS received the highest priority of the Legal Committee. This session was virtually devoted to HNS draft conventions submitted by the Netherlands and the United Kingdom. The draft submitted by the Netherlands proposed a liability scheme under which the shipowners would be strictly liable up to a specific level of HNS damage, which the draft left undefined. The draft did not impose a second tier of shipper liability.[54] The draft submitted by the United Kingdom incorporated the Netherlands proposal as its first tier, but also included a second-tier international fund financed by a levy charged on cargo interests. The proposed fund's purpose was to provide supplemental compensation in cases where the first-tier limits were exceeded.[55] The submission favoured a flat liability rate for shipowners, rather than a tapered rate based on vessel tonnage, because one package of a particular HNS cargo might cause as much damage as a 40,000 ton bulk HNS carrier fully loaded with another type of HNS cargo.[56] Levies charged on exports of HNS carried in bulk would finance the second tier. This system would simplify the levy and collection procedure and minimize the number of shipments involved. The scheme, however, would compensate for damage caused by both bulk and packaged substances. The second tier would use the income received from the levies collected from cargo interests to purchase insurance on the international insurance market and to build up a cash "stand-by" fund.[57]

At the 63rd session, the Legal Committee continued in narrowing discussions to the following issues: (1) the acceptability of imposing liability on the operator

---

[50] *Ibid.*
[51] LEG 62/4 Annex 2.
[52] *Ibid.*
[53] *Ibid.*
[54] LEG 62/4/1.
[55] LEG 62/4/2 Annex 2.
[56] LEG 62/4/2.
[57] *Ibid.*

of the ship; (2) the linking of the HNS compensation scheme to the levels set in the 1976 LLMC; (3) the question of whether to establish a two-tiered HNS regime; (4) the definition of "HNS damage"; and (5) the features of the possible second tier.[58]

It was emphasized that the principal objective of the convention should be to provide effective compensation to the victims of an HNS incident without undue delay.[59] Thus, delegations proposed that strict liability for HNS damage should attach to the vessel's operator, not the shipowner or the shipper, because the operator creates the risk of an incident.[60] Only the operator can exercise sufficient proper care to reduce the risk of an HNS incident to an acceptable minimum.[61] Furthermore, placing liability on the operators accounts for situations where the shipowner parts with possession and control of the vessel, such as letting the vessel on a bareboat charter. However, imposition of liability on the ship's operator instead of the shipowner or shipper lacked sufficient support at the session for several reasons.[62] First, international law had generally imposed shipowner liability, thus a precedent had been set for the convention to follow.[63] Second, most delegates thought that ascertaining the shipowner's identity through public documents could be done relatively easily, while the identity of the specific operator at any given time could not always be so easily determined.[64] Moreover, the shipowner, not the operator, would be responsible for obtaining the insurance necessary for any compensation payable by an offending ship.

Shipping interests continued to lobby against the second tier of shipper's liability at the 63rd session. CEFIC argued that a system of shipowner liability would sufficiently compensate all damage claims.[65] To support this, CEFIC asserted that the number of grave accidents connected with the transport of HNS substances is very small, and where such accidents have occurred, adequate compensation has been afforded.[66] Should a serious incident occur, the current marine insurance market would have the necessary coverage capacity and it should rest with the marine insurers to provide shipowners and operators with the possible additional cover.[67] CEFIC was concerned that a system of liability of shipper's liability would become needlessly complicated, cumbersome to implement and costly.[68] CEFIC was also of the opinion that the imposition of shipper's liability was a dangerous break between the operational responsibility of shipowners and civil liability, divesting shipowners of accountability for their conduct.[69]

---

[58]   LEG 63/14.
[59]   LEG 63/3/1.
[60]   LEG 62/7.
[61]   LEG 62/4/5.
[62]   LEG 63/14.
[63]   *Ibid.*
[64]   *Ibid.*
[65]   LEG 63/3/5.
[66]   *Ibid.*
[67]   *Ibid.*
[68]   *Ibid.*
[69]   *Ibid.*

With regard to the definition of "hazardous substances", the use of lists of substances developed for the IMDG Code was recommended.[70] This was thought to be administratively easy and cost-efficient. However, a contrary view argued that the mere inclusion of a previous list would not meet the special requirements of a new HNS regime.[71] The majority argued that a new HNS regime should cover the broadest possible range of substances in order to meet the international community's objectives of protecting the marine environment, pinpointing liability for accidents, identifying the liability for accidents and ensuring compensation to victims.[72] To resolve some of the issues beforehand, the Legal Committee established a "Working Group of Technical Experts."[73] At the end of the 63rd session, the Legal Committee asked the United Kingdom to prepare a draft convention for consideration at the 64th session.

VI. The 1991 Draft

At the 64th session, held in 1991, eleven nations submitted a joint 1991 Draft HNS Convention.[74] In order to incorporate concerns that substances covered by the convention should be broadly defined, the 1991 Draft substituted the term "dangerous goods" for the term "HNS."[75] The type of damage covered by the draft also received broad definition. Consistent with the 1984 Draft and with most of the discussion and debate that has taken place since, the 1991 Draft extended the strict liability for damage caused by dangerous goods at sea to the shipowner.[76] The draft also added a new defence, not available under the 1984 Draft: the shipowner is not liable for the damage if the consignor failed to inform the shipowner that the consignment contained dangerous goods and if the shipowner did not know nor could have known of the goods' dangerous nature.[77] Art. 6 of the draft provided for limitation of liability of the shipowner, but not for specific liability limits, which presumably awaited negotiation. The proposed system also required the shipowner to carry insurance to cover the limit of liability.[78]

The 1991 Draft Convention addressed the need for a second tier of damage liability by establishing an international Dangerous Goods Scheme to provide compensation for damage in connection with the carriage of dangerous goods by sea, to the extent that recovery from the shipowner is inadequate or unavailable.

---

[70] LEG 63/14. para. 32-35.
[71] *Ibid.* para.33.
[72] *Ibid.* para.38.
[73] This group consisted of experts in manufacturing, shipping, carrying, regulating and the risks of HNS. LEG 63/WP.3 Annex 1 para.2.
[74] LEG 64/4.
[75] *Ibid.*
[76] LEG 64/4 Annex Art. 4 para.1
[77] *Ibid.* Art. 4 para.2(d).
[78] *Ibid.* Art. 10.

VII. Reactions to the 1991 Draft

The initial reaction to the 1991 Draft was favourable.[79] Some delegations submitted proposals for revision of some portions of the Convention. CEFIC approved of the 1991 Draft's first-tier shipowner liability and of the second tier, which does not impose liability directly on the shipper.[80] CEFIC continued to question the need for the second layer of compensation, asserting that "the cover of 100 million SDR per incident – possibly supplementing the 1976 LLMC limitations – should permit adequate compensation in the vast majority of cases."[81] If the Legal Committee deemed a second tier necessary, CEFIC would approve of an administrative compensation scheme, as established by the International Dangerous Goods Scheme.[82] However, CEFIC was concerned that the system of collecting contributions was impractical and recommended that shipowners be responsible for incorporating the cost of contributing to the scheme into the freight price and forwarding that levy to the scheme on a quarterly basis.[83] A shipowner who failed to pay contributions in any year would be refused registration until the contributions were paid.[84]

VIII. Subsequent developments

The 66th session was held in 1992. Recommendations by the Group of Technical Experts for inclusion or deletion of matters in the definition of HNS were integrated. The Group recalled that both reports so far called for guidance from the Committee.[85] Some general guidance was given, although the Committee clearly had the problem that many decisions were necessarily interdependent. On other matters, such as the inclusion of bunker fuels, the Committee itself was still divided. The Group met during the session and reported on the continuing and complicated work on the definition of HNS, the threshold for compulsory shipowner insurance and the list of contributing cargoes.[86]

The report of the Group to the 67th session was considered at the 68th session in 1993.[87] More proposals on the mechanism for the collection of contributions

---

[79]   Schuda, "The International Maritime Organization and the Draft Convention on Liability and Compensation in Connection with the Carriage of Hazardous and Noxious Substances by Sea: An Update on Recent Activity" [1992] 46 *U.Miami L.Rev.*1009, 1045 f.

[80]   LEG 65/3/8.

[81]   *Ibid.*

[82]   *Ibid.*

[83]   *Ibid.*

[84]   *Ibid.*

[85]   LEG 64/10.

[86]   LEG 66/WP.5.

[87]   LEG 67/9.

were presented.[88] At the 69th session in September 1993, the Chairman of the Legal Committee asked delegations to consider three options in order to be able to have a target for a diplomatic conference in 1996. The options were: (1) a single-tier Convention with shipowner liability with a scope limited to bulk cargoes only; (2) a two-step solution, as adopted with the 1969 Liability Convention followed by the 1971 Convention, with the understanding that work on the cargo-financed second tier would begin immediately after the conclusion of the shipowner liability convention; and (3) a two-tier system single Convention, including damage caused by packaged cargo, but with the mechanism for collection restricted to bulk goods only. There was support for options (2) and (3). The Committee decided to consider option (3) on a provisional basis.[89] Most delegations favoured a simple contribution system for bulk goods.

The Group of Technical Experts was instructed to consider three questions: (1) whether a pre- or post-event collection scheme for all contributing cargo could be devised in the time available; (2) whether technical criteria for an independent accounting system could be established while avoiding a proliferation of accounts; and (3) whether technical criteria for identifying "bulk plus" could be established.[90] It was reported that the Group had established an informal Working Group which had presented a proposal under which it would be possible to develop a post-event collection and compensation scheme in time for a 1996 diplomatic conference.[91]

At the 70th session in 1994, a paper prepared by Australia, Canada and Norway (the ACN paper) was presented, which was based on the work in the Group of Technical Experts at the 69th session.[92] It proposed a detailed post-event collection second-tier HNS Scheme with a limited number of separate accounts, and with levies being paid by the importer/receiver rather than the shipper.

The Legal Committee unanimously confirmed that there should be two tiers in a single instrument involving the liability of the shipowner and the contribution of the cargo interests respectively. Likewise, there was nearly unanimous support for a post-event contribution system. The informal Working Group established at the 69th session to work on the HNS list reported at the 70th session that a free-standing list could be produced in the order of classes and instruments already referred to in Art. 1.5 of the 1984 draft HNS, with updating by means of the tacit amendment procedure.[93]

At the 71st session in 1994, the Legal Committee attempted to solve as many problems of policy as possible so that the Secretariat could provide a next text for the 72nd session in 1995. The major item was the development of the ACN paper

---

[88]   Japan presented formal proposals for four separate accounts into the Scheme: for oil, LNG, LPG and HNS. LEG 68/4/4.

[89]   LEG 69/11 para.21.

[90]   *Ibid.* at para.41.

[91]   Gaskell, "The Draft Convention on Liability and Compensation for Damage Resulting from the Carriage of Hazardous and Noxious Substances" in *Essays in Honor of Hugo Tiberg* (1996), 225, 242.

[92]   LEG 70/4/10.

[93]   LEG/70/4/7.

concerning the second tier.[94] The approach of the 1996 diplomatic conference notably quickened the pace of deliberations and a large number of thorny questions were raised. A revised ACN paper provided an analysis and definition of the "receiver."[95] It proposed that the definition should identify as "receiver" the person who was responsible for customs clearance or the person who physically receives delivery. This was generally accepted by the Committee.

Following the 71st session, the Secretariat produced a draft text of the HNS Convention in January 1995. This draft was discussed at the 72nd session in 1995. A number of issues were temporarily resolved.[96] Some minor changes were made to the text. The second-tier "HNS Scheme" was renamed the "HNS Fund", so as to mirror the IOPC Fund more closely.[97] At the 72nd session it was accepted that the draft HNS Convention was ready for a diplomatic conference, although there might be need for more work on some issues.[98]

D. The HNS Convention 1996

After long debates, a diplomatic conference was eventually convened in April/May 1996 to consider the new instrument dealing with compensation for accidents involving hazardous and noxious substances. This time the attempt was successful. The International Convention on Liability and Compensation for Damage in Connection with the Carriage of Hazardous and Noxious Substances by Sea (the "HNS Convention") was adopted on 3 May 1996. The regime established by the HNS is largely modelled on the existing regime for oil pollution from tankers set up under the International Convention on Civil Liability for Oil Pollution Damage 1992 (the "CLC") and the International Convention on the Establishment of an International Fund for Compensation for Oil Pollution Damage 1992 (the "Fund Convention"), which covers pollution damage caused by spills of persistent oil from tankers. By contrast, however, the HNS regime is governed by a single instrument.

I. Definition of HNS

As pointed out above, there was a long debate on whether the Convention should include packaged goods or be restricted only to bulk cargo. Eventually it was agreed that the Convention should cover carriage whatever the form and, in principle, whatever the volume.

---

94   LEG 71/3/4.
95   LEG 71/3/4 para.46.
96   For instance, the inclusion of coal.
97   Gaskell, "The Draft Convention on Liability and Compensation for Damage Resulting from the Carriage of Hazardous and Noxious Substances" in *Essays in Honor of Hugo Tiberg* (1996), 225, 245.
98   *Ibid.*

Thereafter, a great deal of time was spent defining HNS. Any lack of clarity in this concept would be unacceptable both from a legal and a political point of view. The question this time was whether the definition should be made by reference to the lists provided for in the existing technical instruments or whether it would be necessary to enumerate all these substances in a free-standing list as part of the HNS Convention.[99] The latter option was favoured, the argument being that this was the only workable solution from a practical perspective, because it would enable those involved in the carriage of HNS to establish beyond a doubt whether a particular substance fell under the scope of the Convention with the consequences of strict liability, compulsory insurance etc.

The other argument pointed out constitutional difficulties: incorporation by reference would effectively mean that a state might become bound by the treaties containing such lists without being party to them.[100] Apart from that, another view drew attention to other conventions such as the Convention on Civil Liability for Damage Caused during Carriage of Dangerous Goods by Road, Rail and Inland Navigation Vessels, (CRTD), which defined their scope of application by reference to other treaty instruments and relied on the procedure for updating the lists provided for in these instruments. This should not be seen as a constitutional problem, but rather as an opportunity to avoid overburdening the HNS Convention with hundreds of pages listing HNS substances. Moreover, this view regarded it as a natural consequence of incorporating by reference to allow the decisions of the relevant bodies responsible for updating the various list to have an automatic effect on the HNS Convention, thereby ensuring that the HNS Convention will keep pace with technical developments and promoting conformity between the technical instruments and the HNS Convention in this regard.

To encumber the HNS Convention with an estimated three hundred pages of substances was referred to as giving birth to a monster too complicated to handle.[101] Consequently, practical considerations have prevailed. Art. 1.5 of the Convention defines HNS as:

i)   oil carried in bulk listed in Appendix I to Marpol 73/78;
ii) noxious liquid substances carried in bulk referred to in Appendix II of Annex II of Marpol 73/78;
iii) dangerous liquid substances carried in bulk listed in Chapter 17 of the IBC Code;
iv) dangerous, hazardous and harmful substances, materials and articles in packaged form covered by the IMDG Code;
v)   liquefied gases as listed in Chapter 19 of the IGC Code;
vi) liquefied substances carried in bulk with a flashpoint not exceeding 60° C; and

---

[99] Göransson, "The HNS Convention" [1997] *Unif. L. Rev. (Rev. dr.Unif)* 249, 252.
[100] *Ibid.* at 253.
[101] *Ibid.*

vii) solid bulk materials possessing chemical hazards covered by Appendix of the BC Code to the extent that these substances are also subject to the IMDG Code when carried in packaged form.

In addition, the definition includes certain substances and gases which have been provisionally included in the parent lists and in regard to which the provisions of the relevant instrument are applied on a preliminary basis.[102] The definition also includes residues from previous bulk.[103]

The definition does not explicitly exclude the carriage of small quantities in packaged form. However, the reference to the IMDG Code and the requirement that material in packaged form be included to the extent that it is covered by the Code have the effect of excluding from the HNS Convention the same small quantities of cargo that are excluded from the scope of application of the IMDG Code.[104] It is estimated that the Convention incorporates more than 6000 substances.

1. Coal and other solid bulk materials

There was a long debate if coal and other solid bulk materials would have been included. Delegates favouring of the inclusion pointed out that the Convention would cover not only pollution damage but also damage caused by fire and explosion and that coal could be liable to spontaneous heating and fire or the emission of flammable gases apt to be ignited by sparks.[105]

Others who opposed the inclusion of coal and other solid bulk cargoes said that there was no justification for the inclusion of coal. Coal had been safely transported for hundreds of years and no damage had gone uncompensated. It was not a pollutant but a natural mineral and any fire would normally be confined to the cargo space on board and not spread beyond the ship. The risk of a major incident, therefore, was very limited. If coal were to be included, there was also the risk that because of the large volumes involved it would have to bear a disproportionate burden of contributions to the Fund under the second tier which could lead to distortion of competition. Hence, any solution which included coal was seen as unacceptable.[106]

Consequently, it was clear that an overwhelming majority was in favour of the complete exclusion of coal. The legal-technical solution, therefore, was to extend this exclusion to the entire MHB class.[107]

---

[102] HNS Convention Art. 1.5.a(ii),(iii), and (v).

[103] HNS Convention Art. 1.5.b.

[104] Göransson, "The HNS Convention" [1997] *Unif. L. Rev. (Rev. dr. Unif)* 249, 254.

[105] Sasamura, "Development of the HNS Convention", in *13th International Symposium on the Transport of Dangerous Goods by Sea and Inland Waterways* (1998), 491, 497.

[106] A compromise proposal to include coal but to waive its funding obligations under the second tier until experience had demonstrated that it should contribute was rejected.

[107] Aluminium dross, charcoal, coal, direct reduced iron, ferrophosphorus, fluorspar, lime, metal sulphide concentrates, petroleum coke, pitch prill, prilled coal tar, pencil pitch, sawdust, silicon manganese, tankage, vanadium ore, woodchips and wood pulp pellets.

2. Fishmeal

A proposal was made to exclude fishmeal. However, it did not find sufficient support. It was recognized that there was a need to review the classification of fishmeal in the IMDG and BC Codes. As a result, the Conference adopted a Resolution which called on the appropriate technical bodies of the Organization to undertake such a review.[108]

After the Conference, the DSC Sub-Committee, in the light of the above Resolution, reviewed the IMDG Code provisions on the aforementioned types of fishmeal and agreed that they may not be treated as dangerous.[109] Consequently, some types of fishmeal are covered the HNS Convention.[110]

3. Waste

The definition of HNS makes no specific reference to waste. Insofar as hazardous waste carried as cargo came within the definition of different categories of HNS, it would also fall under the scope of the Convention.[111] When the HNS Conference convened, there was also ongoing work for a draft protocol to the Basel Convention on a liability regime for damage in connection with the transboundary movement of hazardous wastes. To avert the risk of overlap, a resolution recommending continued operation between the Secretariats of IMO and the United Nations Environment Program was adopted.[112] In the course of preliminary work, the Legal Committee had principally agreed that waste carried on board for the purpose of dumping in accordance with the provisions of the Convention on the Prevention of Marine Pollution by Dumping of Wastes and other Matter 1972 ("London Convention") should be included in the definition of HNS. At the Diplomatic Conference, however, it was suggested that it would be rather unfair to hold the shipowner strictly liable for what he was hired to carry, the dumping of which had been authorized by a competent authority.[113] It was also pointed out that in virtue of amendments to the London Convention the dumping of industrial waste was no longer permitted;[114] thus such materials would no longer be subject to transport by

---

[108] LEG/CONF.10/8/1.

[109] Sasamura, "Development of the HNS Convention," in *13th International Symposium on the Transport of Dangerous Goods by Sea and Inland Waterways* (1998), 491, 499.

[110] Draft Guide to the Implementation of the HNS Convention, Prepared by the Secretariat of the International Oil Pollution Compensation Fund 1992, <www.iopcfund.org/npdf/HNS-guide_e.pdf, para.3.2>. (visited 13.1.2007).

[111] Göransson, "The HNS Convention" [1997] *Unif. L. Rev. (Rev. dr. Unif)* 249, 255; Sasamura, "Development of the HNS Convention," in *13th International Symposium on the Transport of Dangerous Goods by Sea and Inland Waterways* (1998), 491, 498.

[112] Resolution on the Relationship between the HNS Convention and a Prospective Regime on Liability for Damage in Connection with the Resolution on Transboundary Movements of Hazardous Wastes LEG/CONF.10/8/1.

[113] Göransson, "The HNS Convention" [1997] *Unif. L. Rev. (Rev. dr.Unif)* 249, 255.

[114] The 1993 Amendments entered into force on 20 February 1994 and phased out the dumping of industrial wastes by 31 December 1995.

sea. It was also questionable how materials which had been cleared for dumping at sea could be capable of causing any particular damage. Due to these uncertainties, the Conference left the question of dumping and the carriage of materials for that purpose entirely outside the scope of application of the Convention.

4. Excluded cargoes

a) Oil (causing pollution damage)

The HNS Convention excludes pollution damage as defined by the 1969 Convention on Civil Liability for Oil Pollution Damage, as amended. The exclusion applies whether or not compensation is payable under that Convention. This is so as to exclude any pollution damage caused by persistent oil. Otherwise contributors to the IOPC Fund based on the receipt of such oils might have to pay twice for the same risk. However, the HNS Convention excludes only pollution damage caused by persistent oil. Persistent oil falls under the HNS and fire and explosion caused by such cargo come within the scope of the Convention.

One major issue was whether the Convention should also cover bunker fuel oils. It was felt that including bunker fuel would unnecessarily complicate the HNS Convention since it would basically affect any vessel whatever type it is. Hence, it was concluded that the HNS Convention should not deal with damage caused by bunker fuel oils.[115]

b) Radioactive materials

The Convention explicitly excludes any damage caused by radioactive material of Class 7 of the IMDG Code or listed in Appendix B of the BC Code.[116] There was again a long debate between those who wanted the HNS Convention to deal with "nuclear liability" in any form and those who felt that some categories of radioactive materials should come within the scope of the HNS Convention. Reference was made to so-called "excepted matter." This meant matter excluded from the scope of the nuclear liability regimes on the grounds that it was not deemed to pose any significant risk of nuclear damage to third parties or to the environment

---

[115] Sasamura, "Development of the HNS Convention," in *13th International Symposium on the Transport of Dangerous Goods by Sea and Inland Waterways* (1998), 491, 497. However, the issue went back to the Legal Committee as a request by the Marine Environmental Protection Committee to consider various options to ensure adequate compensation in connection with the establishment of an international regime providing for compensation damage caused by pollution from ship's bunkers. There turned out to be considerable sympathy within the Legal Committee with the concern expressed in respect of the substantial damage that could be caused by bunker spills. Consequently, the Committee decided to reinstate it on its future work agenda as a free-standing item for priority discussion once work on the HNS Convention had been concluded. See LEG 72/9 and 73/14. Eventually, International Convention Civil Liability for Bunker Oil Pollution Damage was adopted on 23 March 2001.

[116] HNS Convention Art. 4.3.2.

which would warrant the application of a special liability regime, channelling the liability to the operator of a nuclear installation.[117]

II. Damages covered

The definition of damage shows significant similarities with that of the CLC 1992. However, there are some differences due to the fact that the HNS Convention not only covers pollution damage but also fire and explosion damage.[118] According to Art. 1.6 of the HNS Convention, "damage" means:

(a) loss of life or personal injury on board or outside the ship carrying the hazardous and noxious substances caused by those substances ;
(b) loss of or damage to property outside the ship carrying the hazardous and noxious substances caused by those substances;
(c) loss or damage by contamination of the environment caused by the hazardous and noxious substances, provided that compensation for impairment of the environment other than loss of profit from such impairment shall be limited to costs of reasonable measures of reinstatement actually undertaken or to be undertaken;
(d) the costs of preventive measures and further loss or damage caused by preventive measures.

As seen, the definition of environment damage closely follows the definition adopted by the CLC 1992.[119] It was important to ensure that compensation for

---

[117] "Excepted matter" is low radioactive material used for medicinal purposes, in watchmaking etc. Such material has found no place in the regimes dealing with the nuclear fuel cycle. Why should the operator of a nuclear installation be liable for damage caused by this particular type of radioactive material? As the Conference excluded completely from the scope of HNS Convention, it was obvious that there was the same lack of justification for making the shipowner liable. Thus, it was excluded from the scope of the Convention. Göransson, "The HNS Convention" [1997] *Unif. L. Rev. (Rev. dr.Unif)* 249, 256 f. It is asserted that the appropriateness of the exclusion of radioactive materials from the HNS Convention is questionable. It is because a comprehensive liability compensation system for the carriage of nuclear materials by sea is lacking and it may take time to have one. Therefore, in view of the aim to achieve a broad coverage of hazardous transports, the HNS Convention should have included all radioactive materials which are not covered by the nuclear liability conventions. Wetterstein, "Carriage of Hazardous Cargoes by Sea- The HNS Convention" [1997] 26 *Ga.J.Int'l & Comp. L.* 595, 601.

[118] Sasamura, "Development of the HNS Convention," in *13th International Symposium on the Transport of Dangerous Goods by Sea and Inland Waterways* (1998), 491, 500.

[119] CLC 1992 Art. 6. "Pollution damage" means (a) loss or damage caused outside of the ship by contamination resulting from the escape or discharge of oil from the ship, wherever such escape or discharge may occur, provided that compensation for impairment of the environment other than loss of profit from such impairment shall be limited to costs of reasonable measures of reinstatement actually undertaken or to be under-

such damage would be available irrespective of what substance, oil or HNS caused the damage. This will also ensure that once the HNS Convention comes into force, it will benefit from the vast experience and jurisprudence of the IOPC Fund.

Where it is not reasonably possible to separate damage caused by the hazardous and noxious substances from that caused by other factors, all such damage shall be deemed to have been caused by the hazardous and noxious substances.[120] However, this rule does not apply if and to the extent that the damage caused by other factors falls under the definition of oil-pollution damage as formulated in the Civil Liability Convention and to radioactive materials, as they are excluded.[121]

III. Liability

1. Strict liability of the shipowner (the first tier)

In terms of the liability established as the first tier, the HNS Convention is parallel to the CLC 1992. Under the HNS Convention, the shipowner is strictly liable for damage caused, unless the circumstances fall within one of the stated exceptions.[122] That means that liability does not depend on the fault of the owner or any other person. The fact that damage has been caused by substances on board the ship is sufficient to establish the shipowner's liability.[123]

The liability is channelled to the shipowner and shall not be attached to other persons connected with the operation of the ship unless the damage resulted from their own fault.[124] The reason is to prevent unnecessary litigation.[125]

The shipowner will be denied the right to limitation of liability if it is proved that the damage resulted from the shipowner's personal act or omission committed either with intent, or recklessly, and with the knowledge that damage would probably result, as a result of the owner's personal act or omission.[126] This is only the case in the event of an act or omission by the shipowner, not in the event of an act or omission by the master or crew of the ship.

---

taken; (b) the costs of preventive measures and further loss or damage caused by preventive measures.

[120] HNS Convention Art. 1.6. para.2. CRTD Art. 1. para.10 is the same. "Caused by those substances" means caused by the hazardous or noxious nature of the substances". The last sentence of the Art. 1.6 of the HNS Convention.

[121] *Ibid.*

[122] HNS Convention Art. 7.

[123] McKinley, *The International Convention on Liability and Compensation for the Carriage of Hazardous and Noxious Substances by Sea: Implications for State Parties, the Shipping, Cargo and Insurance Industries* (Diss. Cape Town 2005), 35 f.

[124] HNS Convention Art. 7.5.

[125] [Draft] IMO Guide for Interested Parties on the Workings of the Hazardous and Noxious Substances Convention 1996, (HNS Convention), LEG83/INF.3 para.4.1.

[126] HNS Convention Art. 9.2. For instance, if the shipowner knowingly allows a ship to sail in such an unseaworthy condition that an accident is likely to occur.

2. Defences available to the shipowner for exoneration from liability

To avoid liability, the shipowner may invoke certain defences similar to those in the CLC. Accordingly, no liability shall attach to the owner if the owner proves that:[127]

(a) the damage resulted from an act of war, hostilities, civil war, insurrection or a natural phenomenon of an exceptional, inevitable and irresistible character; or
(b) the damage was wholly caused by an act or omission done with the intent to cause damage by a third party; or
(c) the damage was wholly caused by the negligence or other wrongful act of any Government or other authority responsible for the maintenance of lights or other navigational aids in the exercise of that function; or
(d) the failure of the shipper or any other person to furnish information concerning the hazardous and noxious nature of the substances shipped either
 (i) has caused the damage, wholly or partly; or
 (ii) has led the owner not to obtain insurance in accordance with Art. 12;

provided that neither the owner nor its servants or agents knew or ought to reasonably known of the hazardous and noxious nature of the substances.

3. Unique defence of the shipowner: shipper's failure to furnish information

The defense available under provision (d) is unique to the HNS Convention. Subparagraph (d) applies only if someone has failed to give the shipowner the relevant information or if the shipowner has not obtained the necessary insurance cover as a consequence of this failure to provide information.[128] This defence was attracted extensive attention and caused much debate.

Some delegations felt that this defence should not be available to the shipowner since it would breach the principle of channeling liability.[129] According to this view, the grounds for exoneration should be as limited as possible. This view also argued that this type of exoneration would open the way to litigation and therefore adversely affect the victim's right to prompt and adequate compensation.

A contrary view suggested that it would be rather unfair to hold the shipowner liable in a situation where he had not been given correct information about a consignment.[130] Even in the absence of first-tier liability, the victim would not suffer, since the HNS Fund would cover the amount that would have been available under the first tier.

---

[127] HNS Convention Art. 7.2.
[128] "[Draft] IMO Guide for Interested Parties on the Workings of the Hazardous and Noxious Substances Convention 1996, (HNS Convention)" LEG 83/INF.3 para.4.1.
[129] Göransson, "The HNS Convention" [1997] *Unif. L. Rev. (Rev. dr.Unif)* 249, 261.
[130] *Ibid.* 262.

IV. Limits of liability

1. Small ships

It is clear that small ships may potentially create much greater risks than their tonnage would indicate. While the Conference appeared to have been unanimous in its view that the Convention should introduce a minimum threshold for small ships so as to avoid the HNS Fund having to intervene in minor cases, it proved more difficult to agree on the actual amount. The LLMC and the CLC recognized this factor to the same extent in the way that the limitation funds under the Conventions are to be calculated. Therefore, there had been suggestions that the minimum limits should be set fairly high for small vessels. If the limits are very low for such vessels, there might be an unacceptably high number of claims against the HNS Fund.[131] This would be inefficient and could affect the viability of the HNS Fund by burdening it with a lot of small claims.

At the 71st session, a paper on the size and number of the fleet was presented by the U.K.[132] Accordingly 46% of the ships were below 2,000 gross tons and 72% were below 10,000 gross tons. Of chemicals tankers used for HNS carriage, 50% were below 2,000 gross tons and 80% below 10,000 gross tons. Under the 1976 LLMC, the minimum tonnage is 500 gross tons. The 1992 Protocol to the 1969 Liability Convention introduced a minimum tonnage limitation of 5,000 gross tons for small ships. It was pointed out that small ships carrying HNS, e.g. in packaged form, could cause more damage then a large cargo ship carrying less harmful products. Under the 1992 CLC, the maximum limit only applies at 140,000 gross tons. An HNS Convention maximum of 100,000 gross tons would mean that only 1% of ships would reach the maximum liability. A maximum reached at 50,000 gross tons would produce a fairly steep progression of limits, but would still apply only to 4% of ships.[133] Consequently, it was concluded that figures indicated that the minimum tonnage figure should be set at 2,000 gross tons with the maximum operating at either 50,000 or 100,000 gross tons.

When it became clear that the Convention would also cover not only international but also domestic voyages, many delegations looked for a rather high minimum liability amount for small ships.[134] The concern was that the levels of insurance cover would mean discontinuing these services, since no insurance would be

---

[131] Gaskell, "The Draft Convention on Liability and Compensation for Damage Resulting from the Carriage of Hazardous and Noxious Substances" in *Essays in Honor of Hugo Tiberg* (1996) 225, 269; Sasamura, "Development of the HNS Convention" in *13th International Symposium on the Transport of Dangerous Goods by Sea and Inland Waterways* (1998), 491, 500.

[132] LEG 71/3/11.

[133] Gaskell, "The Draft Convention on Liability and Compensation for Damage Resulting from the Carriage of Hazardous and Noxious Substances" in *Essays in Honor of Hugo Tiberg* (1996), 225, 269 f.

[134] Attention was particularly drawn to the serious implications this might have for certain states where there was a need to maintain supplies to small island communities.

available for such small ships.[135] This prompted the Conference to agree to allow a state to exclude the application of the Convention in respect of ships not exceeding 200 gross tonnage and carrying HNS in packaged form only while engaged on voyages between ports or facilities of that state.[136]

2. Level of limit

The limit of liability was one of the major stumbling blocks during the Diplomatic Conference.[137] It was clear that before detailed consideration of any second tier was possible, it was necessary to have some idea of the limits which would operate in the first tier. At the 62nd session, the P&I Clubs indicated that it would not be feasible to indicate a figure of more than 100 million Special Drawing Rights ("SDR") as the upper limit.[138] However, it was not possible to ensure that it would ever be possible to develop a special liability cover for HNS alone.[139] International Union of Marine Insurers noted the difficulties of providing cargo liability insurance due to the state of the reinsurance market.[140] Delegations suggested figures ranging between 100-300 million SDR, with the exception of one proposal for 300-500 SDR. 100 million SDR was suggested as an estimate of the likely capacity of the insurance market at the date when the HNS Convention would be likely to come into force. At the 70th session, insurance interests presented a paper indicating the problems of capacity in the insurance and reinsurance markets.[141]

After all these discussions, the limit of shipowners' liability under the Convention was set at an amount calculated on the basis of the units of tonnage of the ship as follows:[142]

(a) 10 million SDR for a ship not exceeding 2,000 units of tonnage
(b) for a ship in excess of 2,000 units of tonnage, the shipowner is entitled to limit his liability to 10 million SDR plus the following amount:

---

[135] Göransson, "The HNS Convention" 1997 *Unif. L. Rev. (Rev. dr. Unif)* 249, 260.
[136] HNS Convention Art. 5.1. A State which wishes to avail itself of this exclusion shall have to make a declaration to that effect. This may be made and withdrawn at any time. Moreover, two neighbouring States may further agree on exclusion on the same conditions for ships engaged on voyages between ports or facilities of those States. Art. 5.2.
[137] In considering the limits of liability under the first tier, there was general agreement that although limits of liability under the first tier should be proportional to the tonnage of the ship, there should be a minimum amount for ships below a certain size in order to avoid the HNS Fund having deal with minor cases.
[138] There was a conflict of interest between shipowners, who are liable under the first tier, and receivers of HNS, such as petroleum and chemical industries, the former wishing to set the limits reasonably low while the latter would like reasonably high ones. Sasamura, "Development of the HNS Convention," in *13th International Symposium on the Transport of Dangerous Goods by Sea and Inland Waterways* (1998) 491, 501.
[139] LEG 62/7 para.18.
[140] LEG 62/7 para.19.
[141] LEG 70/4/2.
[142] HNS Convention Art. 9.1

(i) for each unit of tonnage from 2,001 to 50,000 units of tonnage, 1,500 SDR

(ii) for each unit of tonnage in excess of 50,000 units of tonnage, 360 SDR

In any case, the aggregate amount of the shipowners' liability shall not exceed 100 million SDR.

With regard to the constitution of the limitation fund, two alternative systems were suggested. According to the first alternative, in order for a shipowner to have the right to limit his liability, he was obliged to constitute a "free-standing" limitation fund, i.e. a limitation fund which only covered HNS claims.[143] According to the second alternative, the general limitation fund that was constituted in conformity with the LLMC 1976 Convention or national law should also cover the HNS claim.[144] This system, called "linkage," was crucial to insurers and was therefore supported by the International Group of P&I Association. Accordingly, the need to provide cover for the same type of damage to be compensated by two different funds would result in the reduction of available insurance capacity.[145] The linkage, therefore, would ensure the avoidance of double insurance in respect of risks arising from the same incident.

Considering that non-HNS claimants would be competing with HNS claimants in the general limitation fund, which would result in less compensation for the former group of claimants than what they would get under free-standing funds and other technical difficulties connected with the linkage alternative, a free-standing limitation fund seemed to be more appropriate.[146]

Accordingly, the limitation fund was based on a free-standing fund. In order to be able to limit his liability, the shipowner must establish a limitation fund, in accordance with the determined limit, with a competent court. This is so as to ensure that the limitation amount is actually available for the payment of compensation. Once the limitation fund has been established, no other assets of the owner may be seized, i.e. "legally arrested."[147] Any other person also providing financial

---

[143] Wetterstein, "Carriage of Hazardous Cargoes by Sea – The HNS Convention" [1997] 26 *Ga.J.Int'l & Comp. L.* 595, 608. Namely, when a single occasion resulted in different types of claims, each claim was covered by its own limitation rules.

[144] *Ibid.* Göransson, "The HNS Convention" [1997] *Unif. L. Rev. (Rev. dr.Unif)* 249, 267 f.

[145] Gaskell, "The Draft Convention on Liability and Compensation for Damage Resulting from the Carriage of Hazardous and Noxious Substances" in *Essays in Honor of Hugo Tiberg* (1996), 225, 271.

[146] During the Conference, insurance interests indicated the problems of capacity in the insurance and reinsurance markets. It was maintained that the linkage alternative would facilitate larger insurance capacity, which would reduce the cost of insurance. However, P&I Clubs emphasized that the importance of linkage was not so much a question of cost, rather of capacity: the more independent funds there were, the lower the overall capacity of the market to be able to provide high first-tier limits of liability. LEG 70/4/2.

[147] Any assets that had been seized before the fund was established must be released. Draft IMO guide for interested parties on the workings of the Hazardous and Noxious Substances Convention 1996 (HNS Convention), (IMO Guide), LEG 83/INF.3, 16.

security also has the right to institute a limitation fund, e.g. the shipowner's insurer.[148]

V. Compulsory insurance of the shipowner

The provision on compulsory insurance is closely modelled on Art. 7 of the CLC. According to Art. 12, any ship that carries HNS will have to take out insurance or maintain other acceptable financial security.[149] Depending on the ship's tonnage, the shipowner and the shipowner's insurer will be liable to pay an amount of up to between 10 and 100 million SDR per incident.

The HNS Convention requires the shipowner to provide evidence of insurance cover upon entry into port of any state that is party to the Convention by production of a certificate.[150] This is necessary regardless of whether the state of the ship's registry is party to the Convention.[151] The dominant form of security under the CLC is the so-called blue card issued by P&I Clubs, but other kinds of insurance or security may be utilized.[152] It is often said that such other kinds of insurance will be more common under the HNS Convention than under the CLC.[153]

Each State Party must ensure under its national law that any ship entering or leaving a port within its territory has the required insurance of financial security. The responsibility for issuing insurance certificates will fall upon the state of the ship's registry.[154] Initially, however, many ships may be registered in states that

---

[148] *Ibid.*

[149] Other acceptable security may be the guarantee of a bank or similar institution.

[150] HNS Convention Art. 12.1. The requirement of compulsory insurance applies to all HNS ships. There is no exception for ships carrying HNS below a certain threshold by contrast to CLC. McKinley, *The International Convention on Liability and Compensation for the Carriage of Hazardous and Noxious Substances by Sea: Implications for State Parties, the Shipping, Cargo and Insurance Industries* (Diss. Cape Town 2005), 42 f.

[151] [Draft] IMO Guide for Interested Parties on the Workings of the Hazardous and Noxious Substances Convention 1996, (HNS Convention), LEG 83/INF.3 para.4.4.

[152] When a State Party questions a certificate and when it is to be decided whether a HNS certificate shall be issued on the basis of the blue card the blue card must be evaluated. The question of which blue cards should be accepted as a basis for a HNS certificate must be evaluated under the national law. No clear state practice has formed under the CLCs because almost only P&I Clubs are providers of financial security under these conventions. Røsæg, "HNS Insurers and Insurance Certificates" <http://folk.uio.no/erikro/WWW/HNS/Comp.doc> 2 f. (visited 13.01.2007).

[153] *Ibid.* 1.

[154] According to Art. 12.2, the compulsory insurance certificates issued to ships carrying HNS are required to contain the following particulars:
(a) name of the ship, distinctive number or letters of port of registry;
(b) name and principal place of business of the owner;
(c) IMO ship-identification number;
(d) type and duration of security;

are not parties to the HNS Convention and the burden of issuing the certificates will fall on a few state parties. Ships registered in states not parties to the HNS Convention will be able to seek certificates from states that are parties to the HNS Convention as long as they can satisfy the insurance requirements.[155]

Within this framework, insurance should not be confused with certification. Insurance is underwritten by the insurer, while the certificates are issued by the state against evidence of insurance.[156] Under the terms of the Convention, the relevant authority of the vessel's state of registry issues a certificate for the vessel, subject to conditions of issue, which certifies that the insurance or financial security in place is valid and satisfies the requirements of the Convention.

VI. HNS Fund (the second tier)

During the negotiations for the HNS Convention, a fundamental question was whether it was necessary or possible to have a second tier at all. Although a one-tier Convention, as with the 1989 CRTD Convention, would be a simpler solution, it is probably flawed on the grounds of equitability and insurability. The capacity of the insurance market has always been recognized as having a restrictive effect when establishing the limitation amounts for the shipowner under the first tier. If there was only a system based on the liability of the shipowner, the levels of liability would have to be set extremely high. From the perspective of safeguarding the availability of adequate compensation to victims of HNS damage, it was essential to ensure that there is a supplementary scheme to provide compensation for HNS damage over and above that which the shipowners can insure.

The concept of a two-tier Convention was thus never really a controversial issue, although there was initially considerable resistance on the part of certain chemical industry circles to the idea of mandatory participation in such a system and of paying contributions to it.[157] Because of the problems encountered in revising and amending the Conventions on compensation for oil-pollution damage, a single Convention was favoured.

1. Liability of the HNS Fund

The Convention provides that the HNS Fund has the aim of providing compensation for damage in connection with the carriage of hazardous and noxious sub-

---

(e) name and principal place of business of insurer or other person giving security and, where appropriate, place of business where the insurance or security is established; and
(f) period of validity of the certificate, which shall not be longer than the period of validity of the insurance or other security.

[155] [Draft] IMO Guide for Interested Parties on the Workings of the Hazardous and Noxious Substances Convention 1996, (HNS Convention), LEG 83/INF.3 para.4.1.

[156] Wu, *Pollution from the Carriage of Oil by Sea: Liability and Compensation* (1996), 68.

[157] Göransson, "The HNS Convention" [1997] *Unif. L. Rev. (Rev. dr.Unif)* 249, 262 f.

stances by sea, to the extent that the protection afforded under the first tier is in-adequate or not available. The HNS Fund will apply in the following situations:[158]

(a) because no liability for the damage arises under Chapter II;
(b) because the owner liable for the damage under Chapter II is financially incapable of meeting the obligations under this Convention in full and any financial security that may be provided under Chapter II does not cover or is insufficient to satisfy the claims for compensation for damage;
(c) because the damage exceeds the owner's liability under the terms of Chapter II.

In principle, the same defences available to the IOPC Fund are available to the HNS Fund, i.e. force majeure or contributory negligence.[159] However, there shall be no such exoneration of the Fund with regard to preventive measures.[160] Like the IOPC Fund, the HNS Fund will not be liable to pay compensation for damage caused by unidentified ships unless the claimant proves that the damage resulted from an incident involving one or more ships.[161] Under the HNS Convention, however, this obligation goes no further than the need to prove that there is "a reasonable probability" of this being the case.[162]

The involvement of both tiers in compensation payments will not create any special complications for claimants.[163] Claims will be submitted to the HNS Fund and, providing that claimants can substantiate their losses and the claimants meet the criteria for admissibility, claimants should receive compensation from the shipowner's insurer or the Fund.[164]

The HNS Fund will pay compensation in excess of the shipowner's liability, but with a limit of 250 million SDR, including the amount actually paid by the shipowner or his insurer.[165] If the total amount of the admissible claims is less than

---

[158]   HNS Convention Art. 14.1.
[159]   HNS Convention Art. 14.3. and 4.
[160]   HNS Convention Art. 14.4.
[161]   [Draft] IMO Guide for Interested Parties on the Workings of the Hazardous and Noxious Substances Convention 1996, (HNS Convention), LEG 83/INF.3 para.5.3.
[162]   The choice of expression and the introduction of this additional element followed intensive debate during the preparatory stage, in which divergent views were expressed as to who should carry the burden of proof: the claimant, proving that the damage resulted from an incident involving a ship, or the Fund, proving that it did not. In favour of the latter view, it was suggested that, in order to obtain exoneration, it would be more consistent with the underlying principal objective of providing compensation to victims to place the burden of proving that no ship was involved on the Fund. A majority, however, felt that the HNS Fund should be exonerated unless the claimant could provide some degree of proof that the damage resulted from an incident involving one or more ships. This was seen as representing a fair balance between these opposing views. Göransson, "The HNS Convention" [1997] *Unif. L. Rev. (Rev. dr.Unif)* 249, 264.
[163]   [Draft] IMO Guide for Interested Parties on the Workings of the Hazardous and Noxious Substances Convention 1996, (HNS Convention), LEG 83/INF.3 para.5.3.
[164]   *Ibid.*
[165]   HNS Convention Art. 14.5(a).

the maximum amount available, then all claims will be paid in full. If the total amount of the admissible claims exceeds the maximum amount available for compensation, a pro rata reduction will be made.[166] The HNS Fund's main aim is to provide compensation to the extent that the compensation provided by the shipowner, or the shipowner's insurer or guarantor, is deficient or not available. The Convention provides for claims for loss of life and personal injury to have a certain priority over other claims. Accordingly, similar to LLMC 1976, claims in respect of death or personal injury take priority over other claims, insofar as the aggregate of claims in the former category does not exceed two-thirds of the total amount payable by the Fund,[167] However, unlike LLMC 76, if there are no claims for loss of life and personal injury, under the HNS Convention the total amount of compensation is available for payment of claims for other types of damage.[168]

The HNS Fund shall also pay compensation, even if the owner has not constituted a limitation Fund, if:

– preventive measures to minimise a potential risk were taken by the shipowner.[169]
– the damage resulted from a natural phenomenon of an exceptional, inevitable and irresistible character.[170]

2. Contributions to the HNS Fund

a) Receiver pays

The question "Who pays?" was debated in many of the Legal Committee's reports. As mentioned previously, one of the main difficulties which plagued the 1984 diplomatic conference and negotiations in the Legal Committee thereafter was the basis on which contributions would be made to finance any second tier.[171] A large number of suggestions were presented to the Legal Committee up until 1994, all of which were founded on the concept in the 1984 draft HNS Convention that it would be the shipper which would contribute to the HNS Fund. A proposal in 1994 shifted the burden to the receiver.[172]

It was felt that if the system were to place the obligation to pay contributions to the Fund on the receiver of cargo, like the Fund Conventions, it would make developing countries the major contributors to the system in contrast to the situation with regard to the import of oil.[173] The draft instrument accordingly took as a starting point that it should be up to the exporter to make the contributions in one

---

[166] HNS Convention Art. 14.6.
[167] *Ibid.*
[168] [Draft] IMO Guide for Interested Parties on the Workings of the Hazardous and Noxious Substances Convention 1996, (HNS Convention), LEG 83/INF.3 para.5.3.
[169] HNS Convention Art. 14.2
[170] HNS Convention Art. 14.5.(b).
[171] *Supra* p. 252 ff.
[172] LEG 70/Inf.2
[173] Göransson, "The HNS Convention" [1997] *Unif. L. Rev. (Rev. dr.Unif)* 249, 263.

form or another. Subsequent research and a detailed breakdown of the situation by some of the delegations involved soon revealed, however, that cargo flows were much more complicated than had been assumed and that developing countries were on average "net exporters" of HNS goods.[174] This led the Committee to agree on the principle of "the receiver pays." Some of the complications previously identified vanished forthwith and many other issues automatically fell into place when the Fund Convention model was applied. Accordingly, compensation payments made by the HNS Fund will be financed by levies on receivers of HNS. Levies will be in proportion to the quantities of hazardous and noxious substances received each year. This is a concept similar to that under the Fund Conventions, where receivers of oil – above a minimum threshold – pay contributions.[175]

b) Who is a receiver?

The receiver is the person who physically receives delivery.[176] This definition identifies an actual person (the importer/receiver), avoiding the difficulties of deciding who a shipper is, in law or fact.[177] For commercial reasons, the LNG sector's contributions will be paid by the "title-holder" of the cargo immediately prior to discharge, not the receiver.[178] The concept of physical receipt after the carriage by sea is expressed in the HNS Convention in a similar way as in the 1992 Fund Convention.[179]

The HNS Convention allows a person who receives hazardous and noxious substances on behalf of a third party to designate that third party to the HNS Fund as the receiver for the purposes of the Convention.[180] In such a case, both the person who physically receives the contributing cargo in a port or terminal and the third party must be subject to the jurisdiction of a State Party. The aim of this provision is to meet concerns expressed by the operators of storage facilities, who do not actually own the substances that they receive.[181]

---

[174] *Ibid.*

[175] [Draft] IMO Guide for Interested Parties on the Workings of the Hazardous and Noxious Substances Convention 1996, (HNS Convention), LEG83/INF.3 para.7.1.

[176] HNS Convention Art. 1.4.

[177] Gaskell, "The Draft Convention on Liability and Compensation for Damage Resulting from the Carriage of Hazardous and Noxious Substances" in *Essays in Honor of Hugo Tiberg* (1996), 225, 287.

[178] HNS Convention Art. 19.1(b). This specific designation of liable person was incorporated in the Convention at the request of the LNG industry. As a result concerns have been raised that it will not be possible to collect a high portion of these contributions in respect of LNG. The solution is to make sure that contributions in respect of LNG can be collected. Røsæg, "Non-collectible contributions to the separate LNG account of the HNS Convention" [2007] 13 *JML* 94, 95 ff.

[179] The HNS Fund, therefore, will be able to build on the practice and experience of the IOPC Funds in this regard. Draft Guide to the Implement of the HNS Convention, Prepared by the Secretariat of the International Oil Pollution Compensation Fund 1992, <www.iopcfund.org/npdf/HNS-guide_e.pdf>, June 2005, para.6.1. (visited 13.01.2007).

[180] HNS Convention Art. 4.1(a).

[181] LEG 83/INF.3 Annex, p. 29.

Furthermore, states are allowed to establish their own definition of receiver under national law.[182] Such a definition cannot be used, however, to reduce the overall contributions which that state's receivers would have had paid if the definition in the Convention had been applied. This allows states flexibility in adopting the Convention in conjunction with existing national law, without giving any state the possibility of obtaining an unfair commercial advantage.[183]

With regard to receipts of cargoes carried in domestic traffic, i.e. the trade by sea from one port or terminal to another within the same State Party, states have the option of developing national regimes for the collection of contributions.[184] Any such national regime, however, must also raise the same total contributions for that state as the mechanism specified in the Convention would have done.

c) Contributing cargo

It is obvious that if all cargoes described as HNS were required to contribute to the HNS Fund, it could become unworkable, as there would be millions of potential contributors.[185] Therefore, it is crucial that "contributing cargo" is limited in such a way as to make the HNS Fund workable and financially viable. These complications have eventually been solved by a post-event contribution scheme and a threshold.

A major discussion also took place on whether domestic shipments should also contribute to the second-tier HNS Fund in addition to international voyages. Although most delegations considered that they should contribute, there were practical questions. Clearly, it would be more convenient to have a comprehensive Convention, particularly in those states which do not have, and are unlikely to produce, their own national legislation for domestic carriage. To make no distinction, therefore, between domestic and international carriages was important. On the other hand, to allow contracting states some discretion in the matter was important for archipelagic states.

The definition of "contributing cargo" in the Convention covers domestic traffic, although it is not specifically mentioned.[186] Accordingly, "contributing cargo" means any hazardous and noxious substances which are carried by sea as cargo to a port or terminal in the territory of a State Party and discharged in that state.[187] States then have the option of submitting an annual report showing domestic re-

---

[182] HNS Convention Art. 1.4(b).

[183] LEG 83/INF.3, Annex p. 29.

[184] HNS Convention Art. 21.5.

[185] Gaskell, "The Draft Convention on Liability and Compensation for Damage Resulting from the Carriage of Hazardous and Noxious Substances" in *Essays in Honor of Hugo Tiberg* (1996), 225, 289.

[186] *Ibid.*

[187] HNS Convention Art. 1.10.

ceipts.[188] Transhipped cargo shall be considered as contributing cargo only in respect of receipt at the final destination.[189]

3. The separate accounts and the general account

One of the most controversial questions with the HNS Fund was whether it should operate effectively as one fund or whether there should be separate accounts within it for particular segments of the market. The essential political problem is that certain market sectors, such as LNG, which feel that they have a good safety and environment record, are loath to contribute towards the damage incurred by other less careful industries.[190] By making the separate accounts exclusively liable for damage caused by the substance in question but not for damage caused by other substances, the risk of distortion of trade or cross-subsidization would be eliminated.

In the case of oil, it was preferable for the HNS system to follow that of the IOPC Fund as closely as possible. A receiver of persistent oil would, in principle, have to pay contributions both under the IOPC Fund system for pollution damage and under the HNS Fund for damage caused by fire and explosion.[191] It was felt that it was going too far to ask this same receiver also to contribute to a Fund covering damage caused by chemicals in general, not least given the substantial imbalance in the volume of trade between these two categories of HNS.

Similar arguments were made in favour of establishing a separate account for LNG with specific emphasis on the excellent safety records for the trade, which did not quite fit into a system in which the receiver was liable for contributions to the Fund. The Conference agreed to the establishment of a third separate account in respect of LPG. Initially there had been some doubt as to whether the market was sufficiently stable and of a volume such that a separate LPG account might be regarded as a financially sound option capable of safeguarding victims' interests. These doubts were dispelled at the Conference. Accordingly, when fully operational, the HNS Fund will have four accounts:

(a) oil account
(b) LNG account
(c) LPG account

---

[188] HNS Convention Art. 21.5. On receipt of this, the HNS Fund would provide an invoice to the state, which could then either pay the HNS Fund itself or instruct the HNS Fund to invoice individual receivers in that state.

[189] HNS Convention Art. 1.10.

[190] Japan proposed four separate accounts into the HNS Fund: oil, LNG, LPG and HNS. The danger foreseen by the other delegations was that the more the second tier was segmented, the greater the chance that its financial viability as a whole could be undermined. LEG 68/4/4.

[191] Draft Guide to the Implement of the HNS Convention, Prepared by the Secretariat of the International Oil Pollution Compensation Fund 1992, <www.iopcfund.org/npdf/ HNS-guide_e.pdf>, June 2005, para.7.3. (visited 13.01.2007).

(d) General account
    i. bulk solids
    ii. other HNS

Receivers of the HNS might have to contribute to one or more of the accounts. The levies applying to individual receivers will be calculated according to the quantities of contributing cargo received and, in the case of the general account, according to the regulations in Annex II of the Convention. Liability to contribute to the HNS Fund will arise for a given receiver only when his annual receipts of HNS exceed the following thresholds:[192]

Oil persistent oil	150,000 tonnes
non-persistent oil	20,000 tonnes
LNG	no minimum quantity
LPG	20,000 tonnes
Bulk solids and other HNS	20,000 tonnes

Each account will meet the cost of compensation payments arising from damage caused by substances contributing to that account, i.e. there will be no cross-subsidization.[193] Furthermore, each of the separate accounts will only come into operation when the total quantity of contributing cargo in Member States during the preceding year or another year decided by the Assembly exceeds the following levels:[194]

(a) oil account	350 million tonnes
(b) LNG account	20 million tonnes
(c) LPG account	15 million tonnes

Thus, during the early existence of the HNS Fund, there may not be a sufficient contribution basis in the form of the quantities of HNS received in Member States to set up all the separate accounts. Initially, the HNS Fund may have only two accounts:

a. a separate account for oil; and
b. a general account with four sectors for LNG, LPG, bulk solids and other HNS.

The HNS Convention is based on a system of post-event contributions, i.e. levies are only due after an incident involving the HNS Fund occurs.[195] Levies may be spread over several years in the case of a major incident.[196]

---

[192] HNS Convention Art. 19.1.
[193] [Draft] IMO Guide for Interested Parties on the Workings of the Hazardous and Noxious Substances Convention 1996, (HNS Convention) LEG83/INF.3 para.7.1.
[194] HNS Convention Art. 19.3.
[195] HNS Convention Art. 17.1.
[196] "An Overview of the HNS Convention," <www.imo.org/includes/blastDataOnly.asp/ data_id%3D6505/HNSconventionoverview.pdf> para. 32. (visited 13.01.2007).

4. Administration of the HNS Fund

In a resolution[197] of the Conference which adopted the HNS Convention, the Assembly of the 1992 Fund was invited to assign to the Director of the 1992 Fund, in addition to his functions under the 1992 Fund Convention, the administrative tasks necessary for setting up the HNS Fund in accordance with the HNS Convention. In 1998, the 1992 Fund Assembly instructed the Director to carry out the tasks requested by the HNS Conference.

In 1999, the IMO Legal Committee set up a Correspondence Group to monitor the implementation of the HNS Convention. The Group held a Special Consultative meeting in Ottawa in 2003.[198] At the 90th session of the Legal Committee in March 2005, it was suggested that the 1992 Fund should assume a more active role and work with the IMO regarding the responsibility for co-ordinating the implementation of the HNS Convention.

5. Organization of the HNS Fund

The HNS Fund will operate in a similar way to the 1992 Fund. The HNS Fund will be governed by an Assembly, composed of all Member States.[199] There will, however, be some important differences in the way the HNS Fund will operate compared to the 1992 Fund. The 1992 Fund only deals with claims for pollution damage, whereas the HNS Fund will have to deal with a wider range of potential claims, e.g. death and personal injury. Thus, as already mentioned, the system for contributions to the HNS is much more complicated than that for contributions to the 1992 Fund. To deal with the claims properly, the Assembly will establish a Committee on Claims for Compensation[200]

The Ottawa meeting recommended that the HNS Fund should conclude a Memorandum of Understanding (MOU) with organizations involved in incidents, similar to that between the IOPC Funds and the International Group of P&I Clubs. The meeting also considered that it would be essential for the HNS Fund to conclude a MOU with the 1992 Fund so that, where appropriate, the same experts would be used in the assessment of claims.

6. Secretariat and director of the HNS Fund

According to Art. 24 of the HNS Convention, the HNS Fund will be administered by a Secretariat headed by a Director.

---

[197] LEG/CONF.10/8/1.

[198] LEG 87/11.

[199] HNS Convention Art. 24 and 25. Its first session will be convened by the Secretary-General of the IMO and will be held not more than thirty days after the entry into force of the HNS Convention. Art. 44.

[200] HNS Convention Art. 26 (i). This is similar to the 1992 Fund's Executive Committee.

The IOPC Funds, i.e. the 1971 Fund, the 1992 Fund and the Supplementary Fund, have a joint Secretariat.[201] When the issue of the Secretariat of the HNS Fund was discussed in May 2003 by the 1992 Fund Administrative Council, acting on behalf of the Assembly, a number of delegations expressed the view that the most practical solution would be for the HNS Fund to have a joint Secretariat with the IOPC Funds.[202] The point was made that a joint Secretariat would enable the HNS Fund to benefit from the experience of the IOPC Funds and would reduce the administrative costs for both the IOPC Funds and the HNS Funds.[203] However, one delegation expressed the view that since the HNS Fund would have a different membership to the IOPC Funds, it should have a separate Secretariat so as to ensure that there was a clear delineation of its operation and costs.[204]

The Council recognized that the decision as to the location of the HNS Fund would be taken by the HNS Fund Assembly. However, the Council instructed the Director of the 1992 Fund to continue the preparatory work for the time being on the assumption that the HNS Fund would have a joint Secretariat with the IOPC Funds and would be based in London.[205]

The Ottawa meeting agreed that the issue of the location of the HNS Fund Secretariat required a political decision and requested the Legal Committee of IMO to consider the issue with a view to facilitating a decision by IMO on the location of the HNS Fund. The Ottawa Meeting requested the Legal Committee to recommend that the HNS Fund and the IOPC Funds should have a joint Secretariat, located in London.[206]

VII. Entry into force

The HNS Convention will enter into force eighteen months after both of the following criteria have been fulfilled:[207]

(a) At least 12 states must have expressed their consent to be bound by the Convention. This must include four states each a gross register tonnage of at least 2 million units.

(b) Contributors in the states that have ratified or acceded to the Convention must, between them, receive a minimum of 40 million tonnes of contributing cargo covered by the general account.

---

[201] "Draft Guide to the Implement of the HNS Convention, Prepared by the Secretariat of the International Oil Pollution Compensation Fund 1992" <www.iopcfund.org/npdf/ HNS-guide_e.pdf>, June 2005, para.5.3. (visited 13.01.2007).

[202] 92FUND/AC.1/A/ES.7/7 para. 6.6.

[203] Ibid.

[204] Ibid.

[205] Ibid. para. 6.7.

[206] LEG 87/1.

[207] HNS Convention Art. 46.

The criteria ensure that there will be sufficient quantities of contributing cargo to ensure the financial viability of the HNS Fund when the HNS Convention enters into force, and also a sufficient number of state parties to secure broad international participation.[208]

1. Inconsistency of existing EU legislation with HNS Convention

EU Regulation 44/2001 on jurisdiction and the recognition and enforcement of judgments in civil and commercial matters was adopted on 22 December 2000. By incorporating the Brussels Convention on Jurisdiction and Enforcement of Judgements in Civil and Commercial Matters 1968 into European Community Law, subject matters regulated therein triggered what is known as "exclusive Community competence" not only internally within the EU, but also externally in relation to other states, in line with the so-called AETR doctrine established by the European Court of Justice.[209]

2. EU Regulation 44/2001

Regulation 44/2001 covers all areas of civil and commercial law except for those that are expressly excluded from its scope.[210] It lays down the rules on jurisdiction for a variety of different matters. Recognition of a judgment given in a court of an EU Member State is automatic in another Member State unless contested. When it comes to the relationship between these provisions and international agreements which govern the same subject for particular matters, Regulation 44/2001 represents a small but significant change to the regime applied under the Brussels Convention. While such international conventions were given general precedence in the Brussels Convention, Regulation 44/2001 limits this exception only to conventions to which the Member States are parties at the time of the adoption of the

---

[208] LEG 83/INF.3. As of 31 May 2007, 8 states have ratified the Convention: Angola, Cyprus, Morocco, the Russian Federation, Saint Kitts and Nevis, Samoa, Slovenia and Tonga. Of these, Cyprus and the Russian Federation have tonnage of at least two million units. Only Slovenia declared the total amount of received HNS to be 120,000 tonnes.

[209] In *Commission v. Council* (the AETR judgment) [1973] ECR 263, the Court held that: Each time the Community, with a view to implementing a common policy envisaged by the Treaty, adopts provisions laying down common rules, whatever form these may take, the Member States no longer have the right, acting individually or even collectively, to undertake obligations with third countries which affect those rules... As and when such common rules come into being, the Community alone is in a position to assume and carry out contractual obligations towards third-party countries affecting the whole sphere of application of the community legal system... It follows that to the extent to which Community rules are promulgated for the attainment of the objectives of the treaty, the Member States cannot, outside the framework of the Community institutions, assume obligations which might affect those rules or alter their scope.

[210] Ringbom, "EU Regulation 44/2001 and its Implications for the International Maritime Liability Conventions" [2004] 35 *J.Mar.L.&Com.* 1, 3.

Regulation. Therefore, the adoption of Regulation 44/2001 resulted in a significant set of new rules being brought into the realm of Community law. Of these, one was the resulting exclusive Community competence, which implied significant limitations on the possibility for EU Member States to enter into international obligations on issues that could affect the rules laid down in Regulation 44/2001.[211]

3. Implications of EU Regulation 44/2001 on the HNS Convention

The HNS Convention, like most maritime liability conventions, includes provisions on the competent jurisdiction and on the recognition and enforcement of judgments.[212] In contrast to the multiple grounds for jurisdiction available under Regulation 44/2001, jurisdiction under the maritime liability conventions is, as a rule, in the State Party where the pollution damage occurred.[213] The Conventions further provide that the courts of the state where the shipowner, or the insurer, has constituted a fund in order to benefit from the right to limit the liability shall have exclusive jurisdiction to determine all matters relating to the apportionment and distribution of the fund.[214]

Consequently, there are two issues. One is that there is a maritime approach, stressing the importance of achieving widespread ratification of the maritime liability convention as soon as possible.[215] Therefore, any limitation on the possibility of Member States to ratify the agreements is inadmissible. The other one is that the arguments of Community law are straightforward enough for the proponents of the "Community approach."[216] The European Court of Justice has left member states in no uncertainty about the difficulties they face in concluding international conventions that might affect Community rules or alter their scope. The repeatedly held view of the Court is that Member States are prevented from assuming obligations in those areas outside Community institutions. In addition, the Court, referring to the requirement of unity in the international representation of the Community, has held that, as regards mixed agreements in general, "it is

---

[211] *Ibid.* at 4.

[212] HNS Convention Art. 38-40.

[213] *Ibid.*

[214] *Ibid.* The underlying reasons for limiting the number of jurisdictions include fostering equal treatment of claimants, ensuring a link between the courts involved and the action, limiting forum shopping and avoiding a situation where the same incident is litigated in a variety of courts and jurisdictions at the same time. The maritime conventions seek to avoid complex procedures relating to the recognition of cross-border judgements by requiring the recognition of a judgement that is no longer subject to ordinary forms of review, except where the judgment was obtained by fraud or where the defendant was not given reasonable notice and a fair opportunity to present the case. Judgments shall be enforceable in each State Party as soon as the formalities required in the state where the judgment was given have been satisfied.

[215] Ringbom, "EU Regulation 44/2001 and its Implications for the International Maritime Liability Conventions" [2004] 35 *J.Mar.L.&Com.* 1, 7.

[216] *Ibid.*

important to ensure that there is a close association between institutions of the Community and the Member States both in the process of negotiation and conclusion and in the fulfilment of the obligations entered into" and "the Community institutions and the Member States must take all necessary steps to ensure the best possible cooperation in that regard."[217]

4. Authorizing member states to ratify the conventions with a reservation

The Commission's proposal for a solution with respect to the HNS Convention was presented in 2001. The envisaged solution was a Council decision[218] authorizing the Members States to ratify the Convention subject to making a specific reservation to allow for a continued application of Regulation 44/2001 as between Member States with regard to the recognition and enforcement of judgments. In other words, the rules on the recognition and enforcement of judgments contained in Regulation 44/2001 would, between Member States only, override the corresponding provisions contained in the Convention, when in force. The decision was intended to be a temporary measure until a time when the Convention was amended in order to allow for Community participation.[219]

The Council Decision also requires EU Member States to ratify or accede to the HNS Convention before 30 June 2006, if possible.[220] However, none of the Member States had ratified by that time, although some states have prepared implementation legislation.[221]

---

[217] Ruling 1/78, [1978] E.C.R. 2151.
[218] 2002/971/EC: Council Decision of 18 November 2002, authorizing the Member States, in the interest of the Community, to ratify or accede to the International Convention on Liability and Compensation for Damage in Connection with the Carriage of Hazardous and Noxious Substances by Sea, 1996 (the HNS Convention) OJ L 337 13/12/2002 p. 0055-0056.
[219] Art. 5 of the Decision provides that "Member States shall, at the earliest opportunity, use their best endeavors to ensure that the HNS Convention is amended to allow the Community to become a contracting party to it."
[220] 2002/971/EC Art. 3.1.
[221] <www.hnsconvention.org/legislation.html>. (visited 13.01.2007).

Conclusion

In the carriage of goods by sea, the most common type of damage or loss is loss of or damage to goods. However, there are cases where the carrier suffers loss or damage as a result of the shipment of goods. In this regard, the law relating to "dangerous goods" occupies an important, although often overlooked, aspect of maritime law.

The carriage of dangerous goods involves an inherent risk of danger to those concerned with its care. The best way to eliminate the risk would be either to prohibit altogether the carriage of dangerous goods by sea or to impose such measures as would render their carriage impractical. Either way would be unacceptable, hence policy-makers have searched for a middle path, an "acceptable risk", which lies in between. Regulations have been enacted to reflect the level of "acceptable risk" which translates into practical measures of risk containment. As Part I demonstrates, there are special regulations for all types of dangerous goods. These regulations contain technical and detailed operational procedures. Dangerous goods regulations firstly seek to prescribe rules to ensure the safe carriage of certain goods. Failure to comply with the regulations renders the offending party liable to sanctions in the form of a fine or imprisonment.

Dangerous goods regulations differ in technical details; however, they all have a common aim in providing for risk containment. Firstly, there is a system of documentation which is designed to inform the carrier as to the classification and characteristics of the dangerous goods. Secondly, there are provisions for the proper transmission, and due appreciation, of information relating to the dangerous goods to all concerned. Thirdly, they contain provisions which ensure the proper handling, stowage and carriage of such goods. Additionally, with regard to packaged goods, there must be proper containment of the substance in packages, tanks or receptacles which are sufficient to withstand the ordinary risk of handling and transport by sea with regard to the properties of the goods. In the light of these operational regulations, the actions or conduct of the carrier or shipper may amount to breach of contract. In this regard, operational regulations complement a liability regime. Operational regulations are particularly relevant to the issues of which goods are dangerous, whether adequate notice of the dangerous goods has been given, whether the packaging or container is sufficient and to what extent the carrier knows about the dangerous characteristics of the goods.

As Part 2 shows, not only dangerous goods but any goods may cause damage. Does that mean that a cargo of fish which is stowed together with chocolate or a cargo which has a tendency to infestation is dangerous? Although English common law concept of dangerous goods is broad covering any type of goods causing damage and even goods causing solely delay or detention, this position is hardly justified in modern law of carriage of goods by sea.

In common law, the shipper's liability for the shipment of dangerous goods without notice to the carrier has been long considered as strict. The strict liability of the shipper was confirmed under the Hague/Hague-Visby Rules in English law as well as under U.S. COGSA in American law. German law, on the other hand, expressly provides for strict liability. From the legislative history of the Hague/Hague-Visby Rules, it is clear that Art. IV.6, the special provision on dangerous goods, was aimed at governing a specific group of goods due to safety concerns. Under the Rules, the shipper's strict liability arising from the goods is justified only if the strict liability is restricted to certain groups of dangerous cargo. It is because strict common law liability was modified under the Rules by reducing the duty an obligation to use due diligence. General basis for shipper's liability is fault and strict liability is an exception. Extending the strict liability to any cargo, whatever its nature destroys the said balance between shipper and carrier in favour of the carrier. Therefore, dangerous goods should be understood to be exceptionally dangerous goods, i.e. the risk and potential damage should be significant.

In this regard, there is no doubt that goods contained in the IMDG Code's list well qualify to be within this concept. On the other hand, as we have seen in Part 1, dangerous goods are not restricted to the goods in the IMDG Code, i.e. packaged goods. There are regulations for bulk cargoes. It seems likely that with bulk shipment there is greatest risk of major disasters. On the other hand, the very specialised nature of bulk operations leads naturally to the use of proper equipment and routines. This reduces the risk towards acceptable limits. Hence, a shipowner who specializes in the carriage of a particular cargo is reasonably expected to be aware of the true characteristics of it. Moreover, as it is not packed, there is no element of concealment. However, in any case, shipper must disclose the particular state of cargo. Otherwise liability, strict or fault-based, would be unavoidable.

On the other hand, packaged goods usually offer much less opportunity to cause the major accidents, but they are handled frequently and in close proximity to people. They are mixed freely with a great amount of normally combustible material, namely general cargo, in circumstances with a high potential for minor troubles. However rarely, it is not unheard of that, that packaged dangerous goods cause a major disaster, i.e. loss of the vessel and the death of many people. Since carriers of general cargoes are not specialized as bulk carriers and there are thousands of dangerous goods carried in packaged form in contrast to bulk cargoes, which is limited, the disclosures of the shipper with respect to the nature of the goods and special precautions are of paramount importance.

Furthermore, it is clear from the case law that strict liability of shipper arises if the carrier is unaware of dangerous nature of goods. If a notice of dangerous nature given or if the carrier otherwise knows, the strict liability does not become issue except the fact that particular danger is scientifically unknown. However, this is not to say that shipper is entirely exonerated from liability. This is rather to say that in such a case shipper's liability is fault-based rather than strict.

As the dangerous nature of bulk cargoes is well known, the shipper's strict liability for damages arising from bulk cargoes rarely arises. It is more often the case that the shipper of dangerous bulk cargo often will be liable for misrepresentations, such as statements on the humidity, temperature etc., i.e. generally for negligence.

Moreover, given heightened spectre of terrorism, the security of dangerous goods shipments has become a priority for carriers, shippers, consignees, emergency respondents and governments. As a result security regulations and requirements adopted first in the U.S. and in the process of being adopted in other countries. Due to these security regulations significant losses may arise from delay or detention of the vessel if the particulars furnished by a shipper are inaccurate. Historically neither maritime law nor marine insurance recognized claims for delay. Today much of this changed and damages for delay in cargo delivery are awarded occasionally. It is controversial if damages for delay falls under Hague/Hague-Visby Rules, however, where it falls liability of carrier for delay in delivery is negligence as well as under general law of many national laws. Therefore, it is not justified to make the shipper liable for any delay irrespective of fault by extension the meaning of dangerous to *legally dangerous* cargo. If dangerous cargo causes delay in addition to physical damage, the loss arising from the delay is likely to come within the damage arising from the shipment of dangerous goods.

Historically speaking, tort systems in western countries have vacillated between two ideas: fault and causation. These systems have had to make a choice either to base liability upon fault and make causation a separate question, or to disregard fault and make causation itself the basis of liability. No system is purely causal or purely fault-based, but a system can be classified by the degree to which it makes fault or causation the dominant ground of liability. The younger and more familiar of these two approaches is the fault system. Its first appearance in civil law came centuries before the rise of common law. The older causal systems, which can be found in various ancient primitive legal systems that held sway long before the rise of fault, were based upon strict liability. In modern law, these causal systems have been replaced by fault systems in which strict liability is relegated to certain actions or particular statues that are exceptions to the rule. In the twentieth century, strict liability has been expanding rapidly. Strict liability often applies to inherently dangerous activities and is founded upon a policy that is imposed upon anyone who for his own purposes creates an abnormal risk of harm to his surroundings.

Applying the policy behind strict liability to the transport of dangerous goods means that a shipper who has created the risk, albeit unknowingly, with the likelihood that the harm would be great if it occurred pays for its consequences. The ultimate goal of strict liability is to prevent harm from occurring in the first place by encouraging people to be aware of the nature of the goods with which they deal. One who possesses a dangerous animal, a defective product or dangerous cargo is in the best position to discover its nature and to perhaps control or warn others. Comparatively speaking, dangerous goods are like dangerous activities in terms of the magnitude of harm they can cause. Thus, although a shipper took great care of packing and handling dangerous goods, he would be strictly liable for damage caused by the dangerous cargo.

How does strict liability address safety? If a shipper knows that he will be strictly liable for his dangerous products, he will exercise a high degree of care. Moreover, the shipper is in a much better position than the carrier to learn of the dangerous nature of the goods, especially if the goods are likely to be packed. A carrier relies on the shipper that he is not going to load dangerous goods onto the ship without warning the carrier. At least with the notice of the dangerous nature of the goods, the carrier can make the decision of whether or not to take the risk.

However, due to the harsh nature of strict liability, the strict liability of the shipper would be justified only if it is restricted to goods that posses a high degree of risk and damage therefrom. For solely commercial risks, such as delay, strict liability is a choice which, if it applies, should be equally applied to the parties of a contract. Accordingly, the strict liability of the shipper for the shipment of dangerous goods whose dangers are not known to the carrier is useful for safety reasons; however, it is unfair if it applies to any cargo that causes damage. For the reasons mentioned above, it is suggested that strict liability for the unknown dangers of dangerous goods should be retained in future carriage conventions by restricting it to significant risks and by making clear the distinction between liability arising from regular goods, although innocuous, causing damage and dangerous goods.

The shipper is strictly liable for dangerous goods carried without consent and liable for his fault for damages arising from any type of goods. Moreover, shippers may be liable for damage to the ship during activities such as loading and unloading. A charterer under a charter contract is exposed to even more liabilities. However, neither a limitation under the Hague Rules is available to a shipper nor a global limitation of the shipowner's liability is available to a charterer for damage to the vessel. Limitation of liability evolved from a need to protect investors from huge losses and to promote a fledgling industry. Now shippers and charterers also face extensive losses. The basis of argument for limitation of liability has been, since its inception, largely commercial in nature. It appears that this is still the impetus behind maintaining limitation of liability in the international maritime industry. The focus appears to have shifted from encouraging trade and maintaining the development of shipping to that of capping potential insurance pay-outs. The basis of the argument is that it is simply a matter of insurability and the cost of insurance. Is not this an argument for the insurability of the shipper's or charterer's liability to the carrier or shipowner?

As seen in Part 5, shippers' interests state the need for limitation in future carriage convention and charterers are in need of some protection against their enormous liability exposures to shipowners. It will be interesting to watch developments in this field. Shipper interest has always rejected to be party liable under the third party liability regimes. Indeed, it is the fact under a contract of carriage a shipper may face extensive liabilities. As Parts 4 and 5 demonstrate, shippers and charterers may be in need of insurance protection to cover their liability arising out of the unknown dangers of dangerous goods. World-wide availability of this insurance for liabilities arising from carriage of dangerous goods and where available if prohibitively priced is a question

With regard to liability for third-party damage resulting from the transport of hazardous substances, IMO adopted the International Convention on Liability and Compensation for Damage in Connection with the Carriage of Hazardous and Noxious Substances by Sea (the HNS Convention) in May 1996, to address the need for an international compensation and liability regime governing damage arising from the carriage of hazardous and noxious substances by sea. The HNS Convention was needed to establish an international system for solving problems of compensation linked with the carriage of hazardous and noxious substances. The present rules on compensation do not provide enough safeguards for the interests of claimants in view of the huge damage that may be caused in connection with the carriage of hazardous cargo. HNS incidents pose significant risks due to the immense and growing amount of international trade in HNS and the potential damage resulting from pollution, fire or explosion can be enormous.

The HNS Convention will make it possible for up to 250 million SDR to be paid out to victims of disasters involving HNS. The HNS Convention is based on the two-tier system established under the CLC and Fund Conventions. However it goes further in that it covers not only pollution damage but also risks of fire and explosion, including loss of life or personal injury as well as loss of or damage to property.

Under the HNS Convention liability is channelled to the owner with limited defences. One of these defences is unique to the HNS Convention: "No liability shall attach to the owner if the owner proves that the failure of the shipper or any other person to furnish information concerning the hazardous and noxious substance."[1] Shippers should be cognizant of this fact and should properly disclose the cargo to avoid huge claims.

Contribution system under the HNS Convention is more complicated than under the CLC. However, the HNS Convention will ensure the availability of adequate, prompt and effective compensation for damage resulting from shipping incidents involving the carriage of hazardous and noxious substances by sea.

---

[1]   HNS Convention Art. 7.2.

Appendices

Hague-Visby Rules

Article I

In these Rules the following expressions have the meanings hereby assigned to them respectively, that is to say,

(*a*) "carrier" includes the owner or the charterer who enters into a contract of carriage with a shipper;

(*b*) "contract of carriage" applies only to contracts of carriage covered by a bill of lading or any similar document of title, in so far as such document relates to the carriage of goods by water, including any bill of lading or any similar document as aforesaid issued under or pursuant to a charterparty from the moment at which such bill of lading or similar document of title regulates the relations between a carrier and a holder of the same;

(*c*) "goods" includes goods, wares, merchandise and articles of every kind whatsoever, except live animals and cargo which by the contract of carriage is stated as being carried on deck and is so carried;

(*d*) "ship" means any vessel used for the carriage of goods by water;

(*e*) "carriage of goods" covers the period from the time when the goods are loaded on to the time they are discharged from the ship.

Article II

Subject to the provisions of Article VI, under every contract of carriage of goods by water the carrier, in relation to the loading, handling, stowage, carriage, custody, care and discharge of such goods, shall be subject to the responsibilities and liabilities and entitled to the rights and immunities hereinafter set forth.

Article III

1. The carrier shall be bound, before and at the beginning of the voyage, to exercise due diligence to

    (*a*) make the ship seaworthy;

    (*b*) properly man, equip and supply the ship;

    (*c*) make the holds, refrigerating and cool chambers, and all other parts of the ship in which goods are carried, fit and safe for their reception, carriage and preservation.

2. Subject to the provisions of Article IV, the carrier shall properly and carefully load, handle, stow, carry, keep, care for and discharge the goods carried.

3. After receiving the goods into his charge, the carrier, or the master or agent of the carrier, shall, on demand of the shipper, issue to the shipper a bill of lading showing among other things

(*a*) the leading marks necessary for identification of the goods as the same are furnished in writing by the shipper before the loading of such goods starts, provided such marks are stamped or otherwise shown clearly upon the goods if uncovered, or on the cases or coverings in which such goods are contained, in such a manner as should ordinarily remain legible until the end of the voyage;

(*b*) either the number of packages or pieces, or the quantity, or weight, as the case may be, as furnished in writing by the shipper;

(*c*) the apparent order and condition of the goods:

Provided that no carrier, master or agent of the carrier shall be bound to state or show in the bill of lading any marks, number, quantity, or weight which he has reasonable ground for suspecting not accurately to represent the goods actually received or which he has had no reasonable means of checking.

4. Such a bill of lading shall be *prima facie* evidence of the receipt by the carrier of the goods as therein described in accordance with paragraphs 3(*a*), (*b*) and (*c*).

However, proof to the contrary shall not be admissible when the bill of lading has been transferred to a third party acting in good faith.

5. The shipper shall be deemed to have guaranteed to the carrier the accuracy at the time of shipment of the marks, number, quantity and weight, as furnished by him, and the shipper shall indemnify the carrier against all loss, damages and expenses arising or resulting from inaccuracies in such particulars. The right of the carrier to such indemnity shall in no way limit his responsibility and liability under the contract of carriage to any person other than the shipper.

6. Unless notice of loss or damage and the general nature of such loss or damage be given in writing to the carrier or his agent at the port of discharge before or at the time of the removal of the goods into the custody of the person entitled to delivery thereof under the contract of carriage, or, if the loss or damage be not apparent, within three days, such removal shall be *prima facie* evidence of the delivery by the carrier of the goods as described in the bill of lading.

The notice in writing need not be given if the state of the goods has at the time of their receipt been the subject of joint survey or inspection.

Subject to paragraph 6*bis* the carrier and the ship shall in any event be discharged from all liability whatsoever in respect of the goods, unless suit is brought within one year of their delivery or of the date when they should have been delivered. This period may, however, be extended if the parties so agree after the cause of action has arisen.

In the case of any actual or apprehended loss or damage the carrier and the receiver shall give all reasonable facilities to each other for inspecting and tallying the goods.

6._bis_ An action for indemnity against a third person may be brought even after the expiration of the year provided for in the preceding paragraph if brought within the time allowed by the law of the Court seized of the case. However, the time allowed shall be not less than three months, commencing from the day when the person bringing such action for indemnity has settled the claim or has been served with process in the action against himself.

7. After the goods are loaded the bill of lading to be issued by the carrier, master or agent of the carrier, to the shipper shall, if the shipper so demands, be a "shipped" bill of lading, provided that if the shipper shall have previously taken up any document of title to such goods, he shall surrender the same as against the issue of the "shipped" bill of lading, but at the option of the carrier such document of title may be noted at the port of shipment by the carrier, master, or agent with the name or names of the ship or ships upon which the goods have been shipped and the date or dates of shipment, and when so noted the same shall for the purpose of this Article be deemed to constitute a "shipped" bill of lading.

8. Any clause, covenant or agreement in a contract of carriage relieving the carrier or the ship from liability for loss or damage to or in connection with goods arising from negligence, fault or failure in the duties and obligations provided in this Article or lessening such liability otherwise than as provided in these Rules, shall be null and void and of no effect.

A benefit of insurance or similar clause shall be deemed to be a clause relieving the carrier from liability.

Article IV

1. Neither the carrier nor the ship shall be liable for loss or damage arising or resulting from unseaworthiness unless caused by want of due diligence on the part of the carrier to make the ship seaworthy, and to secure that the ship is properly manned, equipped and supplied, and to make the holds, refrigerating and cool chambers and all other parts of the ship in which goods are carried fit and safe for their reception, carriage and preservation in accordance with the provisions of paragraph 1 of Article III.

Whenever loss or damage has resulted from unseaworthiness, the burden of proving the exercise of due diligence shall be on the carrier or other person claiming exemption under this article.

2. Neither the carrier nor the ship shall be responsible for loss or damage arising or resulting from

(*a*) act, neglect, or default of the master, mariner, pilot or the servants of the carrier in the navigation or in the management of the ship;

(*b*) fire, unless caused by the actual fault or privity of the carrier;

(*c*) perils, dangers and accidents of the sea or other navigable waters;

(*d*) act of God;

(*e*) act of war;

(*f*) act of public enemies;

(g) arrest or restraint of princes, rulers or people, or seizure under legal process;

(h) quarantine restrictions;

(i) act or omission of the shipper or owner of the goods, his agent or representative;

(j) strikes or lock-outs or stoppage or restraint of labour from whatever cause, whether partial or general;

(k) riots and civil commotions;

(l) saving or attempting to save life or property at sea;

(m) wastage in bulk or weight or any other loss or damage arising from inherent defect, quality or vice of the goods;

(n) insufficiency of packing;

(o) insufficiency or inadequacy of marks;

(p) latent defects not discoverable by due diligence;

(q) any other cause arising without the actual fault and privity of the carrier, or without the fault or neglect of the agents or servants of the carrier, but the burden of proof shall be on the person claiming the benefit of this exception to show that neither the actual fault or privity of the carrier nor the fault or neglect of the agents or servants of the carrier contributed to the loss or damage.

3. The shipper shall not be responsible for loss or damage sustained by the carrier or the ship arising or resulting from any cause without the act, fault or neglect of the shipper, his agents or his servants.

4. Any deviation in saving or attempting to save life or property at sea or any reasonable deviation shall not be deemed to be an infringement or breach of these Rules or of the contract of carriage, and the carrier shall not be liable for any loss or damage resulting therefrom.

5. (a) Unless the nature and value of such goods have been declared by the shipper before shipment and inserted in the bill of lading, neither the carrier nor the ship shall in any event be or become liable for any loss or damage to or in connection with the goods in an amount exceeding 666.67 units of account per package or unit or 2 units of account per kilogramme of gross weight of the goods lost or damaged, whichever is the higher.

(b) The total amount recoverable shall be calculated by reference to the value of such goods at the place and time at which the goods are discharged from the ship in accordance with the contract or should have been so discharged.

The value of the goods shall be fixed according to the commodity exchange price, or, if there be no such price, according to the current market price, or, if there be no commodity exchange price or current market price, by reference to the normal value of goods of the same kind and quality.

(c) Where a container, pallet or similar article of transport is used to consolidate goods, the number of packages or units enumerated in the bill of lading as packed in such article of transport shall be deemed the number of packages or units for the purpose of this paragraph as far as these packages or units are concerned. Except as aforesaid such article of transport shall be considered the package or unit.

(*d*) The unit of account mentioned in this Article is the Special Drawing Right as defined by the International Monetary Fund. The amounts mentioned in sub-paragraph (*a*) of this paragraph shall be converted into national currency on the basis of the value of that currency on the date to be determined by the law of the Court seized of the case. The value of the national currency, in terms of the Special Drawing Right, of a State which is a member of the International Monetary Fund, shall be calculated in accordance with the method of valuation applied by the International Monetary Fund in effect at the date in question for its operations and transactions. The value of the national currency, in terms of the Special Drawing Right, of a State which is not a member of the International Monetary Fund, shall be calculated in a manner determined by that State.

Nevertheless, a State which is not a member of the International Monetary Fund and whose law does not permit the application of the provisions of the preceding sentences may, at the time of ratification of the Protocol of 1979 or accession thereto or at any time thereafter, declare that the limits of liability provided for in this Convention to be applied in its territory shall be fixed as follows:

(i) in respect of the amount of 666.67 units of account mentioned in sub-paragraph (*a*) of paragraph 5 of this Article, 10,000 monetary units;
(ii) in respect of the amount of 2 units of account mentioned in sub-paragraph (*a*) of paragraph 5 of this Article, 30 monetary units.

The monetary unit referred to in the preceding sentence corresponds to 65.5 milligrammes of gold of millesimal fineness 900. The conversion of the amounts specified in that sentence into the national currency shall be made according to the law of the State concerned. The calculation and the conversion mentioned in the preceding sentences shall be made in such a manner as to express in the national currency of that State as far as possible the same real value for the amounts in sub-paragraph (*a*) of paragraph 5 of this Article as is expressed there in units of account.

States shall communicate to the depositary the manner of calculation or the result of the conversion as the case may be, when depositing an instrument of ratification of the Protocol of 1979 or of accession thereto and whenever there is a change in either.

(*e*) Neither the carrier nor the ship shall be entitled to the benefit of the limitation of liability provided for in this paragraph if it is proved that the damage resulted from an act or omission of the carrier done with intent to cause damage, or recklessly and with knowledge that damage would probably result.

(*f*) The declaration mentioned in sub-paragraph (*a*) of this paragraph, if embodied in the bill of lading, shall be *prima facie* evidence, but shall not be binding or conclusive on the carrier.

(*g*) By agreement between the carrier, master or agent of the carrier and the shipper other maximum amounts than those mentioned in sub-paragraph (*a*) of this paragraph may be fixed, provided that no maximum amount so fixed shall be less than the appropriate maximum mentioned in that sub-paragraph.

(*h*) Neither the carrier nor the ship shall be responsible in any event for loss or damage to, or in connection with, goods if the nature or value thereof has been knowingly mis-stated by the shipper in the bill of lading.

6. Goods of an inflammable, explosive or dangerous nature to the shipment whereof the carrier, master or agent of the carrier has not consented, with knowledge of their nature and character, may at any time before discharge be landed at any place or destroyed or rendered innocuous by the carrier without compensation, and the shipper of such goods shall be liable for all damages and expenses directly or indirectly arising out of or resulting from such shipment.

If any such goods shipped with such knowledge and consent shall become a danger to the ship or cargo, they may in like manner be landed at any place or destroyed or rendered innocuous by the carrier without liability on the part of the carrier except to general average, if any.

Article IVbis

1. The defences and limits of liability provided for in these Rules shall apply in any action against the carrier in respect of loss or damage to goods covered by a contract of carriage whether the action be founded in contract or in tort.

2. If such an action is brought against a servant or agent of the carrier (such servant or agent not being an independent contractor), such servant or agent shall be entitled to avail himself of the defences and limits of liability which the carrier is entitled to invoke under these Rules.

3. The aggregate of the amounts recoverable from the carrier, and such servants and agents, shall in no case exceed the limit provided for in these Rules.

4. Nevertheless, a servant or agent of the carrier shall not be entitled to avail himself of the provisions of this Article, if it is proved that the damage resulted from an act or omission of the servant or agent done with intent to cause damage or recklessly and with knowledge that damage would probably result.

Article V

A carrier shall be at liberty to surrender in whole or in part all or any of his rights and immunities or to increase any of his responsibilities and liabilities under the Rules contained in any of these Articles, provided such surrender or increase shall be embodied in the bill of lading issued to the shipper.

The provisions of these Rules shall not be applicable to charterparties, but if bills of lading are issued in the case of a ship under a charterparty they shall comply with the terms of these Rules. Nothing in these Rules shall be held to prevent the insertion in a bill of lading of any lawful provision regarding general average.

Article VI

Notwithstanding the provisions of the preceding Articles, a carrier, master or agent of the carrier and a shipper shall in regard to any particular goods be at liberty to enter into any agreement in any terms as to the responsibility and liability of the carrier for such goods, and as to the rights and immunities of the carrier in respect of such goods, or his obligation as to seaworthiness, so far as this stipulation is not contrary to public policy, or the care or diligence of his servants or agents in regard to the loading, handling, stowage, carriage, custody, care and discharge of the goods carried by water, provided that in this case no bill of lading has been or shall be issued and that the terms agreed shall be embodied in a receipt which shall be a non-negotiable document and shall be marked as such.

Any agreement so entered into shall have full legal effect.

Provided that this Article shall not apply to ordinary commercial shipments made in the ordinary course of trade, but only to other shipments where the character or condition of the property to be carried or the circumstances, terms and conditions under which the carriage is to be performed are such as reasonably to justify a special agreement.

Article VII

Nothing herein contained shall prevent a carrier or a shipper from entering into any agreement, stipulation, condition, reservation or exemption as to the responsibility and liability of the carrier or the ship for the loss or damage to, or in connection with the custody and care and handling of goods prior to the loading on and subsequent to the discharge from the ship on which the goods are carried by water.

Article VIII

The provisions of these Rules shall not affect the rights and obligations of the carrier under any statute for the time being in force relating to the limitation of the liability of owners of vessels.

Article IX

These Rules shall not affect the provisions of any international Convention or national law governing liability for nuclear damage.

Article X

The provisions of these Rules shall apply to every bill of lading relating to the carriage of goods between ports in two different States if:
>   (*a*) the bill of lading is issued in a Contracting State, or
>   (*b*) the carriage is from a port in a Contracting State, or
>   (*c*) the contract contained in or evidenced by the bill of lading provides that these Rules or legislation of any State giving effect to them are to govern the contract,

whatever may be the nationality of the ship, the carrier, the shipper, the consignee, or any other interested person.

U.S. Customs 24 Hours Rule Clause for Voyage Charter Parties

(a) If loading cargo destined for the U.S. or passing through U.S. ports in transit, the Charterers shall:

(i) Provide all necessary information, upon request by the Owners, to the Owners and/or their agents to enable them to submit a timely and accurate cargo declaration directly to the U.S. Customs; or

(ii) If permitted by U.S. Customs Regulations (19 CFR 4.7) or any subsequent amendments thereto, submit a cargo declaration directly to the U.S. Customs and provide the Owners with a copy thereof.

In all circumstances, the cargo declaration must be submitted to the U.S. Customs latest 24 hours in advance of loading.

(b) The Charterers assume liability for and shall indemnify, defend and hold harmless the Owners against any loss and/or damage whatsoever (including consequential loss and/or damage) and any expenses, fines, penalties and all other claims of whatsoever nature, including but not limited to legal costs, arising from the Charterers' failure to comply with the provisions of sub-clause (a).

(c) If the Vessel is detained, attached, seized or arrested as a result of the Charterers' failure to comply with the provisions of sub-clause (a), the Charterers shall provide a bond or other security to ensure the prompt release of the Vessel. All time used or lost until the Vessel is free to leave any port of call shall count as laytime or, if the Vessel is already on demurrage, time on demurrage.

U.S. Customs 24 Hours Rule for Time Charter Parties

(a) If loading cargo destined for the U.S. or passing through U.S. ports in transit, the Charterers shall:

(i) Provide all necessary information, upon request by the Owners, to the Owners and/or their agents to enable them to submit a timely and accurate cargo declaration directly to the U.S. Customs; or

(ii) If permitted by U.S. Customs Regulations (19 CFR 4.7) or any subsequent amendments thereto, submit a cargo declaration directly to the U.S. Customs and provide the Owners with a copy thereof.

In all circumstances, the cargo declaration must be submitted to the U.S. Customs latest 24 hours in advance of loading.

(b) The Charterers assume liability for and shall indemnify, defend and hold harmless the Owners against any loss and/or damage whatsoever (including consequential loss and/or damage) and any expenses, fines, penalties and all other claims of whatsoever nature, including but not limited to legal costs, arising from the Charterers' failure to comply with the provisions of sub-clause (a).

(c) If the Vessel is detained, attached, seized or arrested as a result of the Charterers' failure to comply with the provisions of sub-clause (a), the Charterers shall provide a bond or other security to ensure the prompt release of the Vessel. Notwithstanding any other provision in this Charter Party to the contrary, the Vessel shall remain on hire.

International Convention on Liability and Compensation for Damage in Connection with the Carriage of Hazardous and Noxious Substances by Sea, 1996

THE STATES PARTIES TO THE PRESENT CONVENTION,

CONSCIOUS of the dangers posed by the world-wide carriage by sea of hazardous and noxious substances,
CONVINCED of the need to ensure that adequate, prompt and effective compensation is available to persons who suffer damage caused by incidents in connection with the carriage by sea of such substances,
DESIRING to adopt uniform international rules and procedures for determining questions of liability and compensation in respect of such damage,
CONSIDERING that the economic consequences of damage caused by the carriage by sea of hazardous and noxious substances should be shared by the shipping industry and the cargo interests involved,
HAVE AGREED as follows:

Chapter I – GENERAL PROVISIONS

Definitions

Article 1

For the purposes of this Convention:
1 "Ship" means any seagoing vessel and seaborne craft, of any type whatsoever.
2 "Person" means any individual or partnership or any public or private body, whether corporate or not, including a State or any of its constituent subdivisions.
3 "Owner" means the person or persons registered as the owner of the ship or, in the absence of registration, the person or persons owning the ship. However, in the case of a ship owned by a State and operated by a company which in that State is registered as the ship's operator, "owner" shall mean such company.
4 "Receiver" means either:
(a) the person who physically receives contributing cargo discharged in the ports and terminals of a State Party; provided that if at the time of receipt the person who physically receives the cargo acts as an agent for another who is subject to the jurisdiction of any State Party, then the principal shall be deemed to be the receiver, if the agent discloses the principal to the HNS Fund; or
(b) the person in the State Party who in accordance with the national law of that State Party is deemed to be the receiver of contributing cargo discharged in the ports and terminals of a State Party, provided that the total contributing cargo received according to such national law is substantially the same as that which would have been received under (a).
5 "Hazardous and noxious substances" (HNS) means:
(a) any substances, materials and articles carried on board a ship as cargo, referred to in (i) to (vii) below:

(i) oils carried in bulk listed in appendix I of Annex I to the International Convention for the Prevention of Pollution from Ships, 1973, as modified by the Protocol of 1978 relating thereto, as amended;

(ii) noxious liquid substances carried in bulk referred to in appendix II of Annex II to the International Convention for the Prevention of Pollution from Ships, 1973, as modified by the Protocol of 1978 relating thereto, as amended, and those substances and mixtures provisionally categorized as falling in pollution category A, B, C or D in accordance with regulation 3(4) of the said Annex II;

(iii) dangerous liquid substances carried in bulk listed in chapter 17 of the International Code for the Construction and Equipment of Ships Carrying Dangerous Chemicals in Bulk, 1983, as amended, and the dangerous products for which the preliminary suitable conditions for the carriage have been prescribed by the Administration and port administrations involved in accordance with paragraph 1.1.3 of the Code;

(iv) dangerous, hazardous and harmful substances, materials and articles in packaged form covered by the International Maritime Dangerous Goods Code, as amended;

(v) liquefied gases as listed in chapter 19 of the International Code for the Construction and Equipment of Ships Carrying Liquefied Gases in Bulk, 1983, as amended, and the products for which preliminary suitable conditions for the carriage have been prescribed by the Administration and port administrations involved in accordance with paragraph 1.1.6 of the Code;

(vi) liquid substances carried in bulk with a flashpoint not exceeding 60°C (measured by a closed cup test);

(vii) solid bulk materials possessing chemical hazards covered by appendix B of the Code of Safe Practice for Solid Bulk Cargoes, as amended, to the extent that these substances are also subject to the provisions of the International Maritime Dangerous Goods Code when carried in packaged form; and

(b) residues from the previous carriage in bulk of substances referred to in (a)(i) to (iii) and (v) to (vii) above.

6 "Damage" means:

(a) loss of life or personal injury on board or outside the ship carrying the hazardous and noxious substances caused by those substances;

(b) loss of or damage to property outside the ship carrying the hazardous and noxious substances caused by those substances;

(c) loss or damage by contamination of the environment caused by the hazardous and noxious substances, provided that compensation for impairment of the environment other than loss of profit from such impairment shall be limited to costs of reasonable measures of reinstatement actually undertaken or to be undertaken; and

(d) the costs of preventive measures and further loss or damage caused by preventive measures.

Where it is not reasonably possible to separate damage caused by the hazardous and noxious substances from that caused by other factors, all such damage shall be deemed to be caused by the hazardous and noxious substances except if, and to the extent that, the damage caused by other factors is damage of a type referred to in article 4, paragraph 3.

In this paragraph, "caused by those substances" means caused by the hazardous or noxious nature of the substances.

7 "Preventive measures" means any reasonable measures taken by any person after an incident has occurred to prevent or minimize damage.

8 "Incident" means any occurrence or series of occurrences having the same origin, which causes damage or creates a grave and imminent threat of causing damage.

9 "Carriage by sea" means the period from the time when the hazardous and noxious substances enter any part of the ship's equipment, on loading, to the time they cease to be present in any part of the ship's equipment, on discharge. If no ship's equipment is used, the period begins and ends respectively when the hazardous and noxious substances cross the ship's rail.

10 "Contributing cargo" means any hazardous and noxious substances which are carried by sea as cargo to a port or terminal in the territory of a State Party and discharged in that State. Cargo in transit which is transferred directly, or through a port or terminal, from one ship to another, either wholly or in part, in the course of carriage from the port or terminal of original loading to the port or terminal of final destination shall be considered as contributing cargo only in respect of receipt at the final destination.

11 The "HNS Fund" means the International Hazardous and Noxious Substances Fund established under article 13.

12 "Unit of account" means the Special Drawing Right as defined by the International Monetary Fund.

13 "State of the ship's registry" means in relation to a registered ship the State of registration of the ship, and in relation to an unregistered ship the State whose flag the ship is entitled to fly.

14 "Terminal" means any site for the storage of hazardous and noxious substances received from waterborne transportation, including any facility situated off-shore and linked by pipeline or otherwise to such site.

15 "Director" means the Director of the HNS Fund.

16 "Organization" means the International Maritime Organization.

17 "Secretary-General" means the Secretary-General of the Organization.

Annexes

Article 2

The Annexes to this Convention shall constitute an integral part of this Convention.

Scope of application

Article 3

This Convention shall apply exclusively:

(a) to any damage caused in the territory, including the territorial sea, of a State Party;

(b) to damage by contamination of the environment caused in the exclusive economic zone of a State Party, established in accordance with international law, or, if a State Party has not established such a zone, in an area beyond and adjacent to the territorial sea of that State determined by that State in accordance with international law and extending not more than 200 nautical miles from the baselines from which the breadth of its territorial sea is measured;

(c) to damage, other than damage by contamination of the environment, caused outside the territory, including the territorial sea, of any State, if this damage has been caused by a substance carried on board a ship registered in a State Party or, in the case of an unregistered ship, on board a ship entitled to fly the flag of a State Party; and

(d) to preventive measures, wherever taken.

Article 4

1 This Convention shall apply to claims, other than claims arising out of any contract for the carriage of goods and passengers, for damage arising from the carriage of hazardous and noxious substances by sea.

2 This Convention shall not apply to the extent that its provisions are incompatible with those of the applicable law relating to workers' compensation or social security schemes.

3 This Convention shall not apply:

(a) to pollution damage as defined in the International Convention on Civil Liability for Oil Pollution Damage, 1969, as amended, whether or not compensation is payable in respect of it under that Convention; and

(b) to damage caused by a radioactive material of class 7 either in the International Maritime Dangerous Goods Code, as amended, or in appendix B of the Code of Safe Practice for Solid Bulk Cargoes, as amended.

4 Except as provided in paragraph 5, the provisions of this Convention shall not apply to warships, naval auxiliary or other ships owned or operated by a State and used, for the time being, only on Government non-commercial service.

5 A State Party may decide to apply this Convention to its warships or other vessels described in paragraph 4, in which case it shall notify the Secretary-General thereof specifying the terms and conditions of such application.

6 With respect to ships owned by a State Party and used for commercial purposes, each State shall be subject to suit in the jurisdictions set forth in article 38 and shall waive all defences based on its status as a sovereign State.

Article 5

1 A State may, at the time of ratification, acceptance, approval of, or accession to, this Convention, or any time thereafter, declare that this Convention does not apply to ships:

(a) which do not exceed 200 gross tonnage; and

(b) which carry hazardous and noxious substances only in packaged form; and

(c) while they are engaged on voyages between ports or facilities of that State.

2 Where two neighbouring States agree that this Convention does not apply also to ships which are covered by paragraph 1(a) and (b) while engaged on voyages between ports or facilities of those States, the States concerned may declare that the exclusion from the application of this Convention declared under paragraph 1 covers also ships referred to in this paragraph.

3 Any State which has made the declaration under paragraph 1 or 2 may withdraw such declaration at any time.

4 A declaration made under paragraph 1 or 2, and the withdrawal of the declaration made under paragraph 3, shall be deposited with the Secretary-General who shall, after the entry into force of this Convention, communicate it to the Director.

5 Where a State has made a declaration under paragraph 1or 2 and has not withdrawn it, hazardous and noxious substances carried on board ships covered by that paragraph shall not be considered to be contributing cargo for the purpose of application of articles 18, 20, article 21, paragraph 5 and article 43.

6 The HNS Fund is not liable to pay compensation for damage caused by substances carried by a ship to which the Convention does not apply pursuant to a declaration made under paragraph 1or 2, to the extent that:

(a) the damage as defined in article 1, paragraph 6(a), (b) or (c) was caused in:

(i) the territory, including the territorial sea, of the State which has made the declaration, or in the case of neighbouring States which have made a declaration under paragraph 2, of either of them; or

(ii) the exclusive economic zone, or area mentioned in article 3(b), of the State or States referred to in (i);

(b) the damage includes measures taken to prevent or minimize such damage .

Duties of State Parties

Article 6

Each State Party shall ensure that any obligation arising under this Convention is fulfilled and shall take appropriate measures under its law including the imposing of sanctions as it may deem necessary, with a view to the effective execution of any such obligation.

Chapter II – *Liability*

Liability of the owner

Article 7

1 Except as provided in paragraphs 2 and 3, the owner at the time of an incident shall be liable for damage caused by any hazardous and noxious substances in connection with their carriage by sea on board the ship, provided that if an incident consists of a series of occurrences having the same origin the liability shall attach to the owner at the time of the first of such occurrences.

2 No liability shall attach to the owner if the owner proves that:

(a) the damage resulted from an act of war, hostilities, civil war, insurrection or a natural phenomenon of an exceptional, inevitable and irresistible character; or

(b) the damage was wholly caused by an act or omission done with the intent to cause damage by a third party; or

(c) the damage was wholly caused by the negligence or other wrongful act of any Government or other authority responsible for the maintenance of lights or other navigational aids in the exercise of that function; or

(d) the failure of the shipper or any other person to furnish information concerning the hazardous and noxious nature of the substances shipped either

(i) has caused the damage, wholly or partly; or

(ii) has led the owner not to obtain insurance in accordance with article 12;

provided that neither the owner nor its servants or agents knew or ought reasonably to have known of the hazardous and noxious nature of the substances shipped.

3 If the owner proves that the damage resulted wholly or partly either from an act or omission done with intent to cause damage by the person who suffered the damage or from the negligence of that person, the owner may be exonerated wholly or partially from liability to such person.

4 No claim for compensation for damage shall be made against the owner otherwise than in accordance with this Convention.

5 Subject to paragraph 6, no claim for compensation for damage under this Convention or otherwise may be made against:

(a) the servants or agents of the owner or the members of the crew;

(b) the pilot or any other person who, without being a member of the crew, performs services for the ship;

(c) any charterer (howsoever described, including a bareboat charterer), manager or operator of the ship;

(d) any person performing salvage operations with the consent of the owner or on the instructions of a competent public authority;

(e) any person taking preventive measures; and

(f) the servants or agents of persons mentioned in (c), (d) and (e);

unless the damage resulted from their personal act or omission, committed with the intent to cause such damage, or recklessly and with knowledge that such damage would probably result.

6 Nothing in this Convention shall prejudice any existing right of recourse of the owner against any third party, including, but not limited to, the shipper or the receiver of the substance causing the damage, or the persons indicated in paragraph 5.

Incidents involving two or more ships

Article 8

1 Whenever damage has resulted from an incident involving two or more ships each of which is carrying hazardous and noxious substances, each owner, unless exonerated under article 7, shall be liable for the damage. The owners shall be jointly and severally liable for all such damage which is not reasonably separable.

2 However, owners shall be entitled to the limits of liability applicable to each of them under article 9.

3 Nothing in this article shall prejudice any right of recourse of an owner against any other owner.

Limitation of liability

Article 9

1 The owner of a ship shall be entitled to limit liability under this Convention in respect of any one incident to an aggregate amount calculated as follows:

(a) 10 million units of account for a ship not exceeding 2,000 units of tonnage; and

(b) for a ship with a tonnage in excess thereof, the following amount in addition to that mentioned in (a):

for each unit of tonnage from 2,001 to 50,000 units of tonnage, 1,500 units of account

for each unit of tonnage in excess of 50,000 units of tonnage, 360 units of account

provided, however, that this aggregate amount shall not in any event exceed 100 million units of account.

2 The owner shall not be entitled to limit liability under this Convention if it is proved that the damage resulted from the personal act or omission of the owner, committed with the intent to cause such damage, or recklessly and with knowledge that such damage would probably result.

3 The owner shall, for the purpose of benefitting from the limitation provided for in paragraph 1, constitute a fund for the total sum representing the limit of liability established in accordance with paragraph 1 with the court or other competent authority of any one of the States Parties in which action is brought under article 38 or, if no action is brought, with any court or other competent authority in any one of the States Parties in which an action can be brought under article 38. The fund can be constituted either by depositing the sum or by producing a bank guarantee or other guarantee, acceptable under the law of the State Party where the

fund is constituted, and considered to be adequate by the court or other competent authority.

4 Subject to the provisions of article 11, the fund shall be distributed among the claimants in proportion to the amounts of their established claims.

5 If before the fund is distributed the owner or any of the servants or agents of the owner or any person providing to the owner insurance or other financial security has as a result of the incident in question, paid compensation for damage, such person shall, up to the amount that person has paid, acquire by subrogation the rights which the person so compensated would have enjoyed under this Convention.

6 The right of subrogation provided for in paragraph 5 may also be exercised by a person other than those mentioned therein in respect of any amount of compensation for damage which such person may have paid but only to the extent that such subrogation is permitted under the applicable national law.

7 Where owners or other persons establish that they may be compelled to pay at a later date in whole or in part any such amount of compensation, with regard to which the right of subrogation would have been enjoyed under paragraphs 5 or 6 had the compensation been paid before the fund was distributed, the court or other competent authority of the State where the fund has been constituted may order that a sufficient sum shall be provisionally set aside to enable such person at such later date to enforce the claim against the fund.

8 Claims in respect of expenses reasonably incurred or sacrifices reasonably made by the owner voluntarily to prevent or minimize damage shall rank equally with other claims against the fund.

9 (a) The amounts mentioned in paragraph 1 shall be converted into national currency on the basis of the value of that currency by reference to the Special Drawing Right on the date of the constitution of the fund referred to in paragraph 3. The value of the national currency, in terms of the Special Drawing Right, of a State Party which is a member of the International Monetary Fund, shall be calculated in accordance with the method of valuation applied by the International Monetary Fund in effect on the date in question for its operations and transactions. The value of the national currency, in terms of the Special Drawing Right, of a State Party which is not a member of the International Monetary Fund, shall be calculated in a manner determined by that State.

(b) Nevertheless, a State Party which is not a member of the International Monetary Fund and whose law does not permit the application of the provisions of paragraph 9(a) may, at the time of ratification, acceptance, approval of or accession to this Convention or at any time thereafter, declare that the unit of account referred to in paragraph 9(a) shall be equal to 15 gold francs. The gold franc referred to in this paragraph corresponds to sixty-five-and-a-half milligrammes of gold of millesimal fineness nine hundred. The conversion of the gold franc into the national currency shall be made according to the law of the State concerned.

(c) The calculation mentioned in the last sentence of paragraph 9(a) and the conversion mentioned in paragraph 9(b) shall be made in such manner as to express in the national currency of the State Party as far as possible the same real value for the amounts in paragraph 1 as would result from the application of the first two sentences of paragraph 9(a). States Parties shall communicate to the Secretary-General the manner of calculation pursuant to paragraph 9(a), or the result of the conversion in paragraph 9(b) as the case may be, when depositing an instrument of ratification, acceptance, approval of or accession to this Convention and whenever there is a change in either.

10 For the purpose of this article the ship's tonnage shall be the gross tonnage calculated in accordance with the tonnage measurement regulations contained in Annex I of the International Convention on Tonnage Measurement of Ships, 1969.

11 The insurer or other person providing financial security shall be entitled to constitute a fund in accordance with this article on the same conditions and having the same effect as if it were constituted by the owner. Such a fund may be constituted even if, under the provisions of paragraph 2, the owner is not entitled to limitation of liability, but its constitution shall in that case not prejudice the rights of any claimant against the owner.

Article 10

1 Where the owner, after an incident, has constituted a fund in accordance with article 9 and is entitled to limit liability:

(a) no person having a claim for damage arising out of that incident shall be entitled to exercise any right against any other assets of the owner in respect of such claim; and

(b) the court or other competent authority of any State Party shall order the release of any ship or other property belonging to the owner which has been arrested in respect of a claim for damage arising out of that incident, and shall similarly release any bail or other security furnished to avoid such arrest.

2 The foregoing shall, however, only apply if the claimant has access to the court administering the fund and the fund is actually available in respect of the claim.

Death and injury

Article 11

Claims in respect of death or personal injury have priority over other claims save to the extent that the aggregate of such claims exceeds two-thirds of the total amount established in accordance with article 9, paragraph 1.

Compulsory insurance of the owner

Article 12

1 The owner of a ship registered in a State Party and actually carrying hazardous and noxious substances shall be required to maintain insurance or other financial security, such as the guarantee of a bank or similar financial institution, in the sums fixed by applying the limits of liability prescribed in article 9, paragraph 1, to cover liability for damage under this Convention.

2 A compulsory insurance certificate attesting that insurance or other financial security is in force in accordance with the provisions of this Convention shall be

issued to each ship after the appropriate authority of a State Party has determined that the requirements of paragraph 1 have been complied with. With respect to a ship registered in a State Party such compulsory insurance certificate shall be issued or certified by the appropriate authority of the State of the ship's registry; with respect to a ship not registered in a State Party it may be issued or certified by the appropriate authority of any State Party. This compulsory insurance certificate shall be in the form of the model set out in Annex I and shall contain the following particulars:

(a) name of the ship, distinctive number or letters and port of registry;

(b) name and principal place of business of the owner;

(c) IMO ship identification number;

(d) type and duration of security;

(e) name and principal place of business of insurer or other person giving security and, where appropriate, place of business where the insurance or security is established; and

(f) period of validity of certificate, which shall not be longer than the period of validity of the insurance or other security.

3 The compulsory insurance certificate shall be in the official language or languages of the issuing State. If the language used is neither English, nor French nor Spanish, the text shall include a translation into one of these languages.

4 The compulsory insurance certificate shall be carried on board the ship and a copy shall be deposited with the authorities who keep the record of the ship's registry or, if the ship is not registered in a State Party, with the authority of the State issuing or certifying the certificate.

5 An insurance or other financial security shall not satisfy the requirements of this article if it can cease, for reasons other than the expiry of the period of validity of the insurance or security specified in the certificate under paragraph 2, before three months have elapsed from the date on which notice of its termination is given to the authorities referred to in paragraph 4, unless the compulsory insurance certificate has been issued within the said period. The foregoing provisions shall similarly apply to any modification which results in the insurance or security no longer satisfying the requirements of this article.

6 The State of the ship's registry shall, subject to the provisions of this article, determine the conditions of issue and validity of the compulsory insurance certificate.

7 Compulsory insurance certificates issued or certified under the authority of a State Party in accordance with paragraph 2 shall be accepted by other States Parties for the purposes of this Convention and shall be regarded by other States Parties as having the same force as compulsory insurance certificates issued or certified by them even if issued or certified in respect of a ship not registered in a State Party. A State Party may at any time request consultation with the issuing or certifying State should it believe that the insurer or guarantor named in the compulsory insurance certificate is not financially capable of meeting the obligations imposed by this Convention.

8 Any claim for compensation for damage may be brought directly against the insurer or other person providing financial security for the owner's liability for

damage. In such case the defendant may, even if the owner is not entitled to limitation of liability, benefit from the limit of liability prescribed in accordance with paragraph 1. The defendant may further invoke the defences (other than the bankruptcy or winding up of the owner) which the owner would have been entitled to invoke. Furthermore, the defendant may invoke the defence that the damage resulted from the wilful misconduct of the owner, but the defendant shall not invoke any other defence which the defendant might have been entitled to invoke in proceedings brought by the owner against the defendant. The defendant shall in any event have the right to require the owner to be joined in the proceedings.

9 Any sums provided by insurance or by other financial security maintained in accordance with paragraph 1 shall be available exclusively for the satisfaction of claims under this Convention.

10 A State Party shall not permit a ship under its flag to which this article applies to trade unless a certificate has been issued under paragraph 2 or 12.

11 Subject to the provisions of this article, each State Party shall ensure, under its national law, that insurance or other security in the sums specified in paragraph 1 is in force in respect of any ship, wherever registered, entering or leaving a port in its territory, or arriving at or leaving an offshore facility in its territorial sea.

12 If insurance or other financial security is not maintained in respect of a ship owned by a State Party, the provisions of this article relating thereto shall not be applicable to such ship, but the ship shall carry a compulsory insurance certificate issued by the appropriate authorities of the State of the ship's registry stating that the ship is owned by that State and that the ship's liability is covered within the limit prescribed in accordance with paragraph 1. Such a compulsory insurance certificate shall follow as closely as possible the model prescribed by paragraph 2.

Chapter III – *Compensation by the International Hazardous and Noxious Substances Fund (HNS Fund)*

Establishment of the HNS Fund

Article 13

1 The International Hazardous and Noxious Substances Fund (HNS Fund) is hereby established with the following aims:
> (a) to provide compensation for damage in connection with the carriage of hazardous and noxious substances by sea, to the extent that the protection afforded by chapter II is inadequate or not available; and
> (b) to give effect to the related tasks set out in article 15.

2 The HNS Fund shall in each State Party be recognized as a legal person capable under the laws of that State of assuming rights and obligations and of being a party in legal proceedings before the courts of that State. Each State Party shall recognize the Director as the legal representative of the HNS Fund.

Compensation

Article 14

1 For the purpose of fulfilling its function under article 13, paragraph 1(a), the HNS Fund shall pay compensation to any person suffering damage if such person has been unable to obtain full and adequate compensation for the damage under the terms of chapter II:

(a) because no liability for the damage arises under chapter II;

(b) because the owner liable for the damage under chapter II is financially incapable of meeting the obligations under this Convention in full and any financial security that may be provided under chapter II does not cover or is insufficient to satisfy the claims for compensation for damage; an owner being treated as financially incapable of meeting these obligations and a financial security being treated as insufficient if the person suffering the damage has been unable to obtain full satisfaction of the amount of compensation due under chapter II after having taken all reasonable steps to pursue the available legal remedies;

(c) because the damage exceeds the owner's liability under the terms of chapter II.

2 Expenses reasonably incurred or sacrifices reasonably made by the owner voluntarily to prevent or minimize damage shall be treated as damage for the purposes of this article.

3 The HNS Fund shall incur no obligation under the preceding paragraphs if:

(a) it proves that the damage resulted from an act of war, hostilities, civil war or insurrection or was caused by hazardous and noxious substances which had escaped or been discharged from a warship or other ship owned or operated by a State and used, at the time of the incident, only on Government non-commercial service; or

(b) the claimant cannot prove that there is a reasonable probability that the damage resulted from an incident involving one or more ships.

4 If the HNS Fund proves that the damage resulted wholly or partly either from an act or omission done with intent to cause damage by the person who suffered the damage or from the negligence of that person, the HNS Fund may be exonerated wholly or partially from its obligation to pay compensation to such person. The HNS Fund shall in any event be exonerated to the extent that the owner may have been exonerated under article 7, paragraph 3. However, there shall be no such exoneration of the HNS Fund with regard to preventive measures.

5 (a) Except as otherwise provided in subparagraph (b), the aggregate amount of compensation payable by the HNS Fund under this article shall in respect of any one incident be limited, so that the total sum of that amount and any amount of compensation actually paid under chapter II for damage within the scope of application of this Convention as defined in article 3 shall not exceed 250 million units of account.

(b) The aggregate amount of compensation payable by the HNS Fund under this article for damage resulting from a natural phenomenon of an exceptional, inevitable and irresistible character shall not exceed 250 million units of account.

(c) Interest accrued on a fund constituted in accordance with article 9, paragraph 3, if any, shall not be taken into account for the computation of the maximum compensation payable by the HNS Fund under this article.

(d) The amounts mentioned in this article shall be converted into national currency on the basis of the value of that currency with reference to the Special Drawing Right on the date of the decision of the Assembly of the HNS Fund as to the first date of payment of compensation.

6 Where the amount of established claims against the HNS Fund exceeds the aggregate amount of compensation payable under paragraph 5, the amount available shall be distributed in such a manner that the proportion between any established claim and the amount of compensation actually recovered by the claimant under this Convention shall be the same for all claimants. Claims in respect of death or personal injury shall have priority over other claims, however, save to the extent that the aggregate of such claims exceeds two-thirds of the total amount established in accordance with paragraph 5.

7 The Assembly of the HNS Fund may decide that, in exceptional cases, compensation in accordance with this Convention can be paid even if the owner has not constituted a fund in accordance with chapter II. In such cases paragraph 5(d) applies accordingly.

Related tasks of the HNS Fund

Article 15

For the purpose of fulfilling its function under article 13, paragraph 1(a), the HNS Fund shall have the following tasks:

(a) to consider claims made against the HNS Fund;

(b) to prepare an estimate in the form of a budget for each calendar year of:
Expenditure:
(i) costs and expenses of the administration of the HNS Fund in the relevant year and any deficit from operations in the preceding years; and
(ii) payments to be made by the HNS Fund in the relevant year;
(iii) surplus funds from operations in preceding years, including any interest;
(iv) initial contributions to be paid in the course of the year;
(v) annual contributions if required to balance the budget; and
(vi) any other income;

(c) to use at the request of a State Party its good offices as necessary to assist that State to secure promptly such personnel, material and services as are necessary to enable the State to take measures to prevent or mitigate damage arising from an incident in respect of which the HNS Fund may be called upon to pay compensation under this Convention; and

(d) to provide, on conditions laid down in the internal regulations, credit facilities with a view to the taking of preventive measures against damage arising from a particular incident in respect of which the HNS Fund may be called upon to pay compensation under this Convention.

General provisions on contributions

Article 16

1 The HNS Fund shall have a general account, which shall be divided into sectors.

2 The HNS Fund shall, subject to article 19, paragraphs 3 and 4, also have separate accounts in respect of:

(a) oil as defined in article 1, paragraph 5(a)(i) (oil account);

(b) liquefied natural gases of light hydrocarbons with methane as the main constituent (LNG) (LNG account); and

(c) liquefied petroleum gases of light hydrocarbons with propane and butane as the main constituents (LPG) (LPG account).

3 There shall be initial contributions and, as required, annual contributions to the HNS Fund.

4 Contributions to the HNS Fund shall be made into the general account in accordance with article 18, to separate accounts in accordance with article 19 and to either the general account or separate accounts in accordance with article 20 or article 21, paragraph 5. Subject to article 19, paragraph 6, the general account shall be available to compensate damage caused by hazardous and noxious substances covered by that account, and a separate account shall be available to compensate damage caused by a hazardous and noxious substance covered by that account.

5 For the purposes of article 18, article 19, paragraph 1(a)(i), paragraph 1(a)(ii) and paragraph 1(c), article 20 and article 21, paragraph 5, where the quantity of a given type of contributing cargo received in the territory of a State Party by any person in a calendar year when aggregated with the quantities of the same type of cargo received in the same State Party in that year by any associated person or persons exceeds the limit specified in the respective subparagraphs, such a person shall pay contributions in respect of the actual quantity received by that person notwithstanding that that quantity did not exceed the respective limit.

6 "Associated person" means any subsidiary or commonly controlled entity. The question whether a person comes within this definition shall be determined by the national law of the State concerned.

General provisions on annual contributions

Article 17

1 Annual contributions to the general account and to each separate account shall be levied only as required to make payments by the account in question.

2 Annual contributions payable pursuant to articles 18, 19 and article 21, paragraph 5 shall be determined by the Assembly and shall be calculated in accordance with those articles on the basis of the units of contributing cargo received or, in respect of cargoes referred to in article 19, paragraph 1(b), discharged during the preceding calendar year or such other year as the Assembly may decide.

3 The Assembly shall decide the total amount of annual contributions to be levied to the general account and to each separate account. Following that decision the Director shall, in respect of each State Party, calculate for each person liable to

pay contributions in accordance with article 18, article 19, paragraph 1 and article 21, paragraph 5, the amount of that person's annual contribution to each account, on the basis of a fixed sum for each unit of contributing cargo reported in respect of the person during the preceding calendar year or such other year as the Assembly may decide. For the general account, the above-mentioned fixed sum per unit of contributing cargo for each sector shall be calculated pursuant to the regulations contained in Annex II to this Convention. For each separate account, the fixed sum per unit of contributing cargo referred to above shall be calculated by dividing the total annual contribution to be levied to that account by the total quantity of cargo contributing to that account.

4 The Assembly may also levy annual contributions for administrative costs and decide on the distribution of such costs between the sectors of the general account and the separate accounts.

5 The Assembly shall also decide on the distribution between the relevant accounts and sectors of amounts paid in compensation for damage caused by two or more substances which fall within different accounts or sectors, on the basis of an estimate of the extent to which each of the substances involved contributed to the damage.

Annual contributions to the general account

Article 18

1 Subject to article 16, paragraph 5, annual contributions to the general account shall be made in respect of each State Party by any person who was the receiver in that State in the preceding calendar year, or such other year as the Assembly may decide, of aggregate quantities exceeding 20,000 tonnes of contributing cargo, other than substances referred to in article 19, paragraph 1, which fall within the following sectors:

(a) solid bulk materials referred to in article 1, paragraph 5(a)(vii);

(b) substances referred to in paragraph 2; and

(c) other substances.

2 Annual contributions shall also be payable to the general account by persons who would have been liable to pay contributions to a separate account in accordance with article 19, paragraph 1 had its operation not been postponed or suspended in accordance with article 19. Each separate account the operation of which has been postponed or suspended under article 19 shall form a separate sector within the general account.

Annual contributions to separate accounts

Article 19

1 Subject to article 16, paragraph 5, annual contributions to separate accounts shall be made in respect of each State Party:

(a) in the case of the oil account,

(i) by any person who has received in that State in the preceding calendar year, or such other year as the Assembly may decide, total quantities exceeding 150,000 tonnes of contributing oil as defined in article 1, paragraph 3 of the International Convention on the Establishment of an International Fund for Compensation for Oil Pollution Damage, 1971, as amended, and who is or would be liable to pay contributions to the International Oil Pollution Compensation Fund in accordance with article 10 of that Convention; and

(ii) by any person who was the receiver in that State in the preceding calendar year, or such other year as the Assembly may decide, of total quantities exceeding 20,000 tonnes of other oils carried in bulk listed in appendix I of Annex I to the International Convention for the Prevention of Pollution from Ships, 1973, as modified by the Protocol of 1978 relating thereto, as amended;

(b) in the case of the LNG account, by any person who in the preceding calendar year, or such other year as the Assembly may decide, immediately prior to its discharge, held title to an LNG cargo discharged in a port or terminal of that State;

(c) in the case of the LPG account, by any person who in the preceding calendar year, or such other year as the Assembly may decide, was the receiver in that State of total quantities exceeding 20,000 tonnes of LPG.

2 Subject to paragraph 3, the separate accounts referred to in paragraph 1 above shall become effective at the same time as the general account.

3 The initial operation of a separate account referred to in article 16, paragraph 2 shall be postponed until such time as the quantities of contributing cargo in respect of that account during the preceding calendar year, or such other year as the Assembly may decide, exceed the following levels:

(a) 350 million tonnes of contributing cargo in respect of the oil account;

(b) 20 million tonnes of contributing cargo in respect of the LNG account; and

(c) 15 million tonnes of contributing cargo in respect of the LPG account.

4 The Assembly may suspend the operation of a separate account if:

(a) the quantities of contributing cargo in respect of that account during the preceding calendar year fall below the respective level specified in paragraph 3; or

(b) when six months have elapsed from the date when the contributions were due, the total unpaid contributions to that account exceed ten per cent of the most recent levy to that account in accordance with paragraph 1.

5 The Assembly may reinstate the operation of a separate account which has been suspended in accordance with paragraph 4.

6 Any person who would be liable to pay contributions to a separate account the operation of which has been postponed in accordance with paragraph 3 or suspended in accordance with paragraph 4, shall pay into the general account the contributions due by that person in respect of that separate account. For the purpose of calculating future contributions, the postponed or suspended separate account shall form a new sector in the general account and shall be subject to the HNS points system defined in Annex II.

Initial contributions

Article 20

1 In respect of each State Party, initial contributions shall be made of an amount which shall for each person liable to pay contributions in accordance with article 16, paragraph 5, articles 18, 19 and article 21, paragraph 5 be calculated on the basis of a fixed sum, equal for the general account and each separate account, for each unit of contributing cargo received or, in the case of LNG, discharged in that State, during the calendar year preceding that in which this Convention enters into force for that State.

2 The fixed sum and the units for the different sectors within the general account as well as for each separate account referred to in paragraph 1 shall be determined by the Assembly.

3 Initial contributions shall be paid within three months following the date on which the HNS Fund issues invoices in respect of each State Party to persons liable to pay contributions in accordance with paragraph 1.

Reports

Article 21

1 Each State Party shall ensure that any person liable to pay contributions in accordance with articles 18, 19 or paragraph 5 of this article appears on a list to be established and kept up to date by the Director in accordance with the provisions of this article.

2 For the purposes set out in paragraph 1, each State Party shall communicate to the Director, at a time and in the manner to be prescribed in the internal regulations of the HNS Fund, the name and address of any person who in respect of the State is liable to pay contributions in accordance with articles 18, 19 or paragraph 5 of this article, as well as data on the relevant quantities of contributing cargo for which such a person is liable to contribute in respect of the preceding calendar year.

3 For the purposes of ascertaining who are, at any given time, the persons liable to pay contributions in accordance with articles 18, 19 or paragraph 5 of this article and of establishing, where applicable, the quantities of cargo to be taken into account for any such person when determining the amount of the contribution, the list shall be *prima facie* evidence of the facts stated therein.

4 Where a State Party does not fulfil its obligations to communicate to the Director the information referred to in paragraph 2 and this results in a financial loss for the HNS Fund, that State Party shall be liable to compensate the HNS Fund for such loss. The Assembly shall, on the recommendation of the Director, decide whether such compensation shall be payable by a State Party.

5 In respect of contributing cargo carried from one port or terminal of a State Party to another port or terminal located in the same State and discharged there, States Parties shall have the option of submitting to the HNS Fund a report with an annual aggregate quantity for each account covering all receipts of contributing cargo, including any quantities in respect of which contributions are payable

pursuant to article 16, paragraph 5. The State Party shall, at the time of reporting, either:

(a) notify the HNS Fund that that State will pay the aggregate amount for each account in respect of the relevant year in one lump sum to the HNS Fund; or

(b) instruct the HNS Fund to levy the aggregate amount for each account by invoicing individual receivers or, in the case of LNG, the title holder who discharges within the jurisdiction of that State Party, for the amount payable by each of them. These persons shall be identified in accordance with the national law of the State concerned.

Non-payment of contributions

Article 22

1 The amount of any contribution due under articles 18, 19, 20 or article 21, paragraph 5 and which is in arrears shall bear interest at a rate which shall be determined in accordance with the internal regulations of the HNS Fund, provided that different rates may be fixed for different circumstances.

2 Where a person who is liable to pay contributions in accordance with articles 18, 19, 20 or article 21, paragraph 5 does not fulfil the obligations in respect of any such contribution or any part thereof and is in arrears, the Director shall take all appropriate action, including court action, against such a person on behalf of the HNS Fund with a view to the recovery of the amount due. However, where the defaulting contributor is manifestly insolvent or the circumstances otherwise so warrant, the Assembly may, upon recommendation of the Director, decide that no action shall be taken or continued against the contributor.

Optional liability of States Parties for the payment of contributions

Article 23

1 Without prejudice to article 21, paragraph 5, a State Party may at the time when it deposits its instrument of ratification, acceptance, approval or accession or at any time thereafter declare that it assumes responsibility for obligations imposed by this Convention on any person liable to pay contributions in accordance with articles 18, 19, 20 or article 21, paragraph 5 in respect of hazardous and noxious substances received or discharged in the territory of that State. Such a declaration shall be made in writing and shall specify which obligations are assumed.

2 Where a declaration under paragraph 1 is made prior to the entry into force of this Convention in accordance with article 46, it shall be deposited with the Secretary-General who shall after the entry into force of this Convention communicate the declaration to the Director.

3 A declaration under paragraph 1 which is made after the entry into force of this Convention shall be deposited with the Director.

4 A declaration made in accordance with this article may be withdrawn by the relevant State giving notice thereof in writing to the Director. Such a notification shall take effect three months after the Director's receipt thereof.

5 Any State which is bound by a declaration made under this article shall, in any proceedings brought against it before a competent court in respect of any obliga-

tion specified in the declaration, waive any immunity that it would otherwise be entitled to invoke.

Organization and administration

Article 24

The HNS Fund shall have an Assembly and a Secretariat headed by the Director.

Assembly

Article 25

The Assembly shall consist of all States Parties to this Convention.

Article 26

The functions of the Assembly shall be:

(a) to elect at each regular session its President and two Vice-Presidents who shall hold office until the next regular session;

(b) to determine its own rules of procedure, subject to the provisions of this Convention;

(c) to develop, apply and keep under review internal and financial regulations relating to the aim of the HNS Fund as described in article 13, paragraph 1(a), and the related tasks of the HNS Fund listed in article 15;

(d) to appoint the Director and make provisions for the appointment of such other personnel as may be necessary and determine the terms and conditions of service of the Director and other personnel;

(e) to adopt the annual budget prepared in accordance with article 15(b);

(f) to consider and approve as necessary any recommendation of the Director regarding the scope of definition of contributing cargo;

(g) to appoint auditors and approve the accounts of the HNS Fund;

(h) to approve settlements of claims against the HNS Fund, to take decisions in respect of the distribution among claimants of the available amount of compensation in accordance with article 14 and to determine the terms and conditions according to which provisional payments in respect of claims shall be made with a view to ensuring that victims of damage are compensated as promptly as possible;

(i) to establish a Committee on Claims for Compensation with at least 7 and not more than 15 members and any temporary or permanent subsidiary body it may consider to be necessary, to define its terms of reference and to give it the authority needed to perform the functions entrusted to it; when appointing the members of such body, the Assembly shall endeavour to secure an equitable geographical distribution of members and to ensure that the States Parties are appropriately represented; the Rules of Procedure of the Assembly may be applied, *mutatis mutandis*, for the work of such subsidiary body;

(j) to determine which States not party to this Convention, which Associate Members of the Organization and which intergovernmental and international non-governmental organizations shall be admitted to take part, without voting rights, in meetings of the Assembly and subsidiary bodies;

(k) to give instructions concerning the administration of the HNS Fund to the Director and subsidiary bodies;

(l) to supervise the proper execution of this Convention and of its own decisions;

(m) to review every five years the implementation of this Convention with particular reference to the performance of the system for the calculation of levies and the contribution mechanism for domestic trade; and

(n) to perform such other functions as are allocated to it under this Convention or are otherwise necessary for the proper operation of the HNS Fund.

Article 27

1 Regular sessions of the Assembly shall take place once every calendar year upon convocation by the Director.

2 Extraordinary sessions of the Assembly shall be convened by the Director at the request of at least one-third of the members of the Assembly and may be convened on the Director's own initiative after consultation with the President of the Assembly. The Director shall give members at least thirty days' notice of such sessions.

Article 28

A majority of the members of the Assembly shall constitute a quorum for its meetings.

Secretariat

Article 29

1 The Secretariat shall comprise the Director and such staff as the administration of the HNS Fund may require.

2 The Director shall be the legal representative of the HNS Fund.

Article 30

1 The Director shall be the chief administrative officer of the HNS Fund. Subject to the instructions given by the Assembly, the Director shall perform those functions which are assigned to the Director by this Convention, the internal regulations of the HNS Fund and the Assembly.

2 The Director shall in particular:

(a) appoint the personnel required for the administration of the HNS Fund;

(b) take all appropriate measures with a view to the proper administration of the assets of the HNS Fund;

(c) collect the contributions due under this Convention while observing in particular the provisions of article 22, paragraph 2;

(d) to the extent necessary to deal with claims against the HNS Fund and to carry out the other functions of the HNS Fund, employ the services of legal, financial and other experts;

(e) take all appropriate measures for dealing with claims against the HNS Fund, within the limits and on conditions to be laid down in the internal regulations of the HNS Fund, including the final settlement of claims without the prior approval of the Assembly where these regulations so provide;

(f) prepare and submit to the Assembly the financial statements and budget estimates for each calendar year;

(g) prepare, in consultation with the President of the Assembly, and publish a report on the activities of the HNS Fund during the previous calendar year; and

(h) prepare, collect and circulate the documents and information which may be required for the work of the Assembly and subsidiary bodies.

Article 31

In the performance of their duties the Director and the staff and experts appointed by the Director shall not seek or receive instructions from any Government or from any authority external to the HNS Fund.

They shall refrain from any action which might adversely reflect on their position as international officials. Each State Party on its part undertakes to respect the exclusively international character of the responsibilities of the Director and the staff and experts appointed by the Director, and not to seek to influence them in the discharge of their duties.

Finances

Article 32

1 Each State Party shall bear the salary, travel and other expenses of its own delegation to the Assembly and of its representatives on subsidiary bodies.

2 Any other expenses incurred in the operation of the HNS Fund shall be borne by the HNS Fund.

Voting

Article 33

The following provisions shall apply to voting in the Assembly:

(a) each member shall have one vote;

(b) except as otherwise provided in article 34, decisions of the Assembly shall be made by a majority vote of the members present and voting;

(c) decisions where a two-thirds majority is required shall be a two-thirds majority vote of members present; and

(d) for the purpose of this article the phrase "members present" means "members present at the meeting at the time of the vote", and the phrase "members present and voting" means "members present and casting an affirmative or negative vote". Members who abstain from voting shall be considered as not voting.

Article 34

The following decisions of the Assembly shall require a two-thirds majority:

(a) a decision under article 19, paragraphs 4 or 5 to suspend or reinstate the operation of a separate account;

(b) a decision under article 22, paragraph 2, not to take or continue action against a contributor;

(c) the appointment of the Director under article 26(d);

(d) the establishment of subsidiary bodies, under article 26(i), and matters relating to such establishment; and

(e) a decision under article 51, paragraph 1, that this Convention shall continue to be in force.

Tax exemptions and currency regulations

Article 35

1 The HNS Fund, its assets, income, including contributions, and other property necessary for the exercise of its functions as described in article 13, paragraph 1, shall enjoy in all States Parties exemption from all direct taxation.

2 When the HNS Fund makes substantial purchases of movable or immovable property, or of services which are necessary for the exercise of its official activities in order to achieve its aims as set out in article 13, paragraph 1, the cost of which include indirect taxes or sales taxes, the Governments of the States Parties shall take, whenever possible, appropriate measures for the remission or refund of the amount of such duties and taxes. Goods thus acquired shall not be sold against payment or given away free of charge unless it is done according to conditions approved by the Government of the State having granted or supported the remission or refund.

3 No exemption shall be accorded in the case of duties, taxes or dues which merely constitute payment for public utility services.

4 The HNS Fund shall enjoy exemption from all customs duties, taxes and other related taxes on articles imported or exported by it or on its behalf for its official use. Articles thus imported shall not be transferred either for consideration or gratis on the territory of the country into which they have been imported except on conditions agreed by the Government of that country.

5 Persons contributing to the HNS Fund as well as victims and owners receiving compensation from the HNS Fund shall be subject to the fiscal legislation of the State where they are taxable, no special exemption or other benefit being conferred on them in this respect.

6 Notwithstanding existing or future regulations concerning currency or transfers, States Parties shall authorize the transfer and payment of any contribution to the HNS Fund and of any compensation paid by the HNS Fund without any restriction.

Confidentiality of information

Article 36

Information relating to individual contributors supplied for the purpose of this Convention shall not be divulged outside the HNS Fund except in so far as it may be strictly necessary to enable the HNS Fund to carry out its functions including the bringing and defending of legal proceedings.

Chapter IV – *Claims and actions*

Limitation of actions

Article 37

1 Rights to compensation under chapter II shall be extinguished unless an action is brought thereunder within three years from the date when the person suffering the damage knew or ought reasonably to have known of the damage and of the identity of the owner.

2 Rights to compensation under chapter III shall be extinguished unless an action is brought thereunder or a notification has been made pursuant to article 39, paragraph 7, within three years from the date when the person suffering the damage knew or ought reasonably to have known of the damage.

3 In no case, however, shall an action be brought later than ten years from the date of the incident which caused the damage.

4 Where the incident consists of a series of occurrences, the ten-year period mentioned in paragraph 3 shall run from the date of the last of such occurrences.

Jurisdiction in respect of action against the owner

Article 38

1 Where an incident has caused damage in the territory, including the territorial sea or in an area referred to in article 3(b), of one or more States Parties, or preventive measures have been taken to prevent or minimize damage in such territory including the territorial sea or in such area, actions for compensation may be brought against the owner or other person providing financial security for the owner's liability only in the courts of any such States Parties.

2 Where an incident has caused damage exclusively outside the territory, including the territorial sea, of any State and either the conditions for application of this Convention set out in article 3(c) have been fulfilled or preventive measures to prevent or minimize such damage have been taken, actions for compensation may be brought against the owner or other person providing financial security for the owner's liability only in the courts of:

(a) the State Party where the ship is registered or, in the case of an unregistered ship, the State Party whose flag the ship is entitled to fly; or

(b) the State Party where the owner has habitual residence or where the principal place of business of the owner is established; or

(c) the State Party where a fund has been constituted in accordance with article 9, paragraph 3.

3 Reasonable notice of any action taken under paragraph 1 or 2 shall be given to the defendant.

4 Each State Party shall ensure that its courts have jurisdiction to entertain actions for compensation under this Convention.

5 After a fund under article 9 has been constituted by the owner or by the insurer or other person providing financial security in accordance with article 12, the courts of the State in which such fund is constituted shall have exclusive juris-

diction to determine all matters relating to the apportionment and distribution of the fund.

Jurisdiction in respect of action against the HNS Fund or taken by the HNS Fund

Article 39

1 Subject to the subsequent provisions of this article, any action against the HNS Fund for compensation under article 14 shall be brought only before a court having jurisdiction under article 38 in respect of actions against the owner who is liable for damage caused by the relevant incident or before a court in a State Party which would have been competent if an owner had been liable.

2 In the event that the ship carrying the hazardous or noxious substances which caused the damage has not been identified, the provisions of article 38, paragraph 1, shall apply *mutatis mutandis* to actions against the HNS Fund.

3 Each State Party shall ensure that its courts have jurisdiction to entertain such actions against the HNS Fund as are referred to in paragraph 1.

4 Where an action for compensation for damage has been brought before a court against the owner or the owner's guarantor, such court shall have exclusive jurisdiction over any action against the HNS Fund for compensation under the provisions of article 14 in respect of the same damage.

5 Each State Party shall ensure that the HNS Fund shall have the right to intervene as a party to any legal proceedings instituted in accordance with this Convention before a competent court of that State against the owner or the owner's guarantor.

6 Except as otherwise provided in paragraph 7, the HNS Fund shall not be bound by any judgement or decision in proceedings to which it has not been a party or by any settlement to which it is not a party.

7 Without prejudice to the provisions of paragraph 5, where an action under this Convention for compensation for damage has been brought against an owner or the owner's guarantor before a competent court in a State Party, each party to the proceedings shall be entitled under the national law of that State to notify the HNS Fund of the proceedings. Where such notification has been made in accordance with the formalities required by the law of the court seized and in such time and in such a manner that the HNS Fund has in fact been in a position effectively to intervene as a party to the proceedings, any judgement rendered by the court in such proceedings shall, after it has become final and enforceable in the State where the judgement was given, become binding upon the HNS Fund in the sense that the facts and findings in that judgement may not be disputed by the HNS Fund even if the HNS Fund has not actually intervened in the proceedings.

Recognition and enforcement

Article 40

1 Any judgement given by a court with jurisdiction in accordance with article 38, which is enforceable in the State of origin where it is no longer subject to ordinary forms of review, shall be recognized in any State Party, except:

(a) where the judgement was obtained by fraud; or

(b) where the defendant was not given reasonable notice and a fair opportunity to present the case.

2 A judgement recognized under paragraph 1 shall be enforceable in each State Party as soon as the formalities required in that State have been complied with. The formalities shall not permit the merits of the case to be re-opened.

3 Subject to any decision concerning the distribution referred to in article 14, paragraph 6, any judgement given against the HNS Fund by a court having jurisdiction in accordance with article 39, paragraphs 1 and 3 shall, when it has become enforceable in the State of origin and is in that State no longer subject to ordinary forms of review, be recognized and enforceable in each State Party.

Subrogation and recourse

Article 41

1 The HNS Fund shall, in respect of any amount of compensation for damage paid by the HNS Fund in accordance with article 14, paragraph 1, acquire by subrogation the rights that the person so compensated may enjoy against the owner or the owner's guarantor.

2 Nothing in this Convention shall prejudice any rights of recourse or subrogation of the HNS Fund against any person, including persons referred to in article 7, paragraph 2(d), other than those referred to in the previous paragraph, in so far as they can limit their liability. In any event the right of the HNS Fund to subrogation against such persons shall not be less favourable than that of an insurer of the person to whom compensation has been paid.

3 Without prejudice to any other rights of subrogation or recourse against the HNS Fund which may exist, a State Party or agency thereof which has paid compensation for damage in accordance with provisions of national law shall acquire by subrogation the rights which the person so compensated would have enjoyed under this Convention.

Supersession clause

Article 42

This Convention shall supersede any convention in force or open for signature, ratification or accession at the date on which this Convention is opened for signature, but only to the extent that such convention would be in conflict with it; however, nothing in this article shall affect the obligations of States Parties to States not party to this Convention arising under such convention.

Chapter V – *Transitional provisions*

Information on contributing cargo

Article 43

When depositing an instrument referred to in article 45, paragraph 3, and annually thereafter until this Convention enters into force for a State, that State shall submit to the Secretary-General data on the relevant quantities of contributing cargo received or, in the case of LNG, discharged in that State during the preceding calendar year in respect of the general account and each separate account.

First session of the Assembly

Article 44

The Secretary-General shall convene the first session of the Assembly. This session shall take place as soon as possible after the entry into force of this Convention and, in any case, not more than thirty days after such entry into force.

Chapter VI –*Final Clauses*

Signature, ratification, acceptance, approval and accession

Article 45

1 This Convention shall be open for signature at the Headquarters of the Organization from 1 October 1996 to 30 September 1997 and shall thereafter remain open for accession.

2 States may express their consent to be bound by this Convention by:

(a) signature without reservation as to ratification, acceptance or approval; or

(b) signature subject to ratification, acceptance or approval, followed by ratification, acceptance or approval; or

(c) accession.

3 Ratification, acceptance, approval or accession shall be effected by the deposit of an instrument to that effect with the Secretary-General.

Entry into force

Article 46

1 This Convention shall enter into force eighteen months after the date on which the following conditions are fulfilled:

(a) at least twelve States, including four States each with not less than 2 million units of gross tonnage, have expressed their consent to be bound by it, and

(b) the Secretary-General has received information in accordance with article 43 that those persons in such States who would be liable to contribute pursuant to article 18, paragraphs 1(a) and (c) have received during the preceding calendar year a total quantity of at least 40 million tonnes of cargo contributing to the general account.

2 For a State which expresses its consent to be bound by this Convention after the conditions for entry into force have been met, such consent shall take effect three months after the date of expression of such consent, or on the date on which this Convention enters into force in accordance with paragraph 1, whichever is the later.

Revision and amendment

Article 47

1 A conference for the purpose of revising or amending this Convention may be convened by the Organization.

2 The Secretary-General shall convene a conference of the States Parties to this Convention for revising or amending the Convention, at the request of six States Parties or one-third of the States Parties, whichever is the higher figure.

3 Any consent to be bound by this Convention expressed after the date of entry into force of an amendment to this Convention shall be deemed to apply to the Convention as amended.

Amendment of limits

Article 48

1 Without prejudice to the provisions of article 47, the special procedure in this article shall apply solely for the purposes of amending the limits set out in article 9, paragraph 1 and article 14, paragraph 5.

2 Upon the request of at least one half, but in no case less than six, of the States Parties, any proposal to amend the limits specified in article 9, paragraph 1, and article 14, paragraph 5, shall be circulated by the Secretary-General to all Members of the Organization and to all Contracting States.

3 Any amendment proposed and circulated as above shall be submitted to the Legal Committee of the Organization (the Legal Committee) for consideration at a date at least six months after the date of its circulation.

4 All Contracting States, whether or not Members of the Organization, shall be entitled to participate in the proceedings of the Legal Committee for the consideration and adoption of amendments.

5 Amendments shall be adopted by a two-thirds majority of the Contracting States present and voting in the Legal Committee, expanded as provided in paragraph 4, on condition that at least one half of the Contracting States shall be present at the time of voting.

6 When acting on a proposal to amend the limits, the Legal Committee shall take into account the experience of incidents and, in particular, the amount of damage resulting therefrom, changes in the monetary values and the effect of the proposed amendment on the cost of insurance. It shall also take into account the relationship between the limits established in article 9, paragraph 1, and those in article 14, paragraph 5.

7 (a) No amendment of the limits under this article may be considered less than five years from the date this Convention was opened for signature nor less than five years from the date of entry into force of a previous amendment under this article.

(b) No limit may be increased so as to exceed an amount which corresponds to a limit laid down in this Convention increased by six per cent per year calculated on a compound basis from the date on which this Convention was opened for signature.

(c) No limit may be increased so as to exceed an amount which corresponds to a limit laid down in this Convention multiplied by three.

8 Any amendment adopted in accordance with paragraph 5 shall be notified by the Organization to all Contracting States. The amendment shall be deemed to have been accepted at the end of a period of eighteen months after the date of notification, unless within that period no less than one-fourth of the States which were Contracting States at the time of the adoption of the amendment have communicated to the Secretary-General that they do not accept the amendment, in which case the amendment is rejected and shall have no effect.

9 An amendment deemed to have been accepted in accordance with paragraph 8 shall enter into force eighteen months after its acceptance.

10 All Contracting States shall be bound by the amendment, unless they denounce this Convention in accordance with article 49, paragraphs 1 and 2, at least six months before the amendment enters into force. Such denunciation shall take effect when the amendment enters into force.

11 When an amendment has been adopted but the eighteen month period for its acceptance has not yet expired, a State which becomes a Contracting State during that period shall be bound by the amendment if it enters into force. A State which becomes a Contracting State after that period shall be bound by an amendment which has been accepted in accordance with paragraph 8. In the cases referred to in this paragraph, a State becomes bound by an amendment when that amendment enters into force, or when this Convention enters into force for that State, if later.

Denunciation

Article 49

1 This Convention may be denounced by any State Party at any time after the date on which it enters into force for that State Party.

2 Denunciation shall be effected by the deposit of an instrument of denunciation with the Secretary-General.

3 Denunciation shall take effect twelve months, or such longer period as may be specified in the instrument of denunciation, after its deposit with the Secretary-General.

4 Notwithstanding a denunciation by a State Party pursuant to this article, any provisions of this Convention relating to obligations to make contributions under articles 18, 19 or article 21, paragraph 5 in respect of such payments of compensation as the Assembly may decide relating to an incident which occurs before the denunciation takes effect shall continue to apply.

Extraordinary sessions of the Assembly

Article 50

1 Any State Party may, within ninety days after the deposit of an instrument of denunciation the result of which it considers will significantly increase the level of contributions from the remaining States Parties, request the Director to convene an extraordinary session of the Assembly. The Director shall convene the Assembly to meet not less than sixty days after receipt of the request.

2 The Director may take the initiative to convene an extraordinary session of the Assembly to meet within sixty days after the deposit of any instrument of denunciation, if the Director considers that such denunciation will result in a significant increase in the level of contributions from the remaining States Parties.

3 If the Assembly, at an extraordinary session, convened in accordance with paragraph 1 or 2 decides that the denunciation will result in a significant increase in the level of contributions from the remaining States Parties, any such State may, not later than one hundred and twenty days before the date on which the denunciation takes effect, denounce this Convention with effect from the same date.

Cessation

Article 51

1 This Convention shall cease to be in force:

(a) on the date when the number of States Parties falls below 6; or

(b) twelve months after the date on which data concerning a previous calendar year were to be communicated to the Director in accordance with article 21, if the data shows that the total quantity of contributing cargo to the general account in accordance with article 18, paragraphs 1(a) and (c) received in the States Parties in that preceding calendar year was less than 30 million tonnes.

Notwithstanding (b), if the total quantity of contributing cargo to the general account in accordance with article 18, paragraphs 1(a) and (c) received in the States Parties in the preceding calendar year was less than 30 million tonnes but more than 25 million tonnes, the Assembly may, if it considers that this was due to exceptional circumstances and is not likely to be repeated, decide before the expiry of the above-mentioned twelve month period that the Convention shall continue to be in force. The Assembly may not, however, take such a decision in more than two subsequent years.

2 States which are bound by this Convention on the day before the date it ceases to be in force shall enable the HNS Fund to exercise its functions as described under article 52 and shall, for that purpose only, remain bound by this Convention.

Winding up of the HNS Fund

Article 52

1 If this Convention ceases to be in force, the HNS Fund shall nevertheless:

(a) meet its obligations in respect of any incident occurring before this Convention ceased to be in force; and

(b) be entitled to exercise its rights to contributions to the extent that these contributions are necessary to meet the obligations under (a), including expenses for the administration of the HNS Fund necessary for this purpose.

2 The Assembly shall take all appropriate measures to complete the winding up of the HNS Fund including the distribution in an equitable manner of any remaining assets among those persons who have contributed to the HNS Fund.

3 For the purposes of this article the HNS Fund shall remain a legal person.

Depositary

Article 53

1 This Convention and any amendment adopted under article 48 shall be deposited with the Secretary-General.

2 The Secretary-General shall:

(a) inform all States which have signed this Convention or acceded thereto, and all Members of the Organization, of:

(i) each new signature or deposit of an instrument of ratification, acceptance, approval or accession together with the date thereof;

(ii) the date of entry into force of this Convention;

(iii) any proposal to amend the limits on the amounts of compensation which has been made in accordance with article 48, paragraph 2;

(iv) any amendment which has been adopted in accordance with article 48, paragraph 5;

(v) any amendment deemed to have been accepted under article 48, paragraph 8, together with the date on which that amendment shall enter into force in accordance with paragraphs 9 and 10 of that article;

(vi) the deposit of any instrument of denunciation of this Convention together with the date on which it is received and the date on which the denunciation takes effect; and

(vii) any communication called for by any article in this Convention; and

(b) transmit certified true copies of this Convention to all States which have signed this Convention or acceded thereto.

3 As soon as this Convention enters into force, a certified true copy thereof shall be transmitted by the depositary to the Secretary-General of the United Nations for registration and publication in accordance with Article 102 of the Charter of the United Nations.

Languages

Article 54

This Convention is established in a single original in the Arabic, Chinese, English, French, Russian and Spanish languages, each text being equally authentic.

DONE AT LONDON this third day of May one thousand nine hundred and ninety-six.

IN WITNESS WHEREOF the undersigned, being duly authorized by their respective Governments for that purpose, have signed this Convention.

ANNEX I
CERTIFICATE OF INSURANCE OR OTHER FINANCIAL SECURITY IN RESPECT OF LIABILITY FOR DAMAGE CAUSED BY HAZARDOUS AND NOXIOUS SUBSTANCES (HNS)

Issued in accordance with the provisions of Article 12 of the International Convention on Liability and Compensation for Damage in Connection with the Carriage of Hazardous and Noxious Substances by Sea, 1996

Name of ship

Distinctive numbers or letters

IMO ship identification number

Port of registry

Name and full address of the principal business of the owner

This is to certify that there is in force in respect of the above named ship a policy of insurance or other financial security satisfying the requirements of Article 12 of the International Convention on Liability and Compensation for Damage in Connection with the Carriage of Hazardous and Noxious Substances by Sea, 1996.

Type of security

Duration of security

Name and address of the insurer(s) and/or guarantor(s)

Name

Address

This certificate is valid until

Issued or certified by the Government of (Full designation of the State) At (Place)

On (Date)

(Signature and Title of issuing or certifying official)

Explanatory Notes:

1. If desired, the designation of the State may include a reference to the competent public authority of the country where the certificate is issued.

2. If the total amount of security has been furnished by more than one source, the amount of each of them should be indicated.

3. If security is furnished in several forms, these should be enumerated.

4. The entry "Duration of the Security" must stipulate the date on which such security takes effect.

5. The entry "Address" of the insurer(s) and/or guarantor(s) must indicate the principal place of business of the insurer(s) and/or guarantor(s). If appropriate, the place of business where the insurance or other security is established shall be indicated.

ANNEX II
REGULATIONS FOR THE CALCULATION OF ANNUAL CONTRIBUTIONS TO THE GENERAL ACCOUNT

Regulation 1

1 The fixed sum referred to in article 17, paragraph 3 shall be determined for each sector in accordance with these regulations.

2 When it is necessary to calculate contributions for more than one sector of the general account, a separate fixed sum per unit of contributing cargo shall be calculated for each of the following sectors as may be required:

(a) solid bulk materials referred to in article 1, paragraph 5(a)(vii);

(b) oil, if the operation of the oil account is postponed or suspended;

(c) LNG, if the operation of the LNG account is postponed or suspended;

(d) LPG, if the operation of the LPG account is postponed or suspended;

(e) other substances.

Regulation 2

1 For each sector, the fixed sum per unit of contributing cargo shall be the product of the levy per HNS point and the sector factor for that sector.

2 The levy per HNS point shall be the total annual contributions to be levied to the general account divided by the total HNS points for all sectors.

3 The total HNS points for each sector shall be the product of the total volume, measured in metric tonnes, of contributing cargo for that sector and the corresponding sector factor.

4 A sector factor shall be calculated as the weighted arithmetic average of the claims/volume ratio for that sector for the relevant year and the previous nine years, according to this regulation.

5 Except as provided in paragraph 6, the claims/volume ratio for each of these years shall be calculated as follows:

(a) established claims, measured in units of account converted from the claim currency using the rate applicable on the date of the incident in question, for damage caused by substances in respect of which contributions to the HNS Fund are due for the relevant year; divided by

(b) the volume of contributing cargo corresponding to the relevant year.

6 In cases where the information required in paragraphs 5(a) and (b) is not available, the following values shall be used for the claims/volume ratio for each of the missing years:

(a) solid bulk materials referred to in article 1, paragraph 5 (a)(vii) 0

(b) oil, if the operation of the oil account is postponed 0

(c) LNG, if the operation of the LNG account is postponed 0

(d) LPG, if the operation of the LPG account is postponed 0

(e) other substances 0.0001

7 The arithmetic average of the ten years shall be weighted on a decreasing linear scale, so that the ratio of the relevant year shall have a weight of 10, the year prior to the relevant year shall have a weight of 9, the next preceding year shall have a weight of 8, and so on, until the tenth year has a weight of 1.

8 If the operation of a separate account has been suspended, the relevant sector factor shall be calculated in accordance with those provisions of this regulation which the Assembly shall consider appropriate.

Multimodal Dangerous Goods Form

MULTIMODAL DANGEROUS GOODS FORM

This form may be used as a dangerous goods declaration as it meets the requirements of SOLAS 74, chapter VII, regulation 54; MARPOL 79/78, Annex III, regulation 4.

1 Shipper/Consignor/Sender	2 transport document number		
	3 Page 1 of _____ pages		4 Shipper's reference
			5 Freight forwarder's reference
6 Consignee	7 Carrier (to be completed by the Carrier)		
	SHIPPER'S DECLARATION I herby declare that the contents of this consignement are fully and accurately described below by the Proper Shipping Name, and are classified, packaged, marked and labeled/placarded and are in all respects in proper condition for transport according to the applicable international and national government regulations.		
8 This shipment is within the limitations prescribed for: (Delete non-applicable)	9 Additional handling information		

PASSENGER AND CARGO AIRCRAFT	CARGO AIRCRAFT ONLY
10 Vessel/flight no. and date	11 Port/place of loading
12 Port/place of discharge	13 Destination

14 Shipping marks	Number and kind of packages; description of goods	Gross mass (kg)	Net mass (kg)	Cube (m³)

15 Container identification No. Vehicle registration No.	16 Seal number(s)	17 Container/vehicle & type	18 Tare mass (kg)	19 Total gross mass (including tare) (kg)

CONTAINER / VEHICLE PACKING CERTIFICATE I hereby declare that the goods described above have been packed/loaded into the container/vehicle identified above in accordance with the applicable provisions.+ **MUST BE COMPLETED AND SIGNED FOR ALL COUNTAINER/VEHICLE LOADS BY PERSON RESPONSIBLE FOR PACKING/LOADING**	**21 RECEIVING ORGANIZATION RECEIPT** Received the above number of packages/containers/trailers in apparent good order and condition, unless stated hereon: RECEIVING ORGANIZATION REMARKS:	
20 Name of company	Haulier's name	22 Name of company (OF SHIPPER PREPARING THIS NOTE)
Name/status of declarant	Vehicle registration no.	Name/status of declarant
Place and date	Signature and date	Place and date
Signature of declarant	DRIVER'S SIGNATURE	Signature of declarant

DANGEROUS GOODS
You must specify: Proper Shipping Name, hazard class, UN No., packing group, (where assigned) marine pollutant and observe the mandatory requirements under applicable national and international government regulations. For the purposes of the IMDG Code see 5.4.1.1

+For the purpose of the IMDG Code, see 5.4.2

MULTIMODAL DANGEROUS GOODS FORM

1 Shipper/Consignor/Sender			
	2 transport document number		
	3 Page 1 of _____ pages		4 Shipper's reference
			5 Freight forwarder's reference

14 Shipping marks	Number and kind of packages; description of goods	Gross mass (kg)	Net mass (kg)	Cube (m³)

Bibliography

Abbot, Charles, *Treatise of the Law Relative to Ships and Seamen* (5th ed. London: Shaw & Sons, 1901)

Abraham, Hans Jürgen, *Das Seerecht in der Bundesrepublik Deutschland* (4th ed., Berlin: de Gruyter, 1978)

Abraham, Kenneth S., "The Relation between Civil Liability and Environment Regulation: An Analytical Overview" [2002] 41 *Wash. L.J* 379 ff.

Abdul Hamid, Wafi Nazrin, *Loss or Damage from the Shipment of Goods, Rights and Liabilities of the Parties to the Maritime Adventure* (Diss. Southampton 1996)

Anderson, B. Charles/De la Rue, Colin, "Liability of Charterers and Cargo Owners for Pollution from Ships" [2001] 26 *Tul. Mar. L.J* 1 ff.

Anderson, Philip, *ISM Code, A practical Guide to the Legal and Insurance Implications* (London: LLP, 1998)

Andrewartha, Jane/Hayhurst, Penny "English Maritime Law Update: 1998" [1999] 30 *J. Mar. L. & Com.* 457 ff.

Asariotis, Regina, "Main Obligations and Liabilities of the Shipper" [2004] *TranspR* 284 ff.

Ashton, Ralph, "A Comparative Analysis of the Legal Regulation of Carriage of Goods by Sea Under Bills of Lading in Germany" [1999] (14) *MLAANZ* <www.mlaanz.org/docs/99journal7a.html>

Astle, William E., *Shipowner's Cargo Liabilities and Immunities* (3rd. ed, London: Witherby Ltd., 1967)

Astle, William E., *The Hamburg Rules* (London: Fairplay Publications, 1981)

Baram Michael S., "Insurability of Hazardous Materials Activities" [1988] (3) *J. Statist. Sci.* 339 ff.

Baughen, Simon "Sue and be Sued? Dangerous Cargo and the Claimant's Dilemma" [2006] 5(4) *S.&T.L.I* 14 ff.

Baughen, Simon/Campbell, Natalie, "Apportionment of Risk and the Carriage of Dangerous Cargo" [2001] (1) *IntML* 3 ff.

Bederman, David J., "Dead in the Water: International Law, Diplomacy, and Compensation for Chemical Pollution at Sea" [1986] 26 *Va. J. Int'l L.* 485 ff.

Bennion, Francis, *Statutory Interpretation* (3rd. ed., London: Butterworths, 1997)

Berlingieri, Francesco, "Basis of Liability and Exclusions of Liability" [2002] *LMCLQ* 336 ff.

Bishop, Owen, "A 'Secure' Package? Maritime Cargo Container Security after 9/11" [2006] 29 *Transp. L. J.* 313 ff.

Bob, Watt/Burgoyne. J.H. "Know your cargo" [1999] (13) *P&I Int'l* 102 ff.

Bothe, Michael, "Legal and Non-Legal norms: A Meaningful Distinction in International Relations" [1980] (11) *Netherlands YB Int'l L.* 65 ff.

Boyd, Stewart C./Burrows, Andrew S./Foxton, David, *Scrutton on Charterparties* (20th ed., London: Sweat & Maxwell, 1996).

Bremer, Jürgen, *Die Haftung beim Gefahrguttransport* (Karlsruhe: VVW, 1992)

Bristow, David I., "Contributory Fault in Construction Contracts" [1986] 2(4) *Const. L.J.* 252 ff.

Bulk Grain Cargoes, "Hot Stuff from the US, but not Enough from Brazil and Argentina", 1995 (September) No:139 Gard News 15.

Bulow, Lucienne Carasso, "Charter Party Consequences of Maritime Security Initiatives: Potential Disputes and Responsive Clauses" [2006] 37 *J. Mar. L& Com.* 79 ff.

Bulow, Lucienne Carasso, "'Dangerous' Cargoes: The Responsibilities and Liabilities of the Various Parties" [1989] *LMCLQ* 342 ff.

Buglass, Leslie J., "Limitation of Liability from a Marine Insurance Viewpoint" [1979] 53 *Tul. L. Rev.* 1364 ff.

Carriage of goods by sea – Indemnity – Carrier sustaining loss as a result of hazardous cargo – Bills of lading providing for "Merchant" to indemnify carrier regardless of fault- Carrier bringing action in contract and in tort – Whether bill of lading violated COGSA – Whether carrier entitled to claim in negligence, Case Comment, (2007) 708 *LMLN* 3.

"Carrier Liability and Freedom of Contract under The UNCITRAL Draft Instrument on The Carriage of Dangerous Goods [wholly or partly][by Sea], Note by the UNCTAD Secretariat, UNCTAD/SDTE/TLB/2004/2.

Chiang, Yung F., "The Characterization of a Vessel as A Common or Private Carrier" [1985] 60 *Tul. L. Rev.* 299 ff.

Chinkin, C.M, "The Challenge of Soft Law: Development and Change in International Law" [1989] 38 *I.C.L.Q* 850 ff.

Cho, Jae Hyung/Huynh, Eunice Huong K./McKee, Elizabeth MP/Tanner, James/Walsh, Allison, "2002-2003 Survey of International Law in the Second Circuit" [2003] 30 *Syracuse J. Int'l L .& Com.* 75 ff.

Clarke, Malcolm Alistair, *Aspects of the Hague Rules, A Comparative Study in English and French Law*, (The Hague: Nijhoff, 1976)

Colinvaux, Raoul, *Carver Carriage by Sea* (13th ed., London: Stevens & Sons, 1982).

Compton, Mike, "Dangerous Goods" 2004 (January) *Cargo Systems* 34 ff.

"Consideration of a Possible Convention on Liability and Compensation in Connection with The Carriage of Hazardous and Noxious Substances by Sea (HNS)" *Note by Secretariat, IMO Legal Comm.*, LEG 62/4, Annex I, 15 December 1989.

Cooke, Julian/Young, Timothy/Taylor, Andrew, *Voyage Charters* (2nd ed., London: LLP, 2001)

De Bièvre, Aline F.M., "Liability and Compensation for Damage in Connection with The Carriage of Hazardous and Noxious Substances by Sea" [1986] 17 *J. Mar. L. & Com.* 61 ff.

De Wit, Ralph, *Multimodal Transport* (London: LLP, 1995)

Donavan, James J., "The Origins and Development of Limitation of Shipowners' Liability" [1979] 53 *Tul. L. Rev.* 999 ff.

DuClos, Justin, "Liability for Losses Caused by Inherently Dangerous Goods Shipped by Sea and the Determinative Competing Degrees of Knowledge", <www.duclosduclos.org/LiabilityforLossesCausedByInherently.pdf> (visited 13.7.2007) (to be published in *U.S.F. Mar. L.J* Vol. 20 No. 1)

Du Pontavice, Emmanuel "The Victims of Damage Caused by the Ship's Cargo" in Grönfors, Kurt (ed.) *Damage from Goods* (Gothenburg: Akad Foerl, 1978).

Edelman, Paul S., "The Maritime Industry and the ISM Code" [1999] *8-WTR Currents: Int'l Trade L.J* 43 ff.

Edgcomb, John D., "The Trojan Horse Sets Sail: Carrier Defences against Hazmat Cargoes" [2000-2001] 13 *U.S.F.Mar.L.J* 31 ff.

Edgcomb, John, D., "Hazardous Substance Releases from Vessels: Current U.S. Law, the HNS Convention and Its Potential Impact if Ratified" [1997] 10 *U.S.F. Mar. L.J.* 73 ff.

Eyer, Walter W., "Shipowners' Limitation of Liability-New Directions for Old Doctrine" [1964] 16 *Stan. L. Rev.* 370 ff.

Falkanger, Thor/Bull, Hans Jacob/Brautaset, Lasse, *Introduction to Maritime Law* (Oslo: Tanoaschehoug, 1998)

Feldhaus, Henrich, *Zur Entwicklung der Haftung beim Seetransport gefährlicher Güter* (Karlsruhe: Verl. Versicherungs-Wirtschaft, 1985).

Focus on IMO, "Basic Facts about IMO", March 2000, <www.imo.org/includes/blastDataOnly.asp/data_id%3D7983/Basics2000.pdf> (visited 30.01. 2007)

"Focus on IMO, SOLAS: the International Convention for the Safety of Life at Sea, 1974", October 1998, <www.imo.org/includes/blastDataOnly.asp/data_id%3D7992/SOLAS98final.pdf.> (visited 30.01.2007)

"Focus on IMO, IMO and Dangerous Goods at Sea", May 1996, <www.imo.org/includes/blastDataOnly.asp/data_id%3D7999/IMDGdangerousgoodsfocus1997.pdf> (visited 30.01.2007)

"Focus on IMO, MARPOL 25 Years" <www.imo.org/includes/blastDataOnly.asp/data_id%3D7993/MARPOL25years1998.pdf>; (visited 30.01.2007)

"Focus on IMO, MARPOL 73/78", <www.imo.org/includes/blastDataOnly.asp/data_id%3D7575/MARPOL.1998.pdf> (visited 30.01.2007)

Gaskell, Nicholas "Charterer's Liability to Shipowner- Orders, Indemnities and Vessel Damage" in Schelin, Johan (ed.) *Modern Law of Charterparties* (Stockholm: Jure, 2003) 19 ff.

Gaskell, Nicholas, "The Draft Convention on Liability and Compensation for Damage Resulting from The Carriage of Hazardous and Noxious Substances", in Wetterstein, Peter, *Essays in Honor of Hugo Tiberg* (Stockholm: Juristfoerl., 1996) 225 ff.

Gaskell, Nicholas, "Pollution, Limitation and Carriage in The Aegean Sea", in Rose, Francis D, (ed.) *Lex Mercatoria* (London: LLP, 2000) 71 ff.

Gaskell Nicholas/Asariotis Regina/Baatz Yvonne, *Bills of Lading: Law and Contract* (London: LLP, 2000)

Gauci, "Risk Allocation in the Charterparty Relationship: an Analysis English Case Law relating to Cargo and Trading Restrictions" [1997] 28 *J. Mar. L. & Com.* 629 ff.

Giermanski, James/Neipert, David "The Regulations of Freight Forwarders in the USA and its Impact on the USA-Mexico Border" [2000] 9 *WTR Currrents: Int'l Trade L.J.* 11 ff.

Girvin, Stephen D., "Shipper's Liability for the Carriage of Dangerous Goods by Sea" [1996] *LMCLQ* 487 ff.

Glass, David A., *Freight Forwarding and Multimodal Transport Contracts* (London: LLP, 2004)

Gold, Edgar, "Legal Aspects of the transportation of dangerous goods at sea" [1986] (10) *Mar. Pol.* 185 ff.

Gramm Hans, *Das neue deutsche Seefrachtrecht nach den Haager Regeln* (Berlin: Mittler, 1938)

Griggs, Patrick/Williams, Richard/Farr, Jeremy, *Limitation of Liability for Maritime Claims* (4th ed. London: LLP, 2005)

Grönfors, Kurt, "Summarizing a Multi-National Problem", in Grönfors, Kurt (ed), *Damage from Goods* (Gothenburg: Akad Foerl., 1978)

Göransson, Magnus, "The HNS Convention" [1997] *Unif. L. Rev. (Rev. dr.Unif)* 249-270, 252.

Goñi, José L. "Watching from Coastal Shore the Passing through of Vessels Loaded with Crude Oil or Dangerous Merchandises" [1991] 26 *E.T.L* 143 ff.

Gündisch, Nicola, *Die Absenderhaftung im Land- und Seetransportrecht* (Hamburg: Lit, 1999)

Hamblen, Nicholas/Jones, Susannah, "Charterers Orders –"To Obey or Not To Obey" [2001] 26 *Tul. Mar. L. J.* 105 ff.

Hazelwood, Steven J., *P&I Clubs Law and Practice* (London: LLP, 2000).

Henry, Cleopatra Elmira, *The Carriage of Dangerous Goods by Sea, The Role of the International Maritime Organization in International Legislation* (New York: St. Martin's Pr., 1985)

Hill, D.J., *Freight Forwarders* (London: Stevens & Sons, 1972)

Hodges, Susan/Hill, Christopher, *Principles of Maritime Law* (London: LLP, 2001)

Homer, Andrew, "Second Circuit Limits COGSA Strict Liability for Shippers of Dangerous Goods in Contship Containerlines, Ltd. v. PPG Industries, Inc." [2006] 31 *Tul. Mar. L.J.* 199 ff.

"Identification of Priority Hazardous Substances" European Commission Working Document (ENV/191000/01 of 16 January 2001)

"Implications of the United Nations Convention on the Law of the Sea for the International Maritime Organization" (IMO) LEG/MISC/4

Jackson D.C, "Dangerous Cargo – A Legal Overview", in *Maritime Movement of Dangerous Cargoes- Public Regulations Private Liability*, Papers of One Day Seminar, Southampton University, Faculty of Law, 21 Sep. 1981, A1 ff.

Jones, Peter, "Carriers Include Indemnity Undertaking in Dangerous Goods Declaration That They Require Shippers Execute", <www.forwarderlaw.com/library/view.php?article_id=225> (last visited 10.11.2006)

Johnson H.N, "IMCO: The First Four Years (1959-1962)" [1963)] 12 *I.C.L.Q* 31 ff.

Kahn-Freund, Otto, *The Law of Inland Transport* (3rd ed., London: Stevens & Sons, 1956)

Karan, Hakan, *The Carriers Liability under the International Maritime Conventions, The Hague, Hague-Visby and Hamburg Rules* (Lewiston: NY: Mellen Press, 2004).

Kervella O., "IMO Roles on the Transport of Dangerous Goods in Ships and the Work of the International Bodies in the UN system, the Harmonization Issues with Regard to Classification, Criteria, Labelling and Placarding, Data Information, Emergency response and Training" in *The 11th International*

Symposium on the Transport of Dangerous Goods by Sea and Inland Waterways" Tokyo, 1992, pp.74-88.

Kim, Hyun "Shipowners' Limitation of Liability: Comparative Utility and Growth in The United States, Japan and South Korea" [1994] 6 *U.S.F. Mar. L.J.* 357 ff.

Knoynenburg, Fred, "Dangerous Goods – Is the charterer liable when he is not the shipper?", <www.middletonpotts.co.uk/library/default.asp?p=93&c=414> (visited 13.07.2007)

Li, K.X/Ng Jim Ng, "International Maritime Conventions: Seafarers' Safety and Human Rights" [2002] 33 *J. Mar. L. & Com.* 381 ff.

Maloof, David L./Krauzlis, James P, "Shipper's Potential Liabilities in Transit" [1980] 5 *Mar. Law.* 175 ff.

Mankabady, Samir, "Comments on Hamburg Rules", in Mankabady, Samir (ed.) *The Hamburg Rules on The Carriage of Goods by Sea* (Leyden: Sijthoff, 1978) pp. 27-115.

Markesinis, Basil S./Lorenz, Werner/Dannemann, Gerhard, *The German Law of Obligations* (Vol. 1 Oxford: Clarendon Press, 1997)

Martin, Stephen J. "HNS: A P&I Club Perspective", The Transportation of Hazardous Cargoes by Sea: Managing Your Risks and Undertaking the Consequences of the Law, (IBC Legal Studies&Services, 22 Mart 1993, London)

McKinley, Derek, *The International Convention on Liability and Compensation for the Carriage of Hazardous and Noxious Substances by Sea: Implications for State Parties, the Shipping, Cargo and Insurance Industries* (Diss. Cape Town 2005)

Mildoy, David/Scorey, David "Liabilities of Transferees of Bills of Lading" [1992] *IJOSL* 94 ff.

Mustill, Michael J., "Carriers' Liabilities and Insurance" in Grönfors Kurt (ed.) *Damage from* Goods (Gothenburg: Akad Foerl., 1978) pp.69-86.

Mustill, M., "Ships Are Different – or Are They?" [1993] *LMCLQ* 490 ff.

"New IMDG Code 'dangerous' says club" [2000] (14 December) *Fairplay* 7

Nunes, Tony, "Charterer's Liabilities under the Ship Time Charter" [2004] 26 *Hous. J. Int'l. L.* 561 ff.

Ogg, Terry, "IMO's International Safety Management Code (The ISM Code)" [1996] *IJOSL* 143 ff.

Palmer, Vernon, "A General Theory of the Inner Structure of Strict Liability: Common Law, Civil Law, and Comparative Law" [1988] 62 *Tul. L. Rev.* 1303 ff.

Panesar, Sukhnindar, "The Shipment of Dangerous Goods and Strict Liability" [1998] *I.C.C.L.R* 136 ff.

Pawlow, Jonathan R., "Liability for Shipments by Sea of Hazardous and Noxious Substances"[1985] 17 *Law & Pol'y Int't Bus.* 455 ff.

Peermohamed, Faz, "Dangerous Cargo" [2002] (16) *P&I Int'l* July 17 ff.

Pejovic, Caslav "Documents of Title in Carriage of Goods by Sea: Present Status and Possible Future Directions" [2001] *J.B.L.* 2001 461 ff.

Porat, Ariel, "Contributory Negligence in Contract Law: Toward a Principled Approach" [1994] 28 *U.B.C.L.R* 142 ff.

Rabe, Dieter, *Seehandelsrecht* (4th ed., München: Beck, 2000).

"Radioactive Materials Transport, The International Safety Regime, An Overview of Safety Regulations and the Organizations Responsible for their Development", World Nuclear Transport Institute, Review Serious No: 1 (Revised July 2006).

Recommendations on the Transport of Dangerous Goods, Model Regulations (14th ed., New York and Geneva: United Nations, 2005).

Reynolds, F.M.B., "The Carriage of Goods by Sea Act 1992" [1993] *LMCLQ* 436 ff.

Rights of Suits in Respect of Carriage of Goods by Sea Law Com. No. 196

Ringbom, Henrik, "EU Regulation 44/2001 and Its Implications for the International Maritime Liability Conventions" [2004] 35 *J. Mar. L. & Com.* 1 ff.

Roark, Holly "Explosion on the High Seas! The Second Circuit Promotes International Uniformity with Strict Liability for the Shipment of Dangerous Goods: Senator v. Sunway" [2003] 33 *Sw. U. L. Rev.* 139 ff.

Robertson, David W./Sturley Michael F. "Recent Developments in Admiralty and Maritime Law at The National Level and in The Fifth and Eleventh Circuits" [2002-2003] 27 *Tul. Mar. L.J* 495 ff.

Rodriguez, Antonio J./Hubbard, Mary Campbell, "The International Safety Management (ISM) Code: A New Level of Uniformity" [1999] *73 Tul. L. Rev.* 1585 ff.

Rose, F.D., "Cargo Risks: "Dangerous" Goods" [1996] 55 *Cam. L .J.* 601 ff.

Rose, F.D., "Liability for Dangerous Goods" [1998] *LMCLQ* 480 ff.

Roskill, "Book Review, 'The Legislative History of the Carriage of Goods by Sea Act and The Travaux Préparatoires of The Hague Rule'" [1992] 108 *L.Q.R.* 501 ff.

Røsæg, Erik, "HNS Insurers and Insurance Certificates" <http://folk.uio.no/erikro/WWW/HNS/Comp.doc> (visited 13.07.2007)

Røsæg, Erik, "Non-collectible contributions to the separate LNG account of the HNS Convention" [2007] 13 *JML* 94 ff.

Sasamura, Yoshio "Development of The HNS Convention" in *13th International Symposium on the Transport of Dangerous Goods by Sea and Inland Waterways*, Seoul Korea, 26-28 October, 1998, 491 ff.

Schoenbaum, Thomas, *Admiralty and Maritime Law* (4th ed., St. Paul, Minn.: West Publ. Co., 2004)

Schuda, Robert, "The International Maritime Organization and The Draft Convention on Liability and Compensation in Connection with The Carriage of Hazardous and Noxious Substances by Sea: An Update on Recent Activity" [1992] 46 *U. Miami L.Rev.*1009 ff.

Schultsz, "Insurance aspects of shippers' liability" in Grönfors, Kurt (ed.) *Damage from Goods* (Gothenburg: Akad Foerl. 1978) pp. 60-68

Schwampe, Dieter, *Charterers' Liability Insurance* (London, Hamburg: LLP, 1988)

Segolson Mats, *Damage from Goods in Sea Carriage, the Sender's Liability against the Carrier and the Other Owners of the Cargo on Board* (Diss. Stockholm   2001)   <www.juridicum.su.se/transport/Forskning/Uppsatser/ segolson.doc> (visited 31.01.2007)

Selvig, Erling, *Unit Limitation of Carrier's Liability* (Oslo: Oslo Uni. Press, 1961)

Selvig, Erling, "An Introduction to the 1976 Convention," in *Limitation of Shipowners Liability: The New Law* (London: Sweet & Maxwell, 1986)

Seward, R.C., "The Insurance Viewpoint", in *Limitations of Shipowners' Liability: The New Law* (London: Sweet & Maxwell, 1986) pp. 161-186.

Shaw, Malcolm N., *International Law* (4th ed., Cambridge: University Press, 1997)

Sheppard, Mandraka, "Contractual/Uncontractual-Lawful/Unlawful Orders under Charterparties" [1996] *IJOSL* 222 ff.

Simmonds, Kenneth R., *The International Maritime Organization* (London: Simmonds & Hill, 1994)

Steel, Joanna, "Dangerous Beetles? The Hague Rules and the Common Law" [1996] *IJOSL* 229 ff.

Steel, Joanna, "Dangerous Beetles in the House of Lords – Shippers Absolutely Liable" [1998] (2) *IJOSL* 119 f.

Sturley, Michael F., *The Legislative History of the Carriage of Goods by Sea and Travaux Préparatoires of the Hague Rules* (Littleton, Colo.: Fred B.Rothman & Co., 1990)

Tadeusz, Gruchulla-Wesierski, "A Framework for Understanding "Soft Law" [1984-1985] 30 *McGill L.J* 37 ff.

Tetley, William, *Marine Cargo Claims* (3rd. ed., Montreal: Blais, 1998)

Tetley, William, "Articles 9 to 13 of the Hamburg Rules", in Mankabady, Samir (ed.), *The Hamburg Rules on the Carriage of Goods by Sea"* (Boston: A.W. Sijthoff, 1978) pp. 197-207.

Tetley, William, "Responsibility of Freight Forwarders"[1987] 22 *E.T.L* 79 ff.

"The Safe Transport of Dangerous, Hazardous, and Harmful Cargoes by Sea" [1990] (25) *E.T.L* 747 ff.

Thomas, Michael, "British Concept of Limitation of Liability"[1979]) 53 *Tul. L. Rev.* 1205 ff.

Thomas, Rhidian D., "Limitation of Liability – London Convention 1976 – Definition of Charterer – Right to Limit – Limitable Claims – Articles I(2) and 2(1)(a), CMA CGM SA v. Classica Co. Ltd." [2004] (10) *J.I.M.L.* 122 ff.

Tiberg, Hugo, "Legal Survey" in Grönfors, Kurt (ed.) *Damage from Goods* (Gothenburg: Akad Foerl. 1978) pp.9-28.

Tiberg, Hugo "Legal Qualities of Transport Documents" [1998] 23 *Tul. Mar. L. J* 1 ff.

Todd, Paul, *Contract for the Carriage of Goods by Sea* (Oxford: BSP Professional Books, 1988)

Todd, Paul, "ISPS Clauses in Charterparties" [2005] *J.B.L.* 372 ff.

"Tracking Intermodal Shipments of Hazardous Materials Using Intelligent Transportation Systems in a New Security Age", Global Trade, Transportation and Logistics GTTL 502 Term Project, Spring 2002 <http://depts. washington.edu/gttl/StudentPapersAbstracts/2002/GTTL-hazmat%20final.pdf>

Trappe, Johannes, "Haftung beim Transport gefährlicher Güter im Seeverkehr" [1986] *VersR* 942 ff.

Trappe, Johannes, "Transport gefährlicher Güter, Unfallursachen und beteiligte Ladungen" [1988] *TranspR* 396 ff.

Treitel, Guenter H./Reynolds F.M.B, *Carver on Bills of Lading* (2nd ed., London: Sweat & Maxwell, 2005)

Tugman, Richard, "US and International Hazardous Material Regulations" [1995] (9) *P&I Int'l* 154 ff.

"UN Model Regulations the Transport of Dangerous Goods", <http://hazmat. dot.gov/regs/intl/untdg.htm> (visited on 16.10.2006)

Vandall, Frank J., *Strict Liability, Legal and Economic Analysis* (New York: Quorum Books, 1989)

Van der Ziel, G.J, "The UNCITRAL/CMI Draft for a New Convention Relating to The Contrat of Carriage by Sea" [2002] *TranspR* 272 ff.

Wardelmann, H. "Transport by Sea of Dangerous, Hazardous, Harmful and Waster Cargoes" [1991] 26 *E.T.L* 116 ff.

Webster, Andrew, "Managing Risk" [2004] 18(9) *M.R.I.* 15

Werro, Franz/Palmer, Valentine (ed), *The Boundaries of Strict Liability in European Tort Law* (Bern: Staempfli, 2004)

Wetterstein, Peter, "Carriage of Hazardous Cargoes by Sea- The HNS Convention" [1997] 26 *Ga. J. Int'l & Comp. L.* 595 ff.

Wilford, Michael/Coghlin, Terence/Kimball, John D., *Time Charters* (4th ed., London: LLP, 2003)

Williams, Charles, "The Implications of Shipping Dangerous Cargo", in *Pursuit and Defence of Cargo Claims*, 10th & 11th May 1999, London.

Williams, Peter John, "The Implications of the ISM Code for the Transport of Packaged Dangerous Goods by Sea", in *International Symposium on the Transport of Dangerous Goods by Sea and Inland Waterways*, Seoul, 26-28 October, 1998, pp. 117-122.

Willinger, Johann, "Rechtliche Grundlagen für die Verpackungen gefährlicher Güter" [1981] *TranspR* 81 ff.

Wilson D.J./Cooke, J.H.S, *General Average and York-Antwerp Rules* (12th ed., London: Sweet & Maxwell, 1997)

Wilson, John F., *Carriage of Goods by Sea* (4th ed. Munich: Longman, 2001)

Wilson, John F. "Basic Carrier Liability and The Right of Limitation", in Mankabady, Samir *The Hamburg Rules on the Carriage of Goods by Sea* (Leyden: Sijthoff, 1978) pp. 138-150.

Wohlfeld, Anne "The Senator Linie: Shipper's Strict Liability for Inherently Dangerous Goods" [2003] 27 *Tul. Mar. L.J.* 663 ff.

Wong K.K, James, "Packing Dangerous Goods" [1977)] 8 *J. Mar L. & Com.* 387 ff.

Woods, Henry, *Comparative Fault* (2nd ed., Rochester, N.Y: Lawyers Co-Operative Publishing Co., 1978)

Wu, Chao, *Pollution from the Carriage of Oil by Sea: Liability and Compensation* (London: Kluwer, 1996)

Wu, Chao, "What are the key charterers' risks?", *Club Cover* CLC01/07 <www.ukpandi.com/ukpandi/resource.nsf/Files/charterers%20brochurejan2007/$FILE/charterers+brochurejan2007.pdf> (visited 13.07.2007)

Zunarelli, Stefano "The Liability of the Shipper" [2002] *LMCLQ* 350 ff.

Documents

UNCITRAL

A/CN.9/WG.III/WP.32	A/CN.9/WG.III/WP.39
A/CN.9/WG.III/WP.55	A/CN.9/WG.III/WP.56
A/CN. 9/WG.III/WP.64	A/CN.9/552
A/CN.9/591	A/CN.9/594

UNECE

Trans/WP.15/2001/17/Add.2.
Trans/WP.15/2001/17/Add.4.

IMO

LEG XXXII/3	LEG.XXXIII/5	LEG. XXX.IV/7
LEG XXXIX/WP.1	LEG.44/7	LEG.47/7
LEG 60/3/3	LEG 60/3/4	LEG 62/4
LEG 62/4/2	LEG 62/4/2 Annex 2	LEG 62/4/5
LEG 62/7	LEG 63/3/5	LEG 63/3/1
LEG 63/3/14	LEG 63/WP.3	LEG 64/4
LEG 64/10	LEG 65/3/8	LEG 66/WP.5
LEG 67/9	LEG 68/4/4	LEG 69/11
LEG 70/4/2	LEG 70/4/10	LEG 70/Inf.2
LEG 71/3/4	LEG 72/9	LEG 73/14
LEG 83/INF.3	LEG 87/1	LEG 87/11
LEG/CONF.6/3	LEG/CONF.6/22	LEGCONF.6/C.1/WP.22
LEG/CONF.6/C.1/WP.24	LEG/CONF.6/C.1/WP.25	LEG./CONF.6/C.1/5
LEG/CONF.10/8/1	92FUND/AC.1/A/ES.7/7	

Index

About the International Max Planck Research School for Maritime Affairs at the University of Hamburg

The International Max Planck Research School for Maritime Affairs at the University of Hamburg was established by the Max Planck Society for the Advancement of Science, in cooperation with the Max Planck Institute for Comparative and International Private Law (Hamburg), the Max Planck Institute for Comparative Foreign Public Law and International Law (Heidelberg), the Max Planck Institute for Meteorology (Hamburg) and the University of Hamburg. The School's research is focused on the legal, economic, and geophysical aspects of the use, protection, and organization of the oceans. Its researchers work in the fields of law, economics, and natural sciences. The School provides extensive research capacities as well as its own teaching curriculum. Currently, the School has 15 Directors who determine the general work of the School, act as supervisors for dissertations, elect applicants for the School's PhD-grants, and are the editors of this book series:

Prof. Dr. Dr. h.c. Jürgen Basedow is Director of the Max Planck Institute for Foreign Private Law and Private International Law; *Prof. Dr. Peter Ehlers* is the Director of the German Federal Maritime and Hydrographic Agency; *Prof. Dr. Dr. h.c. Hartmut Graßl* is Director of the Max Planck Institute for Meteorology; *Prof. Dr. Hans-Joachim Koch* is Managing Director of the Seminar of Environmental Law of the Faculty of Law at the University of Hamburg; *Prof. Dr. Rainer Lagoni* is Managing Director of the Institute of Maritime Law and the Law of the Sea at the University of Hamburg; *PD Dr. Gerhard Lammel* is Senior Scientist at the Max Planck Institute for Meteorology; *Prof. Dr. Ulrich Magnus* is Managing Director of the Seminar of Foreign Law and Private International Law at the University of Hamburg; *Prof. Dr. Peter Mankowski* is Director of the Seminar of Foreign and Private International Law at the University of Hamburg; *Prof. Dr. Marian Paschke* is Director of the Institute of Maritime Law and the Law of the Sea at the University of Hamburg; *PD. Dr. Thomas Pohlmann* is Senior Scientist at the Centre for Marine and Climate Research and Member of the Institute of Oceanography at the University of Hamburg; *Dr. Uwe Schneider* is Assistant Professor at the Research Unit Sustainability and Global Change, Centre for Marine and Climate Research, Departments of Geosciences and Economics at the University of Hamburg; *Prof. Dr. Jürgen Sündermann* is Director at the Centre for Marine and Climate Research at the University of Hamburg; *Prof. Dr. Richard Tol* is Director of the Research Unit Sustainability and Global Change at the University of Hamburg; *Prof. Dr. Dr. h.c. Rüdiger Wolfrum* is Director at the Max Planck Institute for Comparative Foreign Public Law and International Law and a judge at the International Tribunal for the Law of the Sea; *Prof. Dr. Wilfried Zahel* is professor at the Centre for Marine and Climate Research, Departments of Geosciences and Economics at the University of Hamburg.

At present, *Prof. Dr. Dr. h.c. Jürgen Basedow* and *Prof. Dr. Ulrich Magnus* serve as speakers of the International Max Planck Research School for Maritime Affairs at the University of Hamburg.

Printing: Krips bv, Meppel, The Netherlands
Binding: Stürtz, Würzburg, Germany